Introducing Large Rivers

Introducing Large Rivers

Avijit Gupta
School of Earth, Atmospheric and Life Sciences
University of Wollongong
Australia

With contributions from

Olav Slaymaker
Department of Geography
The University of British Columbia
Vancouver
Canada

Wolfgang J. Junk
National Institute of Science and Technology of Wetlands (INCT-INAU)
Federal University of Mato Grosso (UFMT)
Cuiabá
Brazil

WILEY Blackwell

This edition first published 2020

Registered Office
John Wiley & Sons Ltd, The Atrium, Southern Gate, Chichester, West Sussex, PO19 8SQ, UK

Editorial Office
The Atrium, Southern Gate, Chichester, West Sussex, PO19 8SQ, UK

For details of our global editorial offices, customer services, and more information about Wiley products visit us at www.wiley.com.

Wiley also publishes its books in a variety of electronic formats and by print-on-demand. Some content that appears in standard print versions of this book may not be available in other formats.

Library of Congress Cataloging-in-Publication Data

Name: Gupta, Avijit, author.
Title: Introducing large rivers / by Avijit Gupta.
Description: First edition. | Hoboken, NJ : John Wiley & Sons, Inc., 2020. | Includes bibliographic references and index.
Identifiers: LCCN 2019032032 (print) | LCCN 2019032033 (ebook) | ISBN 9781118451403 (paperback) | ISBN 9781118451427 (adobe pdf) | ISBN 9781118451434 (epub)
Subjects: LCSH: Rivers–Environmental aspects–Research. | Fluvial geomorphology–Environmental aspects–Research.
Classification: LCC GB1205 .G86 2020 (print) | LCC GB1205 (ebook) | DDC 551.48/3–dc23
LC record available at https://lccn.loc.gov/2019032032
LC ebook record available at https://lccn.loc.gov/2019032033

Cover Design: Wiley
Cover Image: © guenterguni/Getty Images

Set in 10/12pt WarnockPro by SPi Global, Chennai, India

10 9 8 7 6 5 4 3 2 1

To Mira and Mae

Contents

Preface *xiii*

1 **Introduction** *1*
1.1 Large Rivers *1*
1.2 A Book on Large Rivers *3*
References *6*

2 **Geological Framework of Large Rivers** *7*
2.1 Introduction *7*
2.2 The Geological Framework: Elevated Land and a Large Catchment *8*
2.3 Smaller Tectonic Movements *9*
2.4 The Subsurface Alluvial Fill of Large Rivers *10*
2.5 Geological History of Large Rivers *12*
2.6 Conclusion *14*
Questions *14*
References *14*

3 **Water and Sediment in Large Rivers** *17*
3.1 Introduction *17*
3.2 Discharge of large Rivers *17*
3.3 Global Pattern of Precipitation *18*
3.4 Large River Discharge: Annual Pattern and Long-Term Variability *21*
3.5 Sediment in Large Rivers *26*
3.6 Conclusion *32*
Questions *32*
References *33*

4 **Morphology of Large Rivers** *35*
4.1 Introduction *35*
4.2 Large Rivers from Source to Sink *35*
4.3 The Amazon River *38*
4.3.1 The Setting *39*
4.3.2 Hydrology *39*
4.3.3 Sediment Load *39*
4.3.4 Morphology *42*
4.4 The Ganga River *44*

4.4.1 The Setting *44*
4.4.2 Hydrology *46*
4.4.3 Sediment Load *46*
4.4.4 Morphology *47*
4.5 Morphology of Large Rivers: Commonality and Variations *48*
4.6 Conclusion *52*
Questions *52*
References *52*

5 Large Rivers and their Floodplains: Structures, Functions, Evolutionary Traits and Management with Special Reference to the Brazilian Rivers *55*
Wolfgang J. Junk, Florian Wittmann, Jochen Schöngart, Maria Teresa F. Piedade and Catia Nunes da Cunha
5.1 Introduction *55*
5.2 Origin and Age of Rivers and Floodplains *57*
5.3 Scientific Concepts and their Implications for Rivers and Floodplains *59*
5.4 Water Chemistry and Hydrology of Major Brazilian Rivers and their Floodplains *60*
5.5 Ecological Characterisation of Floodplains and their Macrohabitats *62*
5.6 Ecological Responses of Organisms to Flood-Pulsing Conditions *64*
5.6.1 Trees *65*
5.6.2 Herbaceous Plants *66*
5.6.3 Invertebrates *66*
5.6.4 Fish *67*
5.6.5 Other Vertebrates *68*
5.7 Biodiversity *68*
5.7.1 Higher Vegetation *69*
5.7.2 Animal Biodiversity *71*
5.8 The Role of Rivers and their Floodplains for Speciation and Species Distribution of Trees *71*
5.9 Biogeochemical Cycles in Floodplains *73*
5.9.1 Biomass and Net Primary Production *73*
5.9.1.1 Algae *73*
5.9.1.2 Herbaceous Plants *74*
5.9.1.3 Trees of the Flooded Forest *75*
5.9.2 Decomposition *76*
5.9.3 The Nitrogen Cycle *77*
5.9.4 Nutrient Transfer Between the Terrestrial and Aquatic Phases *78*
5.9.5 Food Webs *79*
5.10 Management of Amazonian River Floodplains *80*
5.10.1 Amazonian River Floodplains *80*
5.10.2 Savanna Floodplains *82*
5.11 Policies in Brazilian Wetlands *82*
5.12 Discussion and Conclusion *84*
Acknowledgements *89*
References *89*

6 Large River Deltas *103*
6.1 Introduction *103*
6.2 Large River Deltas: The Distribution *104*
6.3 Formation of Deltas *104*
6.4 Delta Morphology and Sediment *110*
6.5 The Ganga-Brahmaputra Delta: An Example of a Major Deltaic Accumulation *112*
6.5.1 The Background *112*
6.5.2 Morphology of the Delta *113*
6.5.3 Late Glacial and Holocene Evolution of the Delta *114*
6.6 Conclusion *115*
Questions *115*
References *116*

7 Geological History of Large River Systems *119*
7.1 The Age of Large Rivers *119*
7.2 Rivers in the Quaternary *121*
7.2.1 The Time Period *121*
7.2.2 The Nature of Geomorphic Changes *123*
7.2.3 The Pleistocene and Large Rivers *124*
7.2.3.1 The Glacial Stage *124*
7.2.3.2 The Transition *125*
7.2.3.3 The Interglacial Stage *127*
7.3 Changes During the Holocene *127*
7.4 Evolution and Development of the Mississippi River *128*
7.5 The Ganga-Brahmaputra System *133*
7.6 Evolution of the Current Amazon *137*
7.7 Evolutionary Adjustment of Large Rivers *141*
Questions *142*
References *142*

8 Anthropogenic Alterations of Large Rivers and Drainage Basins *147*
8.1 Introduction *147*
8.2 Early History of Anthropogenic Alterations *148*
8.3 The Mississippi River: Modifications before Big Dams *149*
8.4 The Arrival of Large Dams *151*
8.5 Evaluating the Impact of Anthropogenic Changes *156*
8.5.1 Land Use and Land Cover Changes *157*
8.5.2 Channel Impoundments *159*
8.6 Effect of Impoundments on Alluvial Rivers *161*
8.7 Effect of Impoundments on Rivers in Rock *163*
8.8 Large-scale Transfer of River Water *166*
8.9 Conclusion *167*
Questions *168*
References *169*

9 **Management of Large Rivers** *173*
9.1 Introduction *173*
9.2 Biophysical Management *177*
9.3 Social and Political Management *178*
9.3.1 Values and Objectives in River Management *179*
9.3.2 International Basin Arrangements *180*
9.4 The Importance of the Channel, Floodplain, and Drainage Basin *180*
9.5 Integrated Water Resources Management *182*
9.6 Techniques for Managing Large River Basins *183*
9.7 Administering the Nile *184*
9.8 Conclusion *188*
Questions *189*
References *190*

10 **The Mekong: A Case Study on Morphology and Management** *193*
10.1 Introduction *193*
10.2 Physical Characteristics of the Mekong Basin *194*
10.2.1 Geology and Landforms *194*
10.2.2 Hydrology *196*
10.2.3 Land Use *197*
10.3 The Mekong: Source to Sea *199*
10.3.1 The Upper Mekong in China *199*
10.3.2 The Lower Mekong South of China *199*
10.4 Erosion, Sediment Storage and Sediment Transfer in the Mekong *202*
10.5 Management of the Mekong and its Basin *204*
10.5.1 Impoundments on the Mekong *204*
10.5.2 Anthropogenic Modification of Erosion and Sedimentation on Slopes *206*
10.5.3 Degradation of the Aquatic Life *207*
10.6 Conclusion *208*
Questions *208*
References *209*

11 **Large Arctic Rivers** *211*
Olav Slaymaker
11.1 Introduction *211*
11.1.1 The Five Largest Arctic River Basins *213*
11.1.2 Climate Change in the Five Large Arctic Basins *213*
11.1.3 River Basin Zones *214*
11.2 Physiography and Quaternary Legacy *216*
11.2.1 Physiographic Regions *216*
11.2.1.1 Active Mountain Belts and Major Mountain Belts with Accreted Terranes (Zone 1) *216*
11.2.1.2 Interior Plains, Lowlands, and Plateaux (Zone 2) *217*
11.2.1.3 Arctic Lowlands (Zone 3) *218*
11.2.2 Ice Sheets and Their Influence on Drainage Rearrangement *218*
11.2.3 Intense Mass Movement on Glacially Over-steepened Slopes *218*
11.3 Hydroclimate and Biomes *220*

11.3.1 Climate Regions *220*
11.3.2 Biomes *220*
11.3.3 Wetlands *224*
11.4 Permafrost *224*
11.4.1 Permafrost Distribution *224*
11.4.2 Permafrost and Surficial Materials *226*
11.4.3 Contemporary Warming *226*
11.5 Anthropogenic Effects *228*
11.5.1 Development and Population *228*
11.5.2 Agriculture and Extractive Industry *228*
11.5.3 Urbanisation: The Case of Siberia *228*
11.6 Discharge of Large Arctic Rivers *229*
11.6.1 Problems in Discharge Measurement *229*
11.6.2 Water Fluxes *229*
11.6.3 Water Budget *231*
11.6.4 Nival River Regime *232*
11.6.5 Lakes and Glaciers *234*
11.6.6 River Ice: Freeze and Break Up *236*
11.6.7 Scale Effects *237*
11.6.8 Effects of River Regulation *238*
11.6.9 Historical Changes *238*
11.7 Sediment Fluxes *239*
11.7.1 Complications in Determining Sediment Fluxes Both Within Arctic Basins and to the Arctic Ocean *239*
11.7.2 Flux of Suspended Sediment and Dissolved Solids *240*
11.7.3 Historical Changes in Water and Sediment Discharge in the Siberian Rivers *240*
11.7.4 Suspended Sediment Sources and Sinks in the Mackenzie Basin *242*
11.7.4.1 Sediment Yield in the Mackenzie Basin *242*
11.7.4.2 West Bank Tributary Sources *243*
11.7.4.3 Bed and Bank Sources *245*
11.8 Nutrients and Contaminants *249*
11.8.1 Supply of Nutrients *249*
11.8.2 Transport of Contaminants *250*
11.9 Mackenzie, Yukon and Lena Deltas *253*
11.9.1 Mackenzie Delta *253*
11.9.2 Lena Delta *253*
11.9.3 Yukon–Kuskokwim Delta *256*
11.10 Significance of Large Arctic Rivers *256*
Acknowledgment *258*
Questions *259*
References *259*

12 Climate Change and Large Rivers *265*
12.1 Introduction *265*
12.2 Global Warming: Basic Concept *266*

12.3 A Summary of Future Changes in Climate *270*
12.4 Impact of Climate Change on Large Rivers *271*
12.5 Climate Change and a Typical Large River of the Future *273*
12.6 Conclusion *277*
Questions *277*
References *278*

Index *281*

Preface

An edited anthology on geomorphology and management of large rivers was published in 2007.[1] The book filled a gap in our knowledge about large rivers as fluvial geomorphology used to be based more on smaller streams of manageable dimensions. We needed to extend our study to big rivers which shape a significant part of the global physiography, carry a high volume of water and sediment to the coastal waters, and support a very large number of people who live on their floodplains and deltas. That was an advanced treatise. This volume is written primarily as a textbook on large rivers, introducing such aspects. A number of line drawings and photographs illustrate the text, and a set of questions at the end of the chapters encourage the reader to explore various issues regarding large rivers.

The book introduces the environmental characteristics of river basins and forms and functions of channels commonly seen among the large rivers of the world. Specific discussions cover their complex geology, water, and sediment. The great lengths of these rivers stretch across a range of different environments. The Mekong, for example, flows on both rock and alluvium with varying form and behaviour. The geological framework of a large river is based primarily on large-scale tectonics commonly derived by plate movements. An uplifted zone, the primary source of sediment in the river, and a nearly subcontinental-scale water catchment area are necessary. A range of morphology exists in large rivers, and the associated floodplains and flood pulses are ecologically important. Large rivers could be geologically long-lived. In future, their forms may change and their functions may alter, following construction of engineering structures and climate change.

The quality of the book has been enhanced by detailed and well-illustrated discussions on two important topics: (i) large rivers and their floodplains: structures, functions, evolutionary traits and management with special reference to the Brazilian rivers by W.J. Junk *et al.* (Chapter 5), and (ii) large arctic rivers by O. Slaymaker (Chapter 11). I am grateful to all of the authors of these two chapters for their in-depth discussion on these topics. Lastly, the book indicates that the existing rivers possibly are undergoing dynamic adjustments in a world with a changing climate. Rivers change with time, and we usually know a large river only at a particular point in its existence.

Completion of the book has been a demanding task and I am grateful to the editorial and production teams of John Wiley & Sons, Ltd for their remarkable patience, editorial assistance, and continuous encouragement. I would like to thank Athira Menon and

1 Gupta, A. (Ed.) (2007). *Large Rivers: Geomorphology and Management*. Wiley: Chichester.

Joseph Vimali for guiding me through the intricacies of book production. Lee Li Kheng has produced many of the diagrams from my rough sketches. I have tremendously benefited from the critical readings by Colin Murray-Wallace of Chapter 7 on past rivers and by Colin Woodroffe of Chapter 6 on large river deltas and a discussion on climate change with John Morrison.

Wollongong, Australia, June 2019 *Avijit Gupta*

1

Introduction

1.1 Large Rivers

We have an intuitive recognition of large rivers although a proper definition is elusive. Even though it is difficult to define a large river, we would probably select the same 15 or 20 rivers as the biggest in the world. Potter identified four characteristic properties of large rivers: they drain big basins; they are very long; they carry a large volume of water; and they transfer a considerable amount of sediment (Potter 1978). It is, however, difficult to attribute quantitative thresholds to these, and not all big rivers exhibit these four characteristics. We associate large rivers with high discharge and sediment transfer, but both water and sediment vary over time and space and their data are difficult to acquire. It is easier to identify large rivers by the size of their drainage basins and their lengths; both are easier to measure.

Based on the areal extent of their drainage basin, Potter (1978) examined 50 of the world's largest rivers, ranked by Inman and Nordstrom (1971), starting with the Amazon. All but one of these rivers are more than 10^3 km long, and the smallest drainage basin is about 10^5 km^2. These 50 rivers collectively drain about 47% of the land mass, excluding Greenland and Antarctica. The Amazon alone drains about 5% of the continental area. These rivers also have modified the physiography of a large part of the world. Table 1.1 lists the top 24 large rivers (Figure 1.1), ranked according to their average annual water discharge. Their ranks would change if the rivers were listed according to any of the other three properties.

There are other lists. Hovius (1998) tabulated the morphometric, climatic, hydrologic, transport, and denudation data for 97 river basins, all of which measured above 2.5×10^4 km^2. Meade (1996) ranked the top 25 rivers twice: first, according to their discharge; and second, according to their suspended sediment load. The two lists do not match well. For example, large rivers such as the Zambezi or Lena carry a large water discharge but a low sediment load. Impoundments too have drastically reduced the once high sediment load of many rivers such as the Mississippi-Missouri. Over approximately the last 100 years, many rivers have been modified by engineering structures such as dams and reservoirs. The Colorado or the Huanghe (Yellow River) at present may not flow to the sea round the year. Such changes have also reduced the amount of sediment that passes from the land to the coastal waters. Large rivers such as the Nile or Indus have been associated with human civilisation for thousands of years and show expected modifications.

Introducing Large Rivers, First Edition. Avijit Gupta.

Table 1.1 Selected characteristics of 24 large rivers.

River	Average annual water discharge (10^6 m^3)	Length (km)	Drainage basin area (km^2)	Current average annual suspended sediment discharge (10^6 t)
1. Amazon	6300	6000	5.9	1000–1300
2. Congo	1250	4370	3.75	43
3. Orinoco	1200	770	1.1	150
4. Ganga-Brahmaputra	970	B-2900 G-2525	1.06 (B-0.63)	900–1200
5. Changjiang	900	6300	1.9	480
6. Yenisey	630	5940	2.62	5
7. Mississippi	530	6000	3.22	210
8. Lena	510	4300	2.49	11
9. Mekong	470	4880	0.79	150–170
10. Paranà-Uruguay	470	3965	2.6	100
11. St. Lawrence	450	3100	1.02	3
12. Irrawaddy	430	2010	0.41	260
13. Ob	400	>5570	2.77	16
14. Amur	325	4060	2.05	52
15. MacKenzie	310	4200	2.00	100
16. Zhujiang	300	2197	0.41	80
17. Salween	300	2820	0.27	About 100
18. Columbia	250	2200	0.66	8
19. Indus	240	3000	0.97	50
20. Magdalena	240	1540	0.26	220
21. Zambezi	220	2575	1.32	20
22. Danube	210	2860	0.82	40
23. Yukon	195	3200	0.83	60
24. Niger	190	4100	2.27	40

These figures vary between sources, although perhaps given the dimensions, such variations are proportionally negligible. Discharge and sediment figures are from Meade (1996) and Gupta (2007) and references therein. Drainage areas are rounded off to 10^6 km to reduce discrepancies between various sources. The Nile is not listed, even though it is 6500 km long. It does not qualify for this table as its water and sediment discharges are relatively low.

The great lengths of these rivers allow them to flow across a range of environments. The Mekong, for example, flows on both rock and alluvium, looking different (Figure 1.2). The end part of the river needs to adjust to all such environmental variations plus the Quaternary changes in sea level.

Fluvial geomorphology generally is based on small and logistically manageable streams. A study of large rivers is necessary, although difficult, for multiple reasons. Large rivers form and modify subcontinental-scale landforms and geomorphological

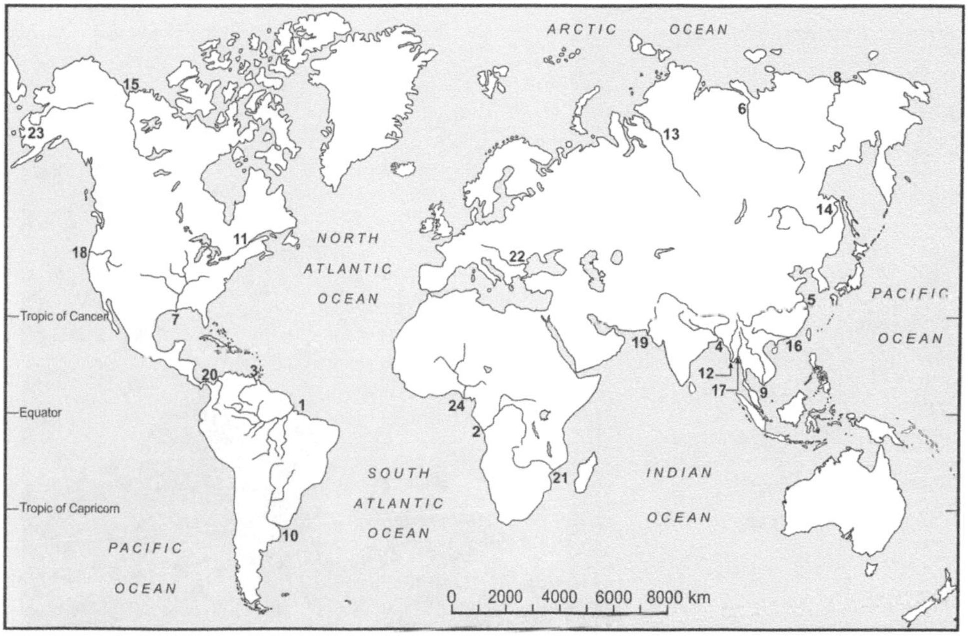

Figure 1.1 A sketch map showing the location of 24 large rivers in the world: 1, Amazon; 2, Congo; 3, Orinoco; 4, Ganga-Brahmaputra; 5, Changjiang; 6, Yenisei; 7, Mississippi; 8, Lena; 9, Mekong; 10, Parana-Uruguay; 11, St. Lawrence; 12, Irrawaddy; 13, Ob; 14, Amur; 15, Mackenzie; 16, Zhujiang; 17, Salween; 18, Columbia; 19, Indus; 20, Magdalena; 21, Zambezi; 22, Danube; 23, Yukon; 24, Niger.

processes. A high number of them convey and discharge a large volume of water and sediment to the coastal seas. An understanding of modern large rivers helps us to explain past sedimentary deposits. Large rivers, such as the Amazon (Mertes and Dunne 2007), and their deposits may reveal basinal and regional tectonics, past and present climate, and sea-level fluctuations. Management of the water resources of a large river is often an essential step toward the supply of water and power to a large number of people. We need to study large rivers for many such reasons.

1.2 A Book on Large Rivers

A number of individual large rivers have been studied and such studies published discretely. A collection of advanced essays on the general characteristics of large rivers, their selected case studies, and their utilisation and management is also available (Gupta 2007). In comparison, this volume is primarily an integrated textbook on large rivers and introduces the reader to the morphology and management of these huge conduits on which both the general physiography of the basins and utilisation of the resources of the rivers depend.

The discussion on large rivers starts with an account of their geological framework (Chapter 2) that determines where they can be located and also what their physical characteristics would be. The geological framework of a large river is based primarily on large-scale tectonics commonly driven by plate movements. An uplifted zone and the

(a)

(b)

Figure 1.2 The Mekong. (a) On rock, downstream of Chiang Saen, northern Thailand. (b) On alluvium near Savanakhet, Lao PDR, photographed from the air. Note the difference in form and behaviour between the two reaches. Large rivers commonly are a combination of a number of similar variations. Source: A. Gupta.

adjoining subcontinental-scale water catchment area are necessary requirements for a big river. Smaller tectonic movements may further modify the basin and the channel and explain their detailed morphological characteristics.

The regional geology should create a drainage basin large enough to accumulate enough precipitation to support and maintain the big river. Chapter 3 discusses the nature of water and sediment in a large river. The discharge in a large river is determined by various climatic criteria depending on its location: annual rainfall, seasonality in rainfall, and high episodic rain from synoptic disturbances such as tropical cyclones. The supply of water to large rivers could be from almost all parts of the watershed but the sediment supply generally is associated selectively with high mountains. For example, the discharge of the Orinoco is collected from most of the basin, irrespective of geology or relief, but its sediment supply is only from the Andes Mountains and the alluvial Llanos plains formed near the Andean foothills. In certain cases, several large rivers flow through arid landscapes without identifiable addition to their discharge but manage to sustain their flow because of the high discharge arriving from the upper non-arid parts of their drainage basins. Sediment in flood moves in large rivers both in downstream and lateral directions if large floodplains are present. The sediment grains travel a long distance to reach the sea and, in the process, become mature and sorted.

Large rivers have been aptly described as massive conveyance systems that move detrital sediment and dissolved matter over transcontinental distances (Meade 2007). Their morphology is dependent on regional geology, discharge and sediment flux, and may change several times between the headwaters and the sea (Chapter 4). Morphologically a large river usually has a channel flanked by bars, floodplain, and terrace fragments. The channel pattern depends on the gradient of the river and the nature of water and sediment it transports, and the pattern varies among different rivers as they adjust to the local physical environment. Floodplains of large rivers are important not only for their origin and age but also for their ecology which supports a wide variety of species, and their economic utilisation by people. The role of flood pulses in the maintenance of the floodplains and its ecology is crucial. This is discussed in detail by Junk *et al.*, in one of the two invited chapters in this book (Chapter 5). The huge discharge of water and sediment that is deposited by a big river in the sea may create a large delta. Deltas are morphologically fragile and change over time (Chapter 6). Deltas of many large rivers support a large population, and hence are of importance.

Large rivers could be geologically long-lived rivers such as the Mississippi or the Nile. A river that exists for a long time has a history. Tectonic processes commonly influence the origin, geographical location, and modification of major rivers. Understanding of such rivers requires knowledge of their history as rivers have changed episodically through tectonic movements, and especially through climate and sea-level changes in the Quaternary (Chapter 7).

Large rivers are a useful resource to people. A proper utilisation (Chapter 8) and management (Chapter 9) of large rivers is important. The land use of their basins and the use of their water have modified the environment over years of human civilisation. This has led to alteration of large rivers and their basins at various levels, especially over the last hundred years. The form and behaviour of many of the present large rivers have been modified mainly due to construction of large dams and reservoirs. The present state of a large river is conditioned by both the original physical environment of the basin and anthropogenic alterations imposed on the channel.

This requires proper management of the rivers so that basinal economic development and environmental degradation can be balanced in a sustainable way. A management procedure which simultaneously allows both economic development and environmental sustenance needs to be chosen. As a large river usually flows across multiple countries, each with different expectations and varying ability of resource utilisation, there is also a political aspect of large river management.

Chapter 10 deals exclusively with the Mekong River as a case study to illustrate the techniques and problems of managing a multistate river in a complex physical environment. It illustrates the reality of river management which involves dealing with the complexity of the physical characteristics of a big river, meeting the different expectancies of multiple stakeholders of the river basin, and maintaining the quality of the river for future generations, all at the same time.

Chapter 11 is on the special case of major rivers in the arctic. It deals mainly with the Lena, Yenisei and Ob in Siberia and the Mackenzie and Yukon in North America. These rivers flow through a unique environment and are expected to go through large changes in the near future due to global warming. This discussion on arctic rivers by Slaymaker is the second invited contribution in this book.

The last chapter deals with the possible modifications of large rivers in the near future. They may undergo significant changes following climate change and construction of large-scale engineering structures. The general tenets of climate change are known and accepted, but we have limited knowledge regarding its impact on large rivers. We, however, need to consider the future for understanding and management of present large rivers, as such changes would impact the lifestyles of a very large number of people, as the rivers of the future are likely to be different.

References

Gupta, A. (ed.) (2007). *Large Rivers: Geomorphology and Management*, 689. Chichester: Wiley.

Hovius, N. (1998). Control of sediment supply by large river. In: *Relative Role of Eustasy, Climate, and Tectonism in Continental Rocks*, vol. 59 (eds. K.W. Shanley and P.C. McCabe), 3–16. Tulsa: Society for Sedimentary Petrology Special Publication.

Inman, D.L. and Nordstrom, C.E. (1971). On the tectonic and morphological classification of coasts. *Journal of Geology* 79: 1–21.

Meade, R.H. (1996). River-sediment inputs to major deltas. In: *Sea-Level Rise and Coastal Subsidence: Causes, Consequences and Strategies* (eds. J.D. Milliman and B.H. Haq), 63–85. Dordrecht: Kluwer.

Meade, R.H. (2007). Transcontinental moving and storage: the Orinoco and Amazon rivers transfer the Andes to the Atlantic. In: *Large Rivers: Geomorphology and Management* (ed. A. Gupta), 45–63. Chichester: Wiley.

Mertes, L. and Dunne, T. (2007). Effects of tectonism, climate change, and sea-level change on the form and behaviour of the modern Amazon River and its floodplain. In: *Large Rivers: Geomorphology and Management* (ed. A. Gupta), 112–144. Chichester: Wiley.

Potter, P.E. (1978). Significance and origin of big rivers. *Journal of Geology* 86: 13–35.

2

Geological Framework of Large Rivers

2.1 Introduction

A large river is a long river which drains an extensive basin, carries a big discharge, and usually, but not always, transports a huge quantity of sediment (Potter 1978). It possesses a suitable three-dimensional geological framework for achieving these characteristics. A linear depression in rock of considerable length commonly lies below the river. A sedimentary fill of varying depth rests on this depressed rock surface, and along with bedrock constitutes the material below the channel of the river. The fill has been deposited by the main river and its ancestors, and some of its sediment is contributed by tributary streams. On the surface, the long trunk river crosses a range of physical environments and changes form and behaviour several times. For example, the Irrawaddy, Narmada and Danube flow in and out of narrow rocky valleys and wide alluvial basins. The basin of a large river commonly is an accumulation of several sub-basins with different character, exhibiting a polyzonal form and behaviour. The end part of the main river needs to adjust to all such variations in the large basin, plus any change in sea level.

The geological framework of a large river is formed primarily by past large-scale tectonics. Its basin should also be big enough to collect sufficient precipitation to form and support a major river system. Conditions vary spatially within the basin of the large river, and different parts of the basin contribute water and sediment in varying fashion to the mainstream. The main river usually receives water from multiple parts of the basin, but almost all of its sediment is usually derived from higher tectonic parts of the catchment, an area of high relief and disintegrated rocks (Meade 2007; Milliman and Syvitski 1992). Usually, the sediment is derived from such areas by glaciation, slope failures, and eroding headstreams of the river.

In brief, the physical characteristics of a large river depend on its structural framework, its geological history, and its pattern of water and sediment supply. Such characteristics form and maintain the river and its basin. Their nature is modified over time following changes in tectonics and climate, and in current times also by anthropogenic alterations of the river and its basin.

Introducing Large Rivers, First Edition. Avijit Gupta.

2.2 The Geological Framework: Elevated Land and a Large Catchment

Many of the existing large rivers start at an elevated orogenic zone, drain a subcontinental-scale area, and flow to the ocean through a major delta. The orogeny is usually created by convergence of two tectonic plates. The course of the river may be determined by a large-scale geofracture which it follows, and its mouth may be positioned by a rock basin at the trailing margin of one of the plates. The Amazon is an excellent example. Another example is that of the present Lower Mississippi, still located over a Cretaceous sub-surface rock embayment near its present confluence with the Ohio. It has been suggested that this embayment underneath the sediment of the river is related to a reactivated rift whose history may have started much earlier in the late Precambrian (Ervin and McGinnis 1975; Potter 1978; Knox 2007).

An uplifted zone, often formed by plate collisions, and an adjoining uplifted sub-continental-scale catchment area are the necessary requirements for a major river (Tandon and Sinha 2007). These conditions exist for a time long enough to create and sustain the river system. The present continental land masses are large enough to support the current big rivers, but the size of the land masses has not always been the same. An existing drainage can be modified over time due to changes in tectonic, geomorphic, and hydrologic systems. Rivers larger than those of the present time probably existed during the supercontinents of the Wilson cycle (Pangea and Rhodinia). In contrast, very early rivers on Earth probably did not become large due to limited hydrologic support and restricted basin area on land (Potter 1978). Tectonic movements and the following geomorphic processes determine the required development of the basin topography marked by elevated boundaries and a regional slope. Continental plate tectonics determine the basin framework and trough of the river, and regional disruptions vary its character, locally.

On fewer occasions, a new topography and a confined river is created by rifting, as shown by the Rio Grande, which is a pull-apart system that began in the early Miocene and developed from separate shallow basins to an integrated system. The failed arm at a plate tectonics triple junction also may give rise to rifting which may carry a big river that occupies the long, narrow, deeply filled depression extending into a craton. At the other end of the linear depression, the river builds a delta at the plate boundary marking the edge of the continent. The Niger is an example. Another example is the Blue Nile which rises within the rift system of Ethiopia to flow through a deep gorge for about half of its length.

A new topography and a modified river may also be formed off a mantle plume (Cox 1989) when it rises to create an extensive domal surface with high elevation, as happened in several parts of the Earth during the Late Cretaceous. Certain areas were uplifted by doming and magmatic underplating, giving rise to topographic highs and new drainage systems. The Orange and part of the Zambezi River are examples.

Not all large rivers fall into such clearly generalised classes. Some do not rise from currently active orogenic belts but drain parts of old mountains that have some elevation or are only slightly active tectonically. Tandon and Sinha (2007) described such rivers as located in cratonic settings as these drain major cratons. Examples of this type are the Mississippi, Yukon, Yenisei, Lena, and Ob. Potter (1978) stated that four major morphological patterns may cover the majority of large rivers. These can be described as:

(1) Most sediment derived from mountains marginal to a large craton (the Amazon).
(2) The river flowing marginal and parallel to a fold belt (the Ganga).
(3) A big river flowing along the strike of a mountain range (the Mekong).
(4) A river superimposed across multiple mountain chains (the Danube).

Potter proposed a fifth possibility which may have occurred in the past: a river on a large craton without bordering mountains (Potter 1978).

2.3 Smaller Tectonic Movements

The basin and the channel of a big river are further modified by smaller tectonic disturbances. For example, the Lower Amazon crosses several transverse structures in its course (Figure 2.1). In the downstream direction these are the Iquitos Arch, Jutai Arch, Purus Arch, Mont Alegre Intrusion and Ridge, and Gurupa Arch. When the Amazon crosses such structural highs, certain geomorphic features characterise the river (Mertes and Dunne 2007). The channel runs straight, its floodplain remains relatively narrow, scroll bars are found only near channel margins, and migration of the channel is limited. In contrast, wide floodplains with significant river movement, scroll bars and anabranches distinguish the river flowing on a low gradient between the upwarps. Even when covered by an alluvial fill, these upwarps affect the river. The morphology of a large river therefore varies along its course.

Large-scale fractures also regionally affect the tributary network in the Amazon Basin. Deep-seated basement fracturing appears to have disturbed the overlying sedimentary rocks that affect the drainage net, oriented in northeast and northwest directions (Potter 1978, Figure 8 and references therein; Mertes and Dunne 2007). The alignment of the Lower Negro, one of the major north bank tributaries of the Amazon, has been interpreted as controlled by a NW-SE tectonic lineament (Franzinelli and Igreja 2002). Here sunken crustal blocks and depressions occur along a half-graben, submerged to approximately 20 m with a width of up to 20 km. This controls the pattern of river islands, bars, and the location of sediment storage in the river.

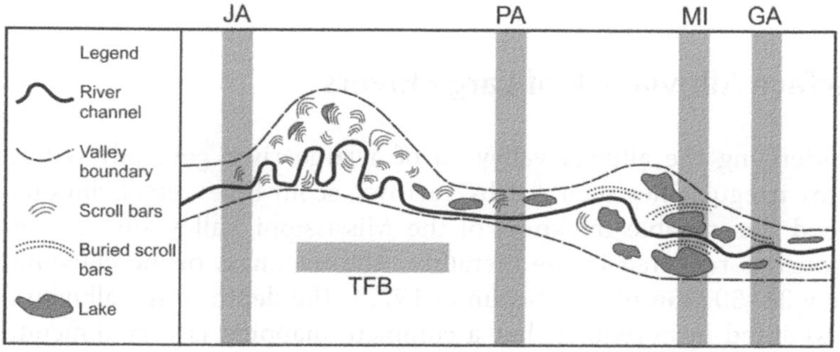

Figure 2.1 Schematic illustration of the relation between structure and morphology, the Lower Amazon Valley. The vertical bars show the approximate location of arches and a tilted block which are structural highs: JA, Jutai Arch; PA, Purus Arch; MI, Monte Allegro Intrusion and Ridge; GA, Gurupa Arch; TFB, approximate location of a tilted fault block. Source: Mertes and Dunne 2007 and references therein.

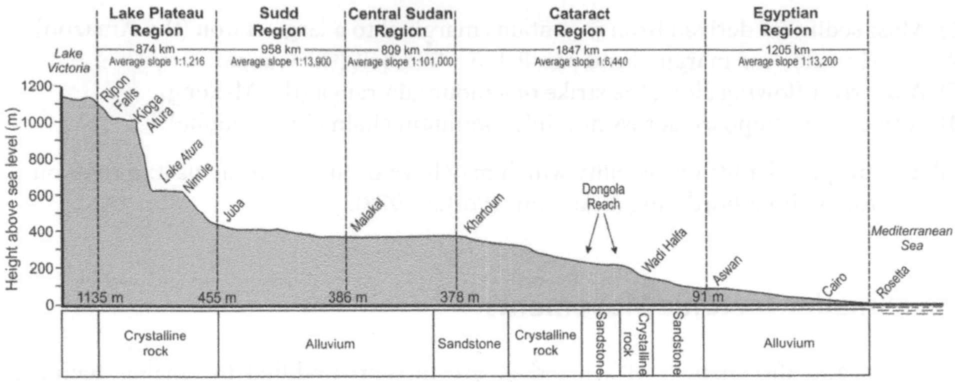

Figure 2.2 Longitudinal profile of the Nile over varying regional structures and lithology. Source: Woodward et al. 2007, from Said 1994.

A number of rivers flow along the structural grains of a folded mountain belt almost to their mouths. The Mekong and Salween are good examples. The Mekong flows for about 80% of its length in a narrow valley flanked by ridges (Figure 1.2a) whereas the Salween flows through a series of gorges. The lower Irrawaddy crosses three gorges in rock separated by wider basins in alluvium, and the morphology and behaviour of the river change between the gorges and basins. The entry and exit to the gorges show adjustment of the river over a short distance. The Irrawaddy finally continues through a flat lowland, discharging into the Gulf of Martaban through a large delta. The entire river is thus a sum of its different parts. The morphology and structural underpinning of large rivers are further discussed in Chapter 4.

Large rivers differ in their form and function, and for many rivers the understanding of such differences is achieved via a history of continental plate tectonics and lesser tectonic movements. The longitudinal profile of large rivers generally reflects a combination of their tectonic settings, basin lithology, and erosional history as their gradient changes (Figure 2.2). The profile of a large river over a long distance thus may be a combination of different sections. Geological structure and lithology tend to control the basic characteristics of the rivers.

2.4 The Subsurface Alluvial Fill of Large Rivers

The rock surface underlying the alluvial valley-fill of a large river generally is not smooth but marked by irregularities such as ribs, furrows, scour holes, etc. Schumm, for example, discussed the variable thickness of the Mississippi Valley alluvium at East St. Louis, Missouri as reported in the literature. The thickness of the alluvium reaches approximately 35–50 m in places (Schumm 1977). The depth of the alluvium and its nature are exhibited in boreholes, but a complete mapping is very difficult. Detailed three-dimensional studies have been carried out on several rivers (Fielding 2007). Construction of dams over large rivers also exposes surface morphology. The section at Hoover Dam on the Colorado River disclosed an inner channel flanked by bedrock terraces. About 200 km of the channel of the Changjiang were mapped between November 1978 and May 1979 near the site chosen later for the construction of the Three Gorges Dam. More than 90 troughs were found at the bottom of the river, cumulatively covering 45% of the total length of the river mapped. All the troughs were more than 40 m below the lowest river level (Yang et al. 2001). The cross-section of the channel of the Upper Mekong is trapezoidal, or has a deep inner channel bounded by rock benches, or a wide scabland-like section with rock ribs and piles rising from the bed (Figure 2.3). Scour pools and rock protrusions occur on the rock benches and the floor of the inner channel (Gupta 2007). The final bed of a big river thus is a composite result of structure and lithology at various scales.

The texture of the river-deposited fill over rock is not uniform. In general, the coarsest sediment is found at the base of the valley-fill and the finest towards the top (Schumm 1977). Sediment also tends to become finer in the downstream direction. However, sedimentation is also likely to indicate: (i) interrupted deposition; (ii) variations along the main channel due to differential contribution by tributaries; and (iii) variations from a general fining-upward sequence probably caused by climate change and tectonics. However, where the available sediment is of similar texture, the fill tends to be texturally rather uniform.

Valleys can be filled longitudinally in three ways. First, basin climate change or uplift of the source area may result in progressive aggradation of relatively coarse material deposited along the valley, downfilling the channel. Secondly, sediment from tributaries joining the trunk stream may vertically fill the valley relatively uniformly over a river

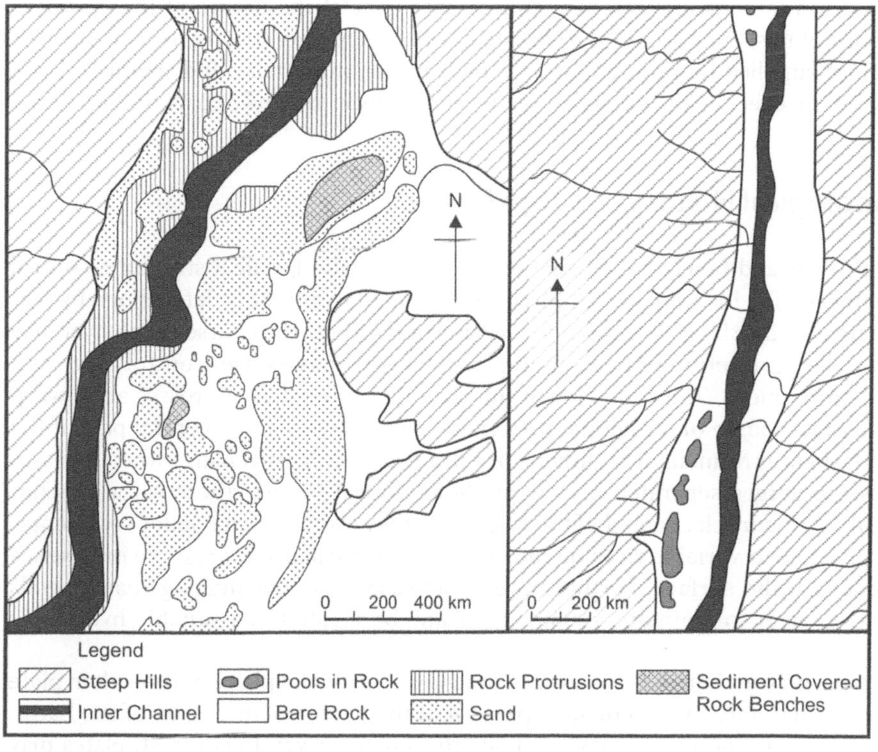

Figure 2.3 Diagrammatic sketches of cross-sections of the Mekong River in rock, Lao PDR. Source: Gupta 2004.

reach. Thirdly, a rise in base level may reduce the gradient over the lower parts of the valley, initiating progressive backfilling (Schumm 1977).

Tandon and Sinha (2007) summarised sedimentation in the Ganga River. Rising from the Himalaya Mountains, the river descends to the plains and flows parallel to the mountains for a considerable part of its course. It then erodes through the igneous Rajmahal Hills; turns south to build part of a major delta; and finally flows into the Bay of Bengal. The upper 20 km of the river downstream below the mountain front carries numerous gravel bars, the riverine alluvium consisting mainly of several metres of gravel. Coarse sand, corresponding to a braided channel belt, replaces gravels downstream (Shukla et al. 2001). This is followed downstream by the meandering and braided middle section of Ganga that deposits mostly fine to very fine sands on medium to coarse sands of older deposits. The fine sands are separated by units of floodplain clay, several metres thick and pedogenically altered, characteristics of a mobile river. Considerable volume of sediment also arrives to the Ganga as fan-deposits of the major Himalayan tributaries such as the Gandak or Kosi. The deposits of these megafans range from the mountain front to the Ganga in zones, from gravelly sand to fine sand and mud. These megafans consist of gravel beds in the upper areas, but most of the fan is made of multistoried sand sheets interbedded with overbank muddy layers (Singh et al. 1993).

Interpretation of both tectonic structures and subsurface fill is required to determine the geological background of a large river and explain its surface configuration. Natural sedimentation in a large river such as the Ganga is controlled primarily by tectonics, climate and sea level. Both tectonics and climate control the nature of the sediment near the mountains, the effect of the climate dominates further downstream along with the sediment supplied by both tributary fans from the mountains and sediment from cratons, and the effect of the changing sea level influences the lower part of the valley.

2.5 Geological History of Large Rivers

The morphology and behaviour of a large river, as discussed, reflect structural control at several levels. The location and origin of the basin are commonly determined by creation of a new surface by plate collision or rifting or doming. The course of the river may follow an older geofracture or be fixed in places where it passes over controls such as a structural embayment. The character of the river changes where it crosses an upwarp transverse to its course or a regional pattern of faults as seen in the central part of the Amazon Basin near Manaus. The dimensions of a major river also change when it passes through steep gorges separated by low-gradient alluvial basins. These variations not only modify the bedrock channel of the rivers, but they also influence valley sedimentation and give rise to variable riverine morphology, bar forms, and presence or absence of floodplains at the surface. Furthermore, in all cases, the new drainage basin needs to be large enough to maintain a major river and be located in a suitable hydrologic environment.

Large rivers vary in their size and age. Mobility of continental plates may result in plate destruction and size reduction or slow passage of the plate to a drier part of the Earth. Such developments modify or may even destroy a major river. In contrast, plates may increase in dimensions or reach more humid areas over time. Potter (1978) suggested that very big rivers may have been formed draining an enormous land mass consisting of welded cratonic blocks which carried at least one system of fold mountains and was well-watered.

The geological history of a major river therefore has a beginning and an end. The Mississippi is recognised to have been in existence since at least the Cretaceous because of a subsurface rock embayment located near its present confluence with the Ohio River mentioned earlier. The embayment is related to later Mississippi River fills and sediment of the Mississippi, dating back to the Jurassic period, has been found in the Gulf of Mexico. The embayment has been dated to late Precambrian (Ervin and McGinnis 1975). A large river therefore has been approximately in a place for nearly 300 million years or about 1/16th of the history of the Earth (Potter 1978). It is an old river, so are several others such as the Nile. The ancestral Nile has been associated with the ancient Pan-African orogenic events of about 550 million years ago or earlier. The river has been modified over time to its present appearance (Woodward et al. 2007 and references therein).

Not all major rivers have a similar long history. The present Amazon came into existence after the rise of the Andes in the Miocene. There was an older drainage system earlier, flowing westward and probably related to the African Plate before the opening of the South Atlantic. The Ganga and Brahmaputra, two major Himalayan rivers, probably came into existence in the Miocene after the formation of the Himalaya Mountains following the collision of the Indian Plate with the Eurasian Plate. There may have been an earlier drainage system, but it has been drastically modified, and the present drainage network has evolved over time. For example, the drainage of the western Himalaya was modified through a set of river captures, changing the direction of flow of the tributaries of the Ganga and Indus (Burbank 1992; Clift and Blusztajn 2005). It is believed that earlier the Sông Hóng (Red River) drainage net included the former upper Changjiang (Yangtze), Mekong, Salween and Tsanpo. The earlier upper drainage of the Sông Hóng then disintegrated to form the headwaters of these separate major rivers (Brookfield 1998). Robinson et al. (2014) have opined that the present Brahmaputra system has changed from the old Yarlung Tsanpo–Brahmaputra–Irrawaddy linkage to the present drainage via river capture and tectonic movements, around 18 Ma ago, and finally was captured by the Bay of Bengal system of the Upper Brahmaputra by the Lohit and newer Himalayan rivers (Licht and Giosan accepted for publication).

A river system may be terminated by other events such as a large-scale marine invasion, a new tectonic deformation, outpouring of lava or continental glaciation (Potter 1978). Tectonic, geomorphic and climatic processes may also modify an earlier drainage. The collision of the Indian and Eurasian Plates resulted in crustal shortening, differential shear and rotation associated with the rise of the Himalaya. An existing river may disappear through river capture, climate change resulting in a drier environment, or deformation due to plate tectonics. The change may affect almost the entire basin or only part of the drainage system. Potter (1978) has described rivers which were an assemblage of parts as having a composite age. Some rivers may persist in spite of major interruptions to experience over a long existence. The major rivers, their deltas, and offshore trenches underwent multiple changes during the Pleistocene. Goodbred Jr (2003) has traced the series of modifications that the Ganga River underwent since Marine Isotope Stage 3 (MIS 3, 58 000 years ago), mainly due to climate change.

2.6 Conclusion

The origin of a large river requires a suitable geological framework that creates (i) a large basin and (ii) a major river that slopes from elevated headwaters to the edge of the continent. Such developments happen primarily via plate tectonics although there could be other explanations. Sufficient precipitation needs to be accumulated in the basin to support and maintain the river. The basic form and function of the river that flows on the surface are spatially modified further by regional and local tectonics.

The size of the basin and the river is determined by plate tectonics and the amount of precipitation received by the area. The size of the river may change because of (i) plate movements which may lead to crustal spread or shortening and (ii) increase or decrease of precipitation. A large river therefore has a beginning and an end, and exists for a length of time. Several rivers such as the Mississippi or the Nile are very old and include parts of an earlier system. Many large rivers of the present are much younger, a number of them coming into existence or being drastically modified after the formation of the young fold mountains such as the Andes or the Himalaya.

Questions

1 What are the characteristics of large rivers? Do all large rivers have the same characteristics?

2 What kind of geological framework is required for a large river to exist? How do such frameworks originate?

3 How old are large rivers?

4 Mertes and Dunne (2007) described the relationship between structure and morphology of the Lower Amazon River. Describe such a relationship for a large river of your choice.

5 Discuss the sedimentary fill below a large river. Does it rest on a smooth rock surface? Give examples in support of your discussion.

6 Do large rivers stay the same in appearance and behaviour over time?

References

Brookfield, M.E. (1998). The evolution of the great river systems of southern Asia, during the Cenozoic India-Asia collision: rivers draining southwards. *Geomorphology* 22: 285–312.

Burbank, D.H. (1992). Causes of recent Himalayan uplift deduced from depositional pattern in the Ganga basin. *Nature* 357: 680–683.

Clift, P.D. and Blusztajn, J.S. (2005). Reorganization of the western Himalayan river system after five million years ago. *Nature* 438: 1001–1003.

Cox, K.G. (1989). The role of mantle plumes in the development of continental drainage patterns. *Nature* 342: 873–877.

Ervin, C.P. and McGinnis, L.D. (1975). Reelfoot rift: reactivated precursor of the Mississippi embayment. *Geological Society of America, Bulletin* 86: 1287–1295.

Fielding, C.R. (2007). Sedimentology and stratigaphy of large river deposits: recognition in the ancient record, and distinction for 'Incised Valley Fills'. In: *Large Rivers: Geomorphology and Management* (ed. A. Gupta), 97–113. Wiley.

Franzinelli, E. and Igreja, H. (2002). Modern sedimentation in the Lower Negro River, Amazonas State, Brazil. *Geomorphology* 44: 259–271.

Goodbred, S.L. Jr. (2003). Response of the Ganges dispersal system to climate change: a source-to-sink view since the last interstade. *Sedimentary Geology* 162: 83–104.

Gupta, A. (2004). The Mekong River: morphology, evolution and palaeoenvironment. *Journal of Geological Society of India* 84: 525–533.

Gupta, A. (2007). The Mekong River: morphology, evolution, management. In: *Large Rivers: Geomorphology and Management* (ed. A. Gupta), 435–455. Wiley.

Knox, J. (2007). The Mississippi River system. In: *Large Rivers: Geomorphology and Management* (ed. A. Gupta), 145–182. Wiley.

Licht, A. and Giosan, L. (accepted for publication). The Ayeyarwady River. In: *Large Rivers: Geomorphology and Management*, 2e (ed. A. Gupta). Wiley.

Meade, R.H. (2007). Transcontinental moving and storage: the Orinoco and Amazon Rivers transfer the Andes to the Atlantic. In: *Large Rivers: Geomorphology and Management* (ed. A. Gupta), 45–63. Wiley.

Mertes, L.A.K. and Dunne, T. (2007). Effects of tectonism, climate change, and sea-level change on the form and behaviour of the modern Amazon River and its floodplain. In: *Large Rivers: Geomorphology and Management* (ed. A. Gupta), 115–144. Wiley.

Milliman, J.D. and Syvitski, J.P.M. (1992). Geomorphic/tectonic control of sediment discharge to the ocean: the importance of small mountainous rivers. *Journal of Geology* 100: 525–544.

Potter, P.E. (1978). Significance and origin of big rivers. *Journal of Geology* 86: 13–33.

Robinson, R.A., Brezina, C.A., Parrish, R.R. et al. (2014). Large rivers and orogens: The evolution of the Yarling-Tsangpo-Irrawaddy system and the eastern Himalayan syntaxis. *Gondwana Research* 26 (1): 112–121.

Said, R. (1994). Origin and evolution of the Nile. In: *The Nile: Sharing a Scarce Resource* (eds. P.P. Howell and J.A. Allan), 17–26. Cambridge: Cambridge University Press.

Schumm, S.A. (1977). *The Fluvial System*. New York: Wiley.

Shukla, U.K., Singh, I.B., Sharma, M., and Sharma, S. (2001). A model of alluvial megafan sedimentation: Ganga megafan. *Sedimentary Geology* 144: 243–262.

Singh, H., Prakash, B., and Gohain, K. (1993). Facies analysis of the Kosi megafan deposits. *Sedimentary Geology* 85: 87–113.

Tandon, S.K. and Sinha, R. (2007). Geology of large river systems. In: *Large Rivers: Geomorphologyn and Management* (ed. A. Gupta), 7–28. Wiley.

Woodward, J.C., Macklin, M.G., Krom, M.D., and Williams, M.A.J. (2007). The Nile: evolution, Quaternary river environments and material fluxes. In: *Large Rivers: Geomorphology and Management* (ed. A. Gupta), 261–291. Wiley.

Yang, D., Li, X., Ke, X. et al. (2001). A note on the troughs in the Three Gorges Channel of the Changjiang River, China. *Geomorphology* 41: 137–142.

3

Water and Sediment in Large Rivers

3.1 Introduction

The potential locations of large river basins are determined mainly by plate tectonics. The existence of a large river is possible in such locations provided a high amount of precipitation fell on the basin and is transformed into river discharge. An integrated network of channels is needed to concentrate the basin runoff into a big main channel. A large river basin also includes actively eroding landforms which provide a significant volume of sediment. This chapter introduces the general sources of water and sediment for such rivers.

3.2 Discharge of large Rivers

River discharge (Q) is computed as the volume of water passing a given point on the river in unit time. It is measured in $m^3\ s^{-1}$, or for annual or long-term periods for large rivers in $km^3\ year^{-1}$. Two rivers can be compared by normalising their discharges into runoff (R) which is discharge per unit area of the basin ($R = Q/A$). The standard expression of runoff is

$$R = P - \sum(AE + S + C)$$

where

R	=	Runoff
P	=	Precipitation (includes snowmelt for certain basins)
AE	=	Actual evapotranspiration
S	=	Surface and subsurface storage of water
C	=	Anthropogenic consumption of water which is not returned to the river

Of all the properties of a large river, a high discharge is the one most expected (Potter 1978). This implies that at least part of the river basin lies in an area of high precipitation, or the area of the river basin is so huge that the cumulative flow in the trunk stream reaches a large volume, or both. The Amazon, the river with the biggest discharge, has a huge basin. It rains and snows heavily over its headwater basins in the Andes Mountains. Large volumes of discharge are contributed by its many tributaries draining the enormous basin. Its neighbour, the Orinoco, exhibits a similar pattern. The Orinoco is

Introducing Large Rivers, First Edition. Avijit Gupta.

fed by drainage from the Andes, the Andean foreland (Llanos), and the Guiana Shield, a craton (Figure 3.1). On average, every square kilometre of its drainage area carries more water than even the basin of the Amazon.

A large river is presumed to have a huge discharge, and large rivers commonly are found in areas of high precipitation, usually determined by the pattern of global climate, and often on the windward slopes of high mountains. In comparison, precipitation over the north-flowing rivers of Eurasia, the Ob, Yenisei, and Lena, are not high but these still have become large rivers because of the collective precipitation falling over their huge catchment areas. The evapotranspiration is low due to prevailing low temperature.

There are variations from such simple explanations. The basin of the Indus is mostly arid, but the water collected in the headwater-mountains from local high annual precipitation maintains its large channel across the dry lower part of the drainage basin. In brief, discharge of a river is derived from both climate and size. By plotting basin area against mean annual discharge for 1100 rivers, Milliman and Farnsworth (2011) demonstrated that 68% of the variance in discharge for the same drainage area can be explained by their climatic characteristics.

We should note that the discharge given for a river refers to the discharge at a specific measurement station. The last gauging station of a large river is located not at its mouth but usually several hundred kilometres upstream, near the end of its tidal limit. The last discharge station on the Amazon is at Óbidos, about 1000 km above its mouth. The last station on the Changjiang is at Datong, 600 km from the sea. Our knowledge about the discharge of water and sediment over the last few hundred kilometres on large rivers is limited.

3.3 Global Pattern of Precipitation

The primary source of pre-precipitation moisture is the atmosphere. Precipitation requires cooling of moist air by upward convection or mixing between two air masses of different temperature. The moist air becomes saturated by cooling and condensation. With further cooling, the moisture falls as rain or snow, depending on the ambient temperature.

Most of the atmospheric water is stored in the troposphere, especially in the warm air of tropical latitudes (Hayden 1988). Evaporation from the seas happens efficiently in the warm climate in the tropics, adding moisture to the atmosphere. Evaporation of more than 60% of water takes places between 30° north and south latitudes. In contrast, only about 5% of total evaporation takes place beyond the 50th parallels. More than 60% of evaporation is from the oceans (Lamb 1972; Hayden 1988). This implies a higher presence of moisture in the tropical air and on the windward side of continents. The stored atmospheric water is condensed before precipitation in two broad ways: barotropic and baroclinic.

Barotropic conditions prevail in the low latitudes. In a *barotropic atmosphere*, the horizontal thermal gradients are small. The condensation is carried out by vertical lifting of the heated air, and where vertical wind shear is low, may give rise to huge convective clouds. Lifting of humid air is accompanied by a continuous production of latent heat by condensation which uplifts air in the barotropic atmosphere in several ways. The uplift commonly happens by:

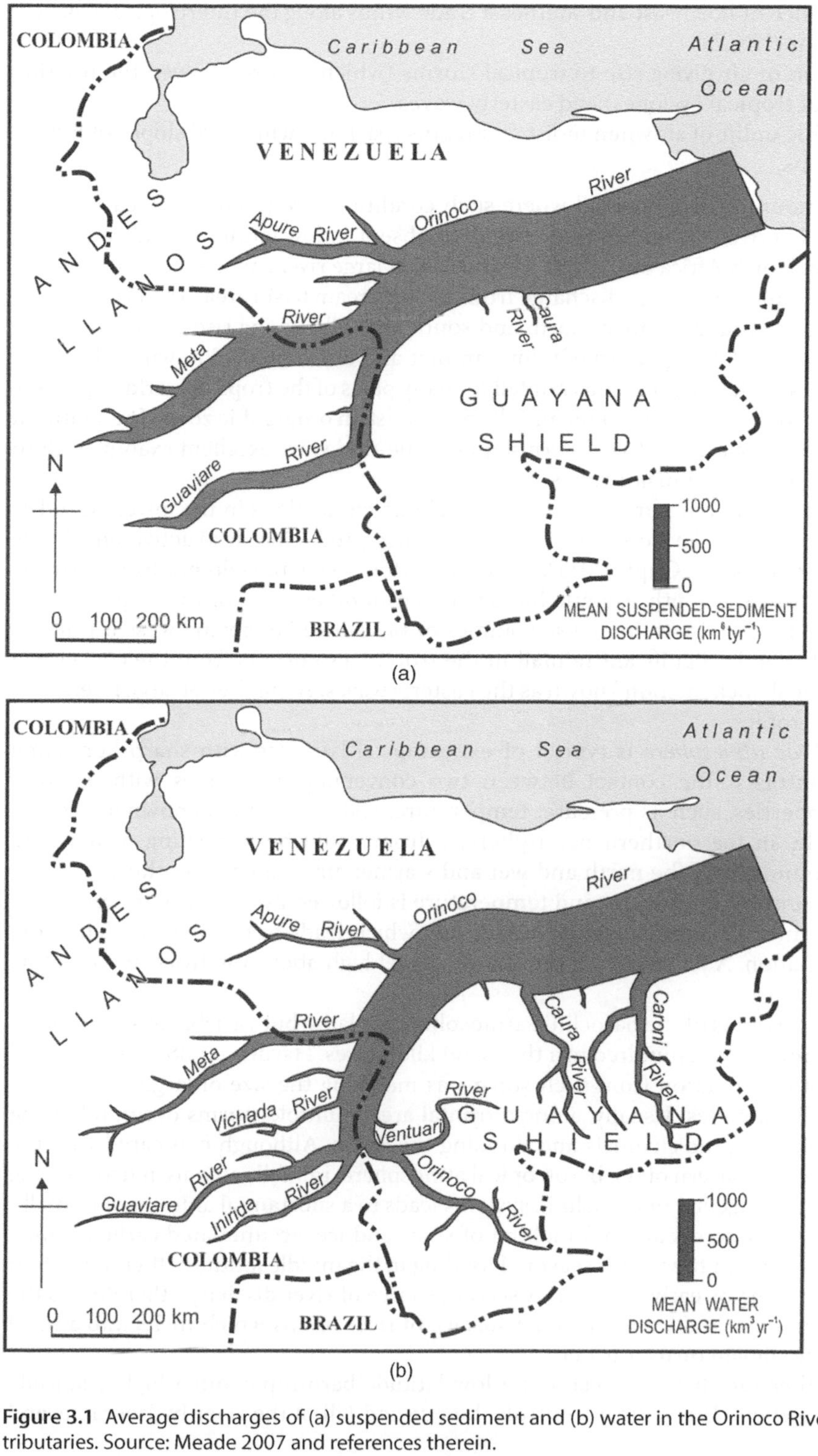

Figure 3.1 Average discharges of (a) suspended sediment and (b) water in the Orinoco River and its tributaries. Source: Meade 2007 and references therein.

- Convergence of northeast and southeast trade winds along the Intertropical Convergence Zone (ITCZ).
- Circulation of air giving rise to tropical storms (which may reach even the rotating velocity of tropical cyclones) and easterly waves.
- Orographic uplift of air when moist airstreams reach the windward slopes of mountain regions.

A high amount of rain may fall where such conditions are fulfilled, increasing river discharge. Where such conditions are weak or absent in low latitudes, arid conditions prevail, as in North Africa or Central Australia, and large rivers are either absent or survive only by importing a high discharge from the upstream basin area. The ITCZ and its associated belt of rainfall moves north and south annually, giving rise to a pronounced seasonality in rainfall. A pattern of rainy summer and dry winter is known as the monsoon system which brings copious rainfall to many parts of the tropical world, especially where the incoming moist summer air is lifted against an orographic zone. The southern slopes of the Himalaya and the eastern slopes of the Andes are excellent examples. Both regions nurture a set of major rivers.

Episodic rainfall occurs from large-scale cyclonic circulations in the lower latitudes, some of which may develop into tropical cyclones producing destructive and heavy rainfall (for details, see Gupta 2011). Tropical cyclones generally do not form near the Equator or over the South Atlantic but are found in other parts of the tropics. These storms tend to give rise to immense volumes of rainfall while moving west within the belt of trade winds. Significant rainfall in the tropics also occurs from the converging meteorological phenomenon known as the easterly waves. A number of large rivers thus exist in the tropics.

A *baroclinic atmosphere* is typical of extratropical latitudes with sharp horizontal thermal contrasts. The contact between two converging air masses with different level of properties, such as pressure, temperature, and moisture, is known as a front. For example, in the northern hemisphere, a front could be a meeting of dry cold polar air coming from the north and wet and warmer air coming from the south. The horizontal contrast in pressure and temperature is followed by a vertical movement of air. The warmer air rises above the colder one which leads to cooling, condensation, and precipitation. A jet stream, if present at a level high above the front, increases its intensity.

The frontal storms of the baroclinic atmosphere are large but variable in size. Diameters range from several hundred to a thousand kilometres. Hayden (1988) has described the areas of precipitation from such storms as matching the size of large river basins of the middle latitudes. Usually, along a frontal area, multiple storms occur, following one another, filling the channels and flooding the rivers. Although compared with the deep convection pattern of the barotropical atmosphere, rainfall rates are much less, the compensating longevity of baroclinic systems leads to a substantial amount of rainfall.

Flooding may also occur from melting of snow and ice, accumulated earlier, from a number of storms in the winter season. Flooding in the middle latitudes therefore often happens in spring or early summer. A second source of river discharge therefore is the accumulated snow and ice on the land surface of river basins which melts into annual floods as the climate turns warmer.

We can therefore have two classes: a low-latitude barotropic and a higher-latitude baroclinic section. This pattern controls the rise and fall of the river hydrographs and floods. Large floods in big rivers occur under specific circumstances. For example, in the tropics, cyclonic circulations give rise to large rain-bearing storms which may develop up to the strength of tropical cyclones. Heavy, intensive, and episodic rainfall from such storms commonly arrives in the middle of the wet season when the river is high and the ground is wet, giving rise to flood discharges, extensive erosion, and sediment transfer (Gabet et al. 2004). In higher latitudes, floods arrive from a series of large-scale frontal storms, often as rain on snow. Rain-bearing tropical cyclones moving towards higher latitudes, may also contribute to floods in major rivers.

Precipitation may vary considerably from year to year (Amarasekera et al. 1997). The basin of a large river may go through a spell of wet or dry years due to various types of climatic shifts such as the short-term El Niño Southern Oscillation (ENSO) or other climate drivers which operate over longer periods. A full ENSO cycle usually runs for five to eight years. It includes particularly dry years (known as El Niño), and wet years (La Niña). The resulting wet or dry climate is found over various parts of the world at the same time explaining variable flows in large river basins. For example, precipitation and river runoff rise during La Niña for the Magdalena, Orinoco, and the northern tributaries of the Amazon across northern South America.

The Pacific Decadal Oscillation (PDO) is similar but lasts for 20–40 years. Rivers of the tropics and subtropics are commonly affected by the short-term ENSO and PDO climate drivers. The North Atlantic Oscillation (NAO) affects rivers flowing to the Atlantic and Arctic Oceans. Its influence on precipitation and runoff can be recognised in the runoff pattern of the northeastern and mid-Atlantic rivers of the United States but rivers in Europe are difficult to interpret. Other long-term climatic oscillations are the Atlantic Multidecadal Oscillation (AMO) and the Southern Annular Mode (SAM). Given that a long series of discharge data is available only for a few rivers, the effects of long-term climate drivers on rivers are difficult to investigate.

Runoff of a large river thus reflects various climatic criteria: annual rainfall, seasonality in rainfall, and episodic rain from synoptic disturbances. Not only are the large rivers thus maintained, but their behaviour is also characterised by the run of changing wet and dry years determined by climate drivers, and episodic storm rainfalls. A high average rainfall or floods of limited recurrence interval is required to maintain the channel of a large river.

3.4 Large River Discharge: Annual Pattern and Long-Term Variability

Discharge of large rivers is derived from (i) the nature of precipitation falling on their basins, (ii) melting of ice and snow in spring and early summer, if present within the basin, and (iii) stepwise contribution of the tributaries to the main streams.

Figure 3.2 demonstrates the annual hydrograph of several major rivers from different climates as examples of basins operating in different climatic zones, starting with the Amazon whose huge basin of nearly 7 million km^2 extends to both sides of the Equator. The precipitation over the basin is dominated by the annual pattern of movement of ITCZ. Precipitation over the Andean regions is affected also by the South Atlantic Convergence Zone (SACZ) which has a seasonal variation. Precipitation is thus generally uniformly distributed over the basin throughout the year, with orographic

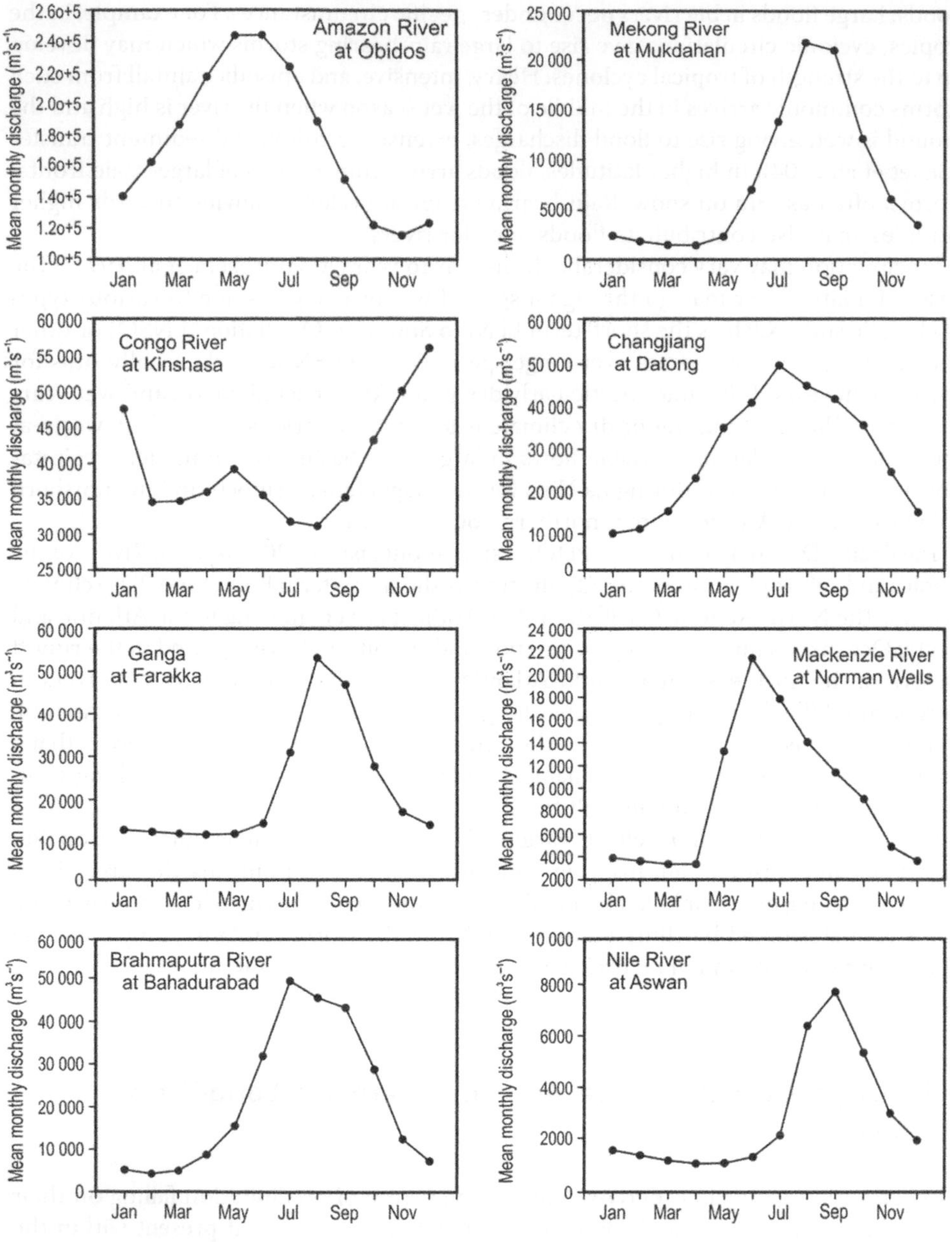

Figure 3.2 Average annual hydrograph of selected large rivers as examples: the Amazon, Congo, Ganga, Brahmaputra, Mekong, Changjiang, Mackenzie, and Nile. Source: Wohl 2007.

enhancement in the west. The average annual precipitation is around 2000–2500 mm of rain, geographically ranging from less than 2000 mm over the northeastern and southern basin to 7000–8000 mm over the lower slopes of the Peruvian Andes (Mertes and Dunne 2007 and references therein).

The rainy season starts in November and December in the southern basin but two months later in the northern part, the pattern followed by the discharge of the Amazon. The peak discharge is attained approximately in May–June, the date depending on the location of the gauging station. Mertes and Dunne (2007) described the annual pattern of discharge as unimodal and damped. The arrival of the rainwater in the Amazon is delayed by the length and size of the drainage network in the basin, and especially by temporary storage in more than 100 000 km^2 of the huge floodplain of the Amazon (see Chapter 5 for a detailed discussion). Temporal offsets also occur between the northern and southern tributaries of the Amazon (Mertes and Dunne 2007; Richey et al. 1989). The ENSO system controls the size of the annual discharge. The El Niño years tend to be drier in the basin, especially over the northern part. In contrast, higher rainfall and greater discharges happen in the La Niña years. Such climatic fluctuations not only affect discharge and inundation along the river but also erosive activities in the headwaters and subsequent sediment load of the Amazon.

The discharge pattern of the Congo, whose huge 3.74 million km^2 drainage basin stretches across the Equatorial region and beyond, illustrates the combined effect of (i) the annual movement of the ITCZ across the basin with (ii) the seasonal wet and dry regime of the headwater tributaries of the Congo coming from the north and south edges of the basin. Flow regimes are different at different locations of the main river. The Congo River receives 1800–2400 mm of rainfall over its middle part with no dry season. A distinct wet and dry distribution, however, characterises the discharge of the streams in the southern regions, as in the Shaba Highlands, which are sub-humid and located in the southern hemisphere. There a peak in discharge occurs between March and May and a conspicuous dryness between September and November. In comparison, a marked high discharge is reached between September and November and the lowest between February and April in the major tributary, the Oubangui River (also known as Ubangi), which drains part of the basin north of the Equator and joins the Congo below its middle section. This derivation from both equatorial and seasonal climates near the source and downstream makes the annual hydrograph of the Congo River complicated, as shown for Kinshasa with a double peaked hydrograph (Figure 3.2). The average annual discharge of the river is 46 200 $m^3\ s^{-1}$ in Kinshasa and the highest recorded discharge is 64 900 $m^3\ s^{-1}$. The variability of flow is swamped by the heavy annual rainfall that regularly falls over the central basin. As expected, variability in flow increases towards the headwaters, away from the Equator (Runge 2007 and references therein).

Moving away from the Equatorial basins, the Ganga, Brahmaputra, Irrawaddy, and Mekong provide excellent examples of monsoon-supported seasonal large rivers in South and Southeast Asia. Most of the annual rainfall arrives in the southwest monsoon with the rain-bearing system moving upstream from the coast. Both the Ganga and Brahmaputra also demonstrate (i) the effect of orographic lifting against the southern slopes of the Himalaya Mountains and (ii) a limited addition of meltwater in summer with melting of snow and ice on slopes.

In the Ganga Basin, the wet southwest monsoon is responsible for 70–80% of the annual rainfall (a figure that rises to 90% in places) between July and September. The average annual rainfall decreases westward over the basin, from about 1000 mm in Bengal to 500 mm in Uttar Pradesh and Haryana, then rising to about 2300 mm on the Himalayan slopes. The rain is seasonal and can be episodic and intense. The Ganga receives its flow from rainfall, subsurface flow, and meltwater from the glacier at the head of the river. Large contributions are added stepwise from various tributaries coming from both the Himalaya Mountain in the north and the peninsular craton to the south. Each square kilometre of the Ganga Basin annually receives 1 million m^3 of water as rainfall. Of this 30% is lost through evaporation, 20% seeps to the subsurface, and the remaining 50% is available as surface runoff (Das Gupta 1984). Annual flooding is typical of all rivers in the plain of the Ganga. Flooding is commonly caused by the arrival of episodic rain-bearing circulations arriving in the wet season when the ground is already wet and the channel full.

The drainage basin of the Brahmaputra is more complicated, being an assemblage of varying climatic and hydrologic zones (Singh 2007). It is cold and arid in Tibet in the rain shadow of the Himalaya, but humid tropical or subtropical in the rest of the basin. Outside Tibet, the main source of the discharge of the river is the wet southwestern monsoon, operating from June to September. The annual rainfall varies from a low of 300 mm in Tibet to about 5000 mm in the eastern basin where the river crosses the Himalaya in a 5075 m deep gorge around the Namcha Barwa Peak. Downstream the annual rainfall varies between 1000 mm and 2000 mm on the southern slopes of the Himalaya, and rises to 3000 mm in the Mishmi Hills of the eastern basin. Apart from the high annual precipitation and discharge with a marked seasonality, floods are very commonly caused in the Brahmaputra. Floods arrive with rainstorm systems in the middle of the wet monsoon season, and also are caused by tectonic disturbances. The enormous earthquakes of 1897 and 1950 (both with magnitude 8.7) partially blocked the river, giving rise to subsequent huge floods (Goswami 1985). Anthropogenic activities such as deforestation on upper slopes and poorly planned floodplain encroachments may also have aggravated the flood situation.

The Mekong, with a basin area of 795 000 km^2, is an example of a large river with seasonal discharge in Southeast Asia. It is a seasonal monsoon river which episodically floods in the rainy season. This 4880 km long river runs on rock through narrow mountainous valleys for the first 3000 km, and flows freely on alluvium only for the last 600 km in a wide lowland that converges to a major delta. The Mekong therefore illustrates the contrasting nature and behaviour of a seasonal large river, on both rock and alluvium. It used to be a natural river but currently is being modified with dams, reservoirs and various other engineering structures in its basin. Such changes are discussed in Chapter 10.

The 6300 km long Changjiang (Yangtze) rises at 6000 m on the snow-covered Tibetan Plateau and flows eastward to the East China Sea. A number of rainfall and gauging stations have recorded the monsoon-driven rainfall and seasonal discharge for this 1.80 million km^2 basin for years. Such information has been important for water management, especially concerning the recent construction of the Three Gorges Dam. A large proportion of rainfall over the basin is due to monsoon-driven precipitation from the warm air from the Pacific and Indian Oceans travelling up the valley between June and October.

Rain over the upper basin plus the snow and glacial melt on the Tibetan Plateau produce about half of the discharge of the river. The rest arrives mainly from the overflow of two lakes (Dongting and Poyang) in the middle Yangtze. Annual rainfall gradually increases downstream, from 400 mm in the upper basin to 1600 mm in the lower. The annual discharge of the river increases downstream: 1.4×10^4 m^3 s^{-1} at Yichang (4300 km from the source); 2.3×10^4 m^3 s^{-1} at Hankou near Wuhan (about 1000 km from Yichang); and 2.8×10^4 m^3 s^{-1} at Datong (about 700 km further east) (Chen et al. 2001 and references therein). The discharge follows the seasonal precipitation but is slightly damped. The wet season floods in the upper Changjiang are caused by the steep rivers of Sichuan. The common sources of floodwater in the middle Changjiang below the three gorges are the Han River from the north joining the Changjiang at Wuhan, and the overflow from the Dongting and Poyang Lakes downstream.

Over the middle latitudes baroclinic conditions prevail. Over the Mississippi Basin this gives rise to frontal storms which result in snowfall in winter. Convective storms also occur, mostly in summer. The huge 3.2 million km^2 Mississippi Basin is located generally in temperate climate but its eastern half is comparatively humid whereas the western part is relatively semiarid. Large differences occur in the basin between the winter and summer temperatures. Annual precipitation decreases east to west and also towards the north. The average annual totals vary: about 1400 mm near the mouth of the river in the south, more than 1000 mm in the east over most of the Ohio River basin, and less than 600 mm to the west in the basins of the Missouri and Arkansas Rivers. This results in a marked east–west reduction in runoff in the west of the Mississippi until the Rocky Mountains are reached. Higher runoff occurs to the east over most of the tributary basin of the Ohio River, even higher in the southern Appalachian Mountains. It is reduced towards the north, even lower towards the northwest basin. The runoff is less on the western Great Plains, but like precipitation, increases abruptly as the Rocky Mountains on the western boundary of the basin are approached. This disproportional distribution of precipitation and runoff is also reflected in episodic flood runoffs of the river (Knox 2007). Other large rivers of the middle latitudes are also maintained by combined flows of frontal rainfall, convectional summer rain, and the melting of glaciers.

The waters of five large arctic rivers (the Ob, Yenisei, and Lena in Eurasia and the Mackenzie and Yukon in North America) flow to the Arctic Ocean. Their runoff ranges between 250 mm and 500 mm and their drainage basins extend over a range of physiographic and bioclimatic zones. Apart from the mountains that occur within their catchment areas, their basins drain variations of tundra, taiga, mid-latitude forests, dry steppe, and semi-desert areas. All these rivers display a highly uneven seasonal pattern of runoff, primarily marked with melting of snow and ice giving rise to large spring floods. In general, the rivers are high from April to June, and low in summer and winter. The large rivers continue to flow through winter but not the smaller tributaries. Some of their discharge comes from rainfall but most of the runoff comes from snowmelt and melting of permafrost. Both the headwaters region in the mountains and the deltaic lowlands in the north, at the two ends of a river, remain frozen for several months. With higher temperature in summer, melting of permafrost, groundwater movement and landslides or bank failures occur in sequence. Permafrost is widespread in the Lena and the northern part of the Yukon Basin, but less in the other three. The annual range of Lena's runoff therefore is impressive, from a minimum of 366 $m^3\ s^{-1}$ to a maximum of 241 000 $m^3\ s^{-1}$ with a mean value of 16 530 $m^3\ s^{-1}$ (see Chapter 11). The range of discharge in these large arctic rivers, however, is commonly reduced by the presence of many lakes and reservoirs in their lower courses.

Significant future increases in discharge are expected on account of current climate change and warming of the temperature in the arctic region. This is happening in the Eurasian rivers despite the construction of a number of dams and reservoirs.

Several large rivers flow through arid landscapes but manage to sustain their flow because of the high discharge arriving from the upper non-arid parts of their drainage basins. The Indus, for example, maintains its long lower course through the arid area of Pakistan by seasonal discharge from snowmelt and orographic monsoon rain that falls in the mountains of its upper course. Other large rivers with a significant part of their drainage basins arid include the Nile, Colorado, Niger, and Murray-Darling. Commonly, water in these rivers is utilised by construction of dams and reservoirs, and as such requires careful management (Chapter 9).

The 6500 km long Nile is a well-known example. It rises as the White Nile in wet Central Africa from Lake Victoria and flows north for about 2700 km through the Sahara Desert without any significant water input. On the other hand, a high rate of evapotranspiration occurs in the wetlands of the Sudd. The annual rainfall decreases from near 2000 mm in the Lake Victoria area to about 175 mm at Khartoum. The White Nile is joined at Khartoum by the Blue Nile and further downstream by the Atbara. Both streams are seasonal and monsoon-fed from the mountains of Ethiopia. About 85% of the annual flow of the Blue Nile is concentrated between July and October. The Atbara is even more seasonal (Woodward et al. 2007). A seasonally flood prone Nile then flows through Egypt to build a fertile delta.

Milliman and Farnsworth (2011) estimated that the rivers of the world altogether discharge about 36 000 km^3 of water to the oceans annually. Given the pattern of global precipitation, rivers of northern South America and South, Southeast and East Asia contribute about half of this amount. Table 1.1 shows that large rivers of these regions provide most of this discharge. The discharge of the Amazon is particularly high, being 6300 km^3 per year, a figure comparable with the total annual discharge of the next eight large rivers.

3.5 Sediment in Large Rivers

Meade (2007) described large rivers as massive conveyance systems for moving clastic sediment and dissolved matter over transcontinental distances. To illustrate, the Amazon and Orinoco are large rivers that transfer sediment for thousands of kilometres from the active margin of South America to its passive edge – from the Andes and its forelands in the west to the low floodplains and deltas of the two rivers on the east coast. The sediment is transferred between the source and the sink in steps, being alternatively transferred down-channel and stored in the valley for years in extensive alluvial plains, huge floodplains, or channel bars.

The mountains and forelands of the Andes are essentially the sources of the sediment of these rivers and such sediment is recycled as it travels in the downstream direction. The source of virtually all the modern sediments in the rivers is the Andes itself. The tributaries of the Amazon which do not originate in the Andes or its foothills, contribute very little sediment to the Amazon. This may also be true for other large rivers that originate in young fold mountains. In their study of global sediment, Milliman and Syvitiski have associated sediment sources with high mountains and stated 'probably the entire tectonic milieu of fractured and brecciated rocks, oversteepened slopes, seismic, and volcanic activity, rather than simple elevation/relief .... promotes the large sediment yields from active orogenic belts.' (Milliman and Syvitiski 1992, p. 540). Meade (2007) discussed (Figure 3.1) the example of the Orinoco River which collects water discharge from most of the basin irrespective of geology, but sediment only from the Andes and

Llanos. Figure 3.1 also illustrates the progressive increase in discharges of water and suspended sediment downstream.

Examples of downstream sediment transfer are common. The Rio Madeira deposits a considerable amount of sediment as it emerges from the Andes Mountains to the lowlands of the Amazon. Some of this sediment is stored in the river but eventually moves downstream (Guyot et al. 1996; Dunne et al. 1998). About half of the sediment eroded from the Andes is deposited in the Andean foreland (Aalto et al. 2006) before removal. Before the Three Gorges Dam was closed, the Changjiang used to deposit nearly 100 million tonnes of sediment annually on floodplains and in lakes and stream channels between Yichang and Datong (Xu et al. 2007). The sediment was likely to have been derived from upstream mountains.

As the sediment emerges from the highlands, individual grains are as likely to be stored in the valleys of many large rivers as to travel downstream. When stored, they may remain at rest for a sufficiently long enough time to decompose in situ to a partly dissolved state which is removed by the river as solution load. The rest of the sediment remains in the solid state in the channel and is transferred downstream during high flows of the river as suspended load, or even as bed load if the grains are still big and the flow is powerful enough. Moving downstream via alternate storage and transfer, the sediment becomes mature in composition and over time may consist of more than 90% quartz, having lost rock fragments and feldspar by abrasion and decomposition. Of its maximum annual channel sediment load of about 1200 million tonnes at Óbidos, virtually all the suspended sediment of the Amazon comes from the Andes either via the main stem Amazon or one of its major tributaries, the Madeira, the headwaters of both starting deep in the Andes. The floods may take the sediment-laden water across the floodplain, deposit the sediment on the floodplain, and then during the falling-water season, return the clear water to the channel. The floodplain of the Amazon within Brazil measures about 90 000 km^2. The floodplains slowly grow vertically in floods and lose area by bank erosion associated with channel movement. These floodplains of Brazil (*várzea)* form a special environment with typical vegetation and animal life, strongly related to flood pulses. Junk et al. provide a detailed account of the physical and ecological characteristics of these floodplains in Chapter 5.

It is not uncommon for large rivers to transfer sediment both as downward flux and also sideways for floodplain-building (Goodbred and Kuehl 1998; Dietrich et al. 1999; Galy and France-Lanord 2001). The volume, however, is enormous for the Amazon River (Figure 3.3). If we consider the entire Amazon, more sediment moves laterally in and out of floodplains than as downstream flux (Meade 2007). Residence time for such sediment between the confluences of the Amazon with the Jutai and Madera has been estimated to be in the range of 1000–2000 years (Mertes et al. 1996). It could be longer, and thus sufficient time to change the proportion of sediment to nearly all quartz grains.

The lowermost gauging station on the Amazon is at Óbidos, about 1000 km above the sea. It is difficult to measure sediment deposition further downstream but apparently much sediment is deposited on floodplains and in partly filled floodplain lakes. Plumes of sediment are seen in satellite images entering the Atlantic from the Amazon mouth (Figure 3.4). Part of that sediment that passes through the sediment mouth settles on the ocean floor but the rest travels northeast along the coast to reach the Orinoco Delta. The outer Orinoco Delta includes more sediment from the Amazon than from the Orinoco.

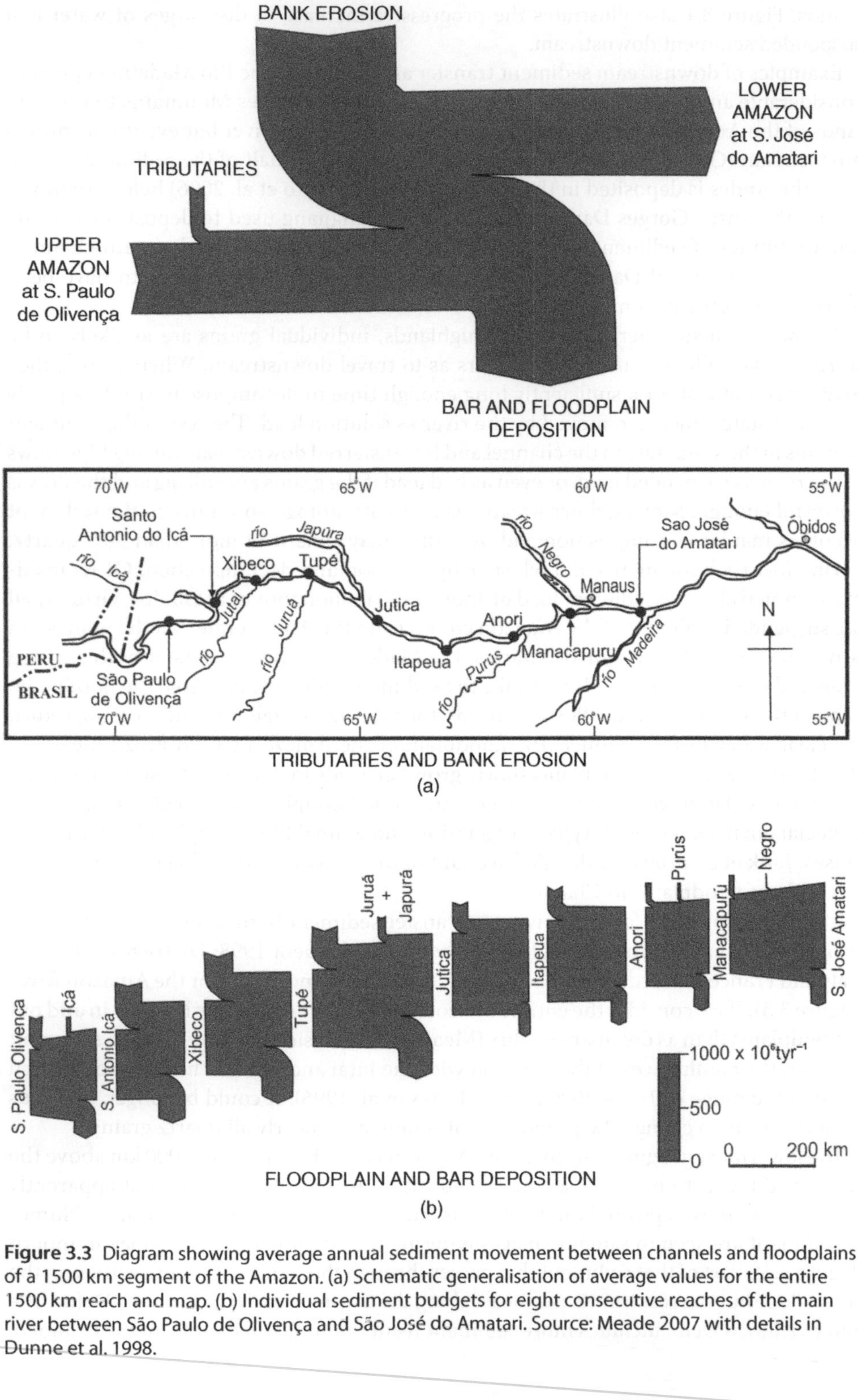

Figure 3.3 Diagram showing average annual sediment movement between channels and floodplains of a 1500 km segment of the Amazon. (a) Schematic generalisation of average values for the entire 1500 km reach and map. (b) Individual sediment budgets for eight consecutive reaches of the main river between São Paulo de Olivença and São José do Amatari. Source: Meade 2007 with details in Dunne et al. 1998.

Figure 3.4 Satellite image showing plumes of sediment entering the Atlantic from the Amazon mouth and then moving northeast along the coast to build part of the Orinoco Delta. Source: NASA worldview application (https://worldview.earthdata.nasa.giv), part of the NASA Earth Observatory System Data and Information System (EOSDIS).

The Amazon probably provides the best example of large-scale sediment transfer but similar processes of sediment transfer and storage occur in other large rivers too.

Depending on texture, fluvial sediment can be transported as dissolved load in solution, in suspension through the water column, and as bed load moving in traction along the river bottom. Dissolved load is finer than 0.62 μm. Suspended load, which is coarser than 0.62 μm, is often described in two parts: wash load (the finer fraction of suspended load that nearly always remains in suspension) and bed material load (the material which is carried by eddies only in high flow). The coarser part of the suspended load can be picked up from the bed and transported in suspension when the power of the river increases. Sand for example can be transported both as suspended and bed load. The coarsest material is usually moved discretely along the bed in very high flows, coming to rest on the bed at other times. Pebbles are dragged and rolled downstream as bed load but have been noted to travel in suspension in very big floods. Very coarse material such as cobbles and boulders are carried as bed load, except in extreme floods. Such catastrophic floods occurred when glacial lakes burst in overflows towards the end of the Pleistocene (Baker 1981, 2007). Very large floods, for example those caused by tropical cyclones, have been known to suspend pebbles and cobbles for short distances. This rarely happens, and boulder, cobbles, and pebbles are carried usually as bed load. Sand usually travels as bed load but in floods is often carried suspended by eddies. Silt and clay are transported as suspended load unless in aggregates. When separated, grains of such fine material may be transported either as suspended or dissolved load.

Data for clastic river sediment often only include total suspended load. Bed load is very difficult to measure. For many meandering rivers it is only a few per cent of the total load and considered to be within the margin of error for sampling (Milliman and Meade 1983 citing C. Nordin). This assumption probably does not work for sediment supplied by small mountainous streams or where large rivers cut through mountain ranges. The Brahmaputra, for example, is considered to be a river which carries a higher percentage (probably 30%) of bed load (Goswami 1985; Best et al. 2007).

The removal (erosion) of particles is a function of stream velocity. The critical erosion velocity of a particle is the velocity at which it starts to move. Once sediment grains are entrained, they can be transported with a velocity lower than that for entrainment. A sediment grain being carried either on bed or in water comes to rest (deposition) when the velocity of the river falls below the value needed to carry it. This is the fall velocity, related directly to grain size. Sand is the easiest to erode from the channel perimeter and bars. Pebbles, cobbles, and boulders require a higher critical erosion velocity but as they are bigger and heavier, requiring high fall velocity, they can be transported only for a short time or distance. Sand is carried longer. The critical erosion velocity of silt and clay, sticky and forming aggregates, is higher than sand, but once suspended in water they are transported for a long time and distance (Hjulström 1939).

This pattern of river transport leads to a sorting of material downstream, finer material travelling longer. In large rivers with room for deposition inside the valley, such sorting also happens across large bars and floodplains (Figure 3.5). The channel of a long river therefore displays pebble, cobbles, and boulders near the mountains and silt and clay near the sea. Sand is ubiquitous, and because of sorting and weathering on bars and floodplains increases in proportion along the channel. The modal size of sand grain, however, decreases downstream. This general pattern persists except where major

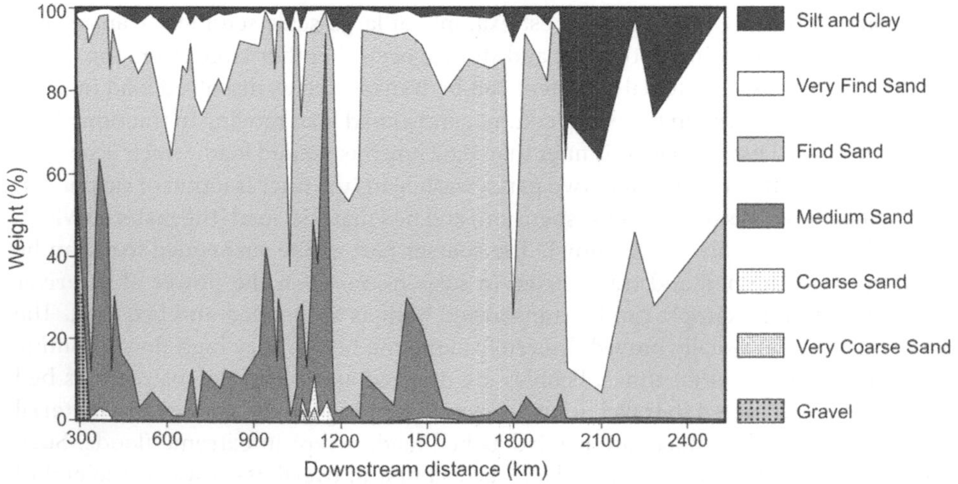

Figure 3.5 The Ganga River. Changes in the grain size of bar material from Hardwar in the Himalayan foothills to Ganga Sagar on the delta. The coarsening of the bars in the middle reach is due to the contribution of southern rivers draining the Indian Peninsular. Source: Singh 2007.

rivers cross erodible fills or are joined by short tributaries bringing coarse sediment. The anomaly is corrected over a stretch below the confluence with the tributary downstream along the main river. The formation and development of floodplains and sorting of sediment grains across them is discussed by Junk et al. in Chapter 5.

The total volume of sediment per unit time is considered as sediment load or sediment discharge. Sediment yield is the total load of the river divided by the upstream basin area. This assumption implies uniform load shedding from all parts of the basin which is incorrect as a very large part of the sediment on large rivers may come from the headwaters with high relief. Furthermore, the eroded sediment in the basin is not always transferred efficiently. Only about 10% of the total eroded sediment in the conterminous United States may reach the ocean (Milliman and Farnsworth 2011, referring to Holeman). Wasson et al. (1996) indicated that only about 1% of the entire eroded soil mass reaches the sea in Australia. Sediment discharge also varies with time, changes in vegetation cover, and anthropogenic alterations of the environment. The question of reliability is more relevant for sediment than water discharge. Milliman and Farnsworth (2011) opined that rounded figures are safer to use, attempted precise measurements are likely to be less accurate.

It is difficult to prioritise all the factors behind erosion and sediment supply to large rivers. Certain factors have been discussed by geomorphologists, such as relief, intensity and amount of rainfall, water discharge, the weathered nature of country rock, etc. (Milliman and Farnsworth 2011 and references therein). Numerical models have been proposed to compare the relative importance of such environmental factors for sediment discharge. For example, Syvitski and Milliman (2007) opined that geological factors explain 65% of the variation in sediment load, whereas climate and anthropogenic factors account for another 30%. The importance of these factors, however, vary among rivers, and anthropogenic modifications can significantly modify the natural pattern. For example, a series of dams have considerably reduced the volume of sediment that used to flow into the Mississippi River from the basin of its west bank tributary, the Missouri (as discussed in Chapter 8). Hovius and Leeder (1998) discussed the difficulty of establishing a reliable universal relationship between certain characteristics of drainage basins and sediment production. The difficulty arises mainly because of the varying importance of a series of tectonic, climatic and geomorphic processes, all three working in an integrated fashion to determine the sediment of a drainage basin.

The data sets used for these conclusions may include measurements from hundreds of rivers but not exclusively from large rivers. We therefore may not only need to prioritise certain basin properties for all rivers but also determine the relative contribution of individual smaller rivers and sum them to construct the total discharge of a specific large river. High young fold mountains, such as the Himalaya or Andes, are directly associated with high sediment discharge because of tectonics. Older ranges such as the Rockies or the Urals produce less sediment because of lack of tectonics and hardness of the older rocks. The orographic effect on precipitation adds to enhanced discharge increasing both runoff and sediment discharge of such streams. Briefly, the sediment discharge of large rivers increases directly with relief, lithology, tectonics, precipitation, and basin area.

According to Milliman and Farnsworth (2011) most large rivers with high dissolved loads have large drainage basins, drain high mountains, and carry a high runoff. Their list of a dozen major rivers with the highest dissolved loads includes six Himalayan rivers:

the Changjiang, Irrawaddy, Ganga, Mekong, Salween, and Brahmaputra. The Amazon, Mississippi, Danube, MacKenzie, Parana, and St. Lawrence complete the list. In contrast, a large river may carry very little dissolved load, given its basin geology and low precipitation. Certain major river basins are dominated by a single lithology. The Zhujiang drains about 80% carbonate rocks whereas more than 80% of the basin of the Yukon is on shale. More than half of the St Lawrence basin is on shield rocks (Amoitte-Suchet et al. 2003). In brief, sediment yield of large rivers depends on several environmental factors: basin elevation, tectonics, lithology, precipitation, and basin area. All these factors determine the nature of sediment load a large river would carry.

3.6 Conclusion

Existence of a large river requires a considerable amount of water and sediment. For this, its drainage basin needs to be big enough to collect a necessary volume of water, large enough to nourish the main stream. At least part of the basin should be in an area of high rainfall so that the main river can be sustained all along its course. The volume of discharge may be annually seasonal, and also may vary over a group of years as directed by climate drivers such as ENSO. The drainage basin may contain sources of high sediment production which commonly comes from a high relief and erosive fold mountain. Sediment grains in long rivers tend to demonstrate size-sorting and quartz enrichment in the downstream direction. Many large rivers transport and store sediment also in a lateral direction, producing dynamic growth and decay of floodplains. Very large volumes of water and sediment are generally needed to sustain large rivers but the patterns of their supply and distribution are different.

Questions

1 Explain the difference between discharge and runoff.

2 What are the sources of big discharges of large rivers? Give examples.

3 Explain the difference between barotropic and barclinic conditions. How do they relate to large rivers?

4 What is a climate driver? How does a short-term climate driver relate to floods and dry conditions?

5 Name five large rivers that flow through arid environments. How do they make it possible?

6 Explain the lateral transport and storage of sediment in large rivers. What effect does it have on floodplains?

7 Describe the changes in sediment of a large river moving downslope.

References

Aalto, R., Dunne, T., and Guyot, J.L. (2006). Geomorphic control on Andean denudation rates. *Journal of Geology* 114: 85–99.

Amarasekera, K.N., Lee, R.F., Williams, E.R., and Eltahir, E.A.B. (1997). ENSO and the natural variability in the flow of tropical rivers. *Journal of Hydrology* 200: 24–39.

Amoitte-Suchet, P., Probst, J.-L., and Ludwig, W. (2003). Worldwide distribution of continental rock lithology: implications for the atmospheric/soil CO_2 uptake by continental weathering and alkalinity river transport to the oceans. *Global Biogeochemical Cycles* 17: 1–13.

Baker, V.R. (1981). *Catastrophic Flooding: The Origin of the Channeled Scabland.* Stroudsburgh, PA: Hutchinson Ross.

Baker, V.R. (2007). Greatest floods and largest rivers. In: *Large Rivers: Geomorphology and Management* (ed. A. Gupta), 65–74. Chichester: Wiley.

Best, J.L., Ashworth, P.J., Sarkar, M.H., and Roden, J.E. (2007). The Brahmaputra-Jamuna River, Bangladesh. In: *Large Rivers: Geomorphology and Management* (ed. A. Gupta), 395–433. Chichester: Wiley.

Chen, Z., Li, J., Shen, H., and Wang, Z. (2001). Yangtze River of China: historical analysis of discharge variability and sediment flux. *Geomorphology* 41: 77–91.

Das Gupta, S.P. (1984). *The Ganga Basin, Part 1.* New Delhi: Central Board for the Prevention and Control of Water Pollution.

Dietrich, W.E., Day, G., and Parker, G. (1999). The Fly River, Papua New Guinea: inferences about rive dynamics, floodplain sedimentation and fate of sediment. In: *Varieties of Fluvial Form* (eds. A.J. Miller and A. Gupta), 346–376. Chichester: Wiley.

Dunne, T., Mertes, L.A.K., Meade, R.H. et al. (1998). Exchanges of sediment between the floodplain and channel of the Amazon River in Brazil. *Geological Society of America Bulletin* 110: 450–470.

Gabet, F.J., Burbank, D.W., and Putkonen, J.K. (2004). Rainfall thresholds for landslides in the Himalayas of Nepal. *Geomorphology* 63: 131–143.

Galy, A. and France-Lanord, C. (2001). Higher erosion rates in the Himalaya: geochemical constrains on riverine fluxes. *Geological Society of America Bulletin* 29: 23–26.

Goodbred, S.L. Jr. and Kuehl, S.A. (1998). Floodplain processes in the Bengal Basin and the storage of Ganges-Brahmaputra River sediment: an accretion study using 137 Cs and 210Pb geochronology. *Sedimentary Geology* 121: 239–258.

Goswami, D.C. (1985). Brahmaputra River, Assam, India: physiography, basin degradation, and channel aggradation. *Water Resources Research* 221: 858–878.

Gupta, A. (2011). *Tropical Geomorphology.* Cambridge: Cambridge University Press.

Guyot, J.L., Filizola, N., Quintanilla, J., and Cortez, J. (1996). Dissolved solids and suspended yields in the Rio Madeira basin from the Bolivian Andes to the Amazon. *IAHS Publication* 236: 55–63.

Hayden, B.P. (1988). Flood climates. In: *Flood Geomorphology* (eds. V.R. Baker, R.C. Kochel and P.C. Patton), 13–26. New York: Wiley.

Hjulström, F. (1939). Transportation of detritus by moving water. In: *Recent Marine Sediments: A Symposium* (ed. P. Trask). Tulsa, PK: American Association of Petroleum Geologists.

Hovius, N. and Leeder, M. (1998). Clastic sediment supply to basins. *Basin Research* 10: 1–5.

Knox, J.C. (2007). The Mississippi River system. In: *Large Rivers: Geomorphology and Management* (ed. A. Gupta), 145–182. Chichester: Wiley.

Lamb, H.H. (1972). *Climate: Past, Present and Future*, vol. 1. London: Methuen.

Meade, R.H. (2007). Transcontinental moving and storage: the Orinoco and Amazon Rivers transfer the Andes to the Atlantic. In: *Large Rivers: Geomorphology and Management* (ed. A. Gupta), 45–63. Chichester: Wiley.

Mertes, L.A.K. and Dunne, T. (2007). Effects of tectonism, climate change, and sea-level change on the form and behaviour of the modern Amazon River and its floodplain. In: *Large Rivers: Geomorphology and Management* (ed. A. Gupta), 115–144. Chichester: Wiley.

Mertes, L.A.K., Dunne, T., and Martinelli, L.A. (1996). Channel-floodplain geomorphology along the Solimoes-Amazon River, Brazil. *Geological Society of America Bulletin* 108: 1088–1107.

Milliman, J.D. and Farnsworth, K.L. (2011). *River Discharge to the Coastal Ocean: A Global Synthesis*. Cambridge: Cambridge University Press.

Milliman, J.D. and Meade, R.H. (1983). World-wide delivery of river sediment to the ocean. *Journal of Geology* 91: 1–21.

Milliman, J.D. and Syvitski, J.P.M. (1992). Geomorphic/tectonic control of sediment discharge to the ocean: the importance of small mountainous rivers. *Journal of Geology* 100: 525–544.

Potter, P.E. (1978). Significance and origin of big rivers. *Journal of Geology* 86: 13–33.

Richey, J.E., Mertes, L.A.K., Dunne, T. et al. (1989). Sources and routing of the Amazon River flood wave. *Global Biogeochemical Cycles* 3: 191–204.

Runge, J. (2007). The Congo River, Central Africa. In: *Large Rivers: Geomorphology and Management* (ed. A. Gupta), 292–309. Chichester: Wiley.

Singh, S.K. (2007). Erosion and weathering in the Brahmaputra River system. In: *Large Rivers: Geomorphology and Management* (ed. A. Gupta), 373–393. Chichester: Wiley.

Syvitski, J.P.M. and Milliman, J.D. (2007). Geology, geography, and humans battle for dominance over the delivery of fluvial sediment to the coastal ocean. *Journal of Geology* 115: 1–19.

Wasson, R.J., Olive, L.L., and Rosewall, C.J. (1996). Rates of erosion and sediment transport in Australia. *IAHS Publication* 236: 139–148.

Wohl, E.E. (2007). Hydrology and discharge. In: *Large Rivers: Geomorphology and Management* (ed. A. Gupta), 29–44. Chichester: Wiley.

Woodward, J.C., Macklin, M.G., Krom, M.D., and Williams, M.A.J. (2007). The Nile: evolution, Quaternary river environments and material fluxes. In: *Large Rivers: Geomorphology and Management* (ed. A. Gupta), 261–292. Chichester: Wiley.

Xu, K.H., Milliman, J.D., Yang, Z., and Xu, H. (2007). Climatic and anthropogenic impacts on water and sediment discharges from the Yangtze River (Changjiang), 1950–2005. In: *Large Rivers: Geomorphology and Management* (ed. A. Gupta), 609–626. Chichester: Wiley.

4

Morphology of Large Rivers

4.1 Introduction

What are the characteristics of a large river? Meade (2007) has described large rivers as 'massive convenience systems for moving detrital sediment and dissolved matter across transcontinental distances'. This implies a very long river, say 1000 km or longer, that carries a huge volume of water and a large sediment load of several types from the land to the sea.

Many smaller rivers have been studied in detail but only a limited number of large rivers. The biggest, the Amazon, has been observed in detail (Junk 1997; Dunne et al. 1998; Mertes and Dunne 2007), and also several others, particularly the Mississippi (Knox 2007 and references therein). Detailed morphological descriptions and relevant hydrologic and sedimentological data, however, are not available for all. We know a lot less about certain rivers such as the Congo or Salween.

This raises several questions. Is there a continuum among rivers of different size, i.e. does a scale invariance in form and behaviour exist between small and large rivers or do large rivers have special properties? Do large rivers, immense in form and complicated in function, record evidence of crustal deformation, hard rock erosion, and water and sediment fluxes over long time periods, as Mertes and Dunne (2007) have shown for the Amazon, which could not be studied in smaller streams? Does the presence of a large river imply a geological history and a climatic condition as described in Chapters 2 and 3 of this book? Do large rivers last longer than streams of smaller dimensions? Do the effects of sea-level changes remain recorded in detail over a long stretch of the lower section of a continental-scale river because the gradient is very low? Do floods and floodplains play an important role regarding the morphology and ecology of large river valleys? Can huge sedimentary alluvial deposits/bars of large rivers such as the Brahmaputra (Figure 4.1) be recognised in later sedimentary rocks (Fielding accepted for publication)? In this chapter we explore these and other queries, and describe the general morphology and behaviour of big rivers.

4.2 Large Rivers from Source to Sink

Imagine a large river that rises in a folded mountain range and flows to the sea and has a gentle offshore gradient (Figure 4.2). The tectonically active mountains of high relief tend to produce large quantities of rock fragments from seismic movements, glaciation,

Introducing Large Rivers, First Edition. Avijit Gupta.

Figure 4.1 At least 10 m high midchannel bar in the Brahmaputra at Sirajganj, Bangladesh; locally known as the Jamuna. A high volume of sand is moved in the wet season and stored during the dry monsoon. Large-scale sedimentary structures are visible. The top of the bar is at several levels. Source: Gupta 2007.

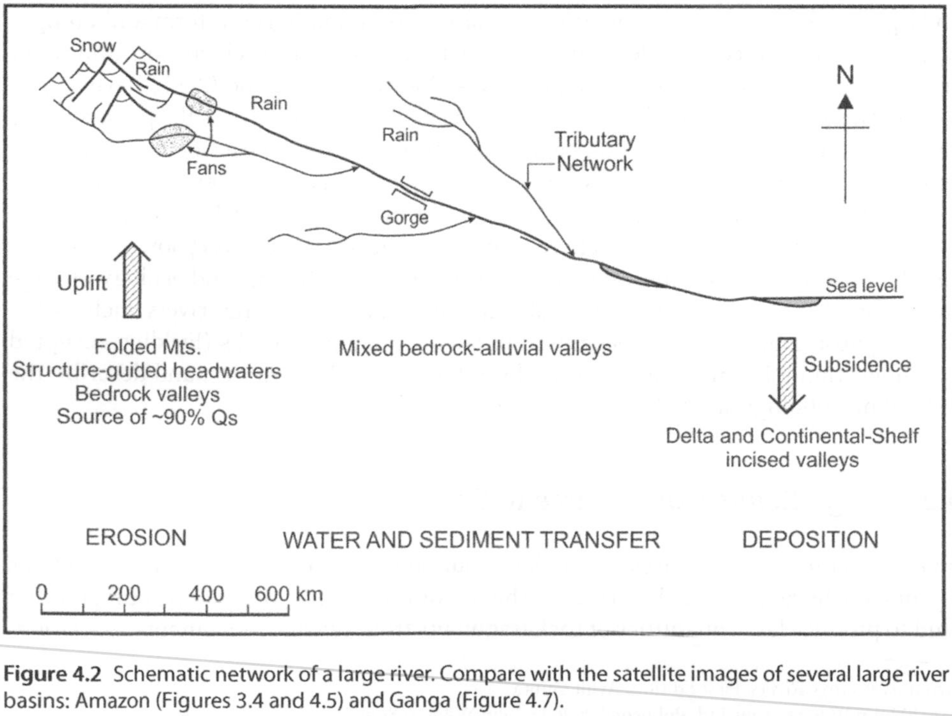

Figure 4.2 Schematic network of a large river. Compare with the satellite images of several large river basins: Amazon (Figures 3.4 and 4.5) and Ganga (Figure 4.7).

slope failures, and erosion by steep headwaters. The mountainous headwaters of the large river tend to flow in a pattern of deep valleys, are often structure-guided, have steep gradients, and their bed and banks are constituted of coarse sediment and rock. The combination of steep gradient and pulses of high discharge associated with orographic precipitation flush a volume of coarse sediment out of the mountains along river channels. As the rivers exit the mountains, their slopes sharply decrease, causing them to deposit sediment and build triangles of alluvial deposits called fans near the highland-lowland contact (Box 4.1). These fans mask the sharp contact between the mountains and the plain and reduce the gradient of the river. The large trunk river continues for a long distance beyond the fan, over a gentler gradient. Tributaries join it at intervals across the plain, increasing its discharge and sediment load. Finally, the river flows through a set of deltaic distributaries into the sea.

This is a short and simplified description. Form, behaviour and sediment of large rivers, however, vary among themselves. Two rivers, the Amazon and Ganga, are described to establish a general picture and several common properties. Several other rivers are then discussed, to highlight an expected assemblage of form and behaviour of large rivers and variations from such expected outcomes. Form and behaviour of a large river is the combined result of such an expected outcome and deviations therefrom.

Box 4.1 Rivers on Alluvial Fans

Sediment yield from fold mountains are characteristically very high (e.g. 10^4 $\text{tkm}^{-2}\ \text{yr}^{-1}$) (Douglas and Guyot 2005). The nature of the sediment depends on the regional geology which in a tectonic mountain may include material from active subduction zones, collision belts, volcanic arcs, granitic plutons, rift valleys and uplifted sedimentary rocks (Scatena and Gupta 2013). A significant amount of the sediment is deposited at the foot of the mountains to form alluvial fans. The upper Amazon and its tributaries have built very large alluvial fans at the foot of the Andes. There the material derived from the mountains after deposition is stored for a long time (10^2–10^3 years), and progressively weathered into a mature sediment enriched in quartz grains before the grains are removed downstream. The sediment of the Amazon River downstream is derived almost entirely from the Andes and these fans. A series of such large alluvial fans extend along fold mountains. The especially large ones are referred to as megafans as they differ from the common alluvial fans by their large size (10^3 km^2 or greater), low gradient (commonly $<0.1°$), typical sorted sedimentary texture (boulders at the apex and predominantly silt and mud at toes), and depositional processes (mostly by wandering channels of a single river) (Leier et al. 2005). Megafans are important source of sediment from tectonically active areas in the seasonal tropics (Figure 4.3).

Emerging from the mountains to the plain, The Ganga and its Himalayan tributaries also have built alluvial megafans. The best known of the Himalayan megafans is that of the Kosi River which covers 154 × 147 km in area, and slope 0.89–0.025 m km^{-1} longitudinally from the Himalayan front to the Ganga River. The climate of the region is affected by the Indian monsoon system. The regional average annual rainfall is between 1300 mm and 1800 mm, 80% of which occurs between June and October. Individual large storms are superimposed on the general rainfall pattern. The Kosi fan consists of large volumes of clastic sediment derived from the Himalaya. On the upper part of the fan, the Kosi River

(Continued)

Box 4.1 (Continued)

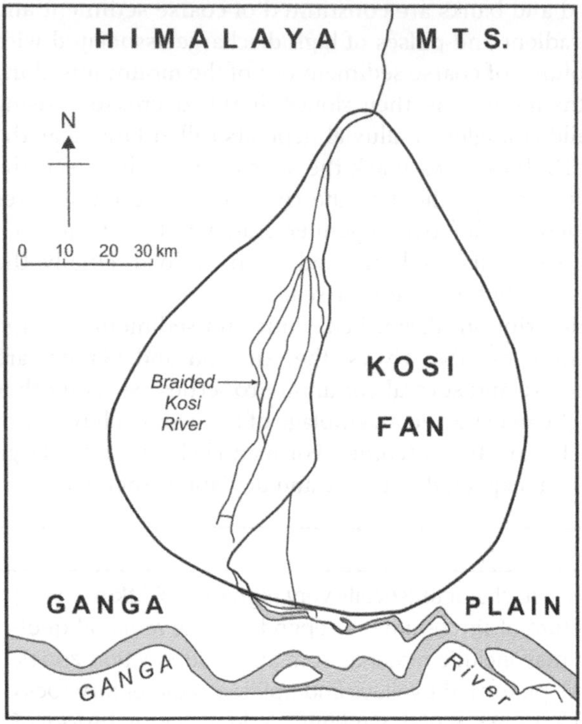

Figure 4.3 Schematic sketch of the megafan of the Kosi River.

braids over boulders and sand; it has a straighter channel over the sand of the middle fan, and a long meandering reach occurs over fine sand and mud of the lower fan. A tropical large river like the Amazon or Ganga or its major tributaries crosses the fan below the mountains in a channel which is not particularly mobile. The entire length of the fan river, however, may avulse in specific tectonic movements or episodic floods. Its course may also shift randomly during high flow, and the fan becomes a principal source of water and sediment to the trunk river skirting the fan below (Chakraborty et al. 2011; Gole and Chitale 1966; Singh et al. 1993; Wells and Dorr 1987).

4.3 The Amazon River

Big rivers of the world have been long utilised as a natural resource, and in the process many of them have been physically modified by anthropogenic activities (Chapter 8). The Amazon, however, remains in near-natural conditions, unmodified by anthropogenic activities. This summary of the geomorphology of the river is based primarily on Dunne et al. (1998), Mertes and Dunne (2007), and Meade (2007). The ecology of its floodplain is discussed in detail by Junk et al. in Chapter 5 of this book.

4.3.1 The Setting

The approximately 6000 km long Amazon and its tributaries drain a river basin of about 7 million km^2 (Figure 4.4). Its headwaters originate in the folded mountain belt of the Andes in the west and its present east-flowing course post-dates the Miocene uplift of the ranges. Descending from the Andes, the Amazon crosses a downwarped foreland basin, and then a vast lowland, before passing between two cratonic highlands (Guyana Shield to the north and Brazil Shield to the south) to reach the low trailing eastern coast of South America on the Atlantic Ocean (Figures 4.4 and 4.5). Plate tectonics control the location and lithologic and topographic frameworks of the Amazon (Potter 1978). The basin was delineated following the Miocene uplift of the tectonically and volcanically active Andes due to the subduction of the Nazca Plate below the South American Plate. Exploration for petroleum has revealed the underlying structure along the channel and floodplain of the Andes (Mertes and Dunne 2007). Evidence from deep cores indicates an east–west crustal sag underneath the basin axis at a depth of 6000 m that links with a graben, the Marajó Rift, roughly located underneath the mouth of the river. It is a very large river with low gradient and limited power (Baker and Costa 1987), even at the average peak discharge (12 W m^{-2}), which generally carries a huge amount of sediment, but finer than about 0.5 mm.

4.3.2 Hydrology

The discharge of the Amazon comes essentially from precipitation although some snowmelt is derived from the Andes. The precipitation pattern of the region is controlled by the annual shifting of the Intertropical Convergence Zone (ITCZ) and the South Atlantic Convergence Zone (SACZ). Average annual basin precipitation is about 2000–2500 mm, which is near-uniformly distributed over most of the basin. The maximum precipitation of 7000–8000 mm falls on the lower eastern slopes of the Peruvian Andes and, in contrast, the extreme northern and southern parts of the basin are relatively dry. The rain arrives first over the southern basin in November to December, and then moves north. The annual hydrograph of the river (Figure 3.2) is unimodal and damped. The river floods regularly (Chapter 5), but the rising of the river stage is slowed by the sheer size of the basin, length of the drainage network, and storage of water in the enormous floodplains which has a cumulative size approaching 100 000 km^2. The precipitation is affected by El Niño Southern Oscillation (ENSO) and so is the discharge. Low flows of the Amazon occur in the El Niño years (Mertes and Dunne 2007). Such climatic fluctuations affect flooding and sedimentation (Aalto et al. 2003) but their effect on morphology and behaviour of the regional smaller rivers is yet to be understood.

4.3.3 Sediment Load

Only 800 000 km^2 of the Andes and Sub-Andes falls within the 7 million km^2 Amazon Basin but this area contributes almost the entire sediment load of the river. Throughout the Late Cenozoic, 500–600 million tonnes of sediment arrived annually from this

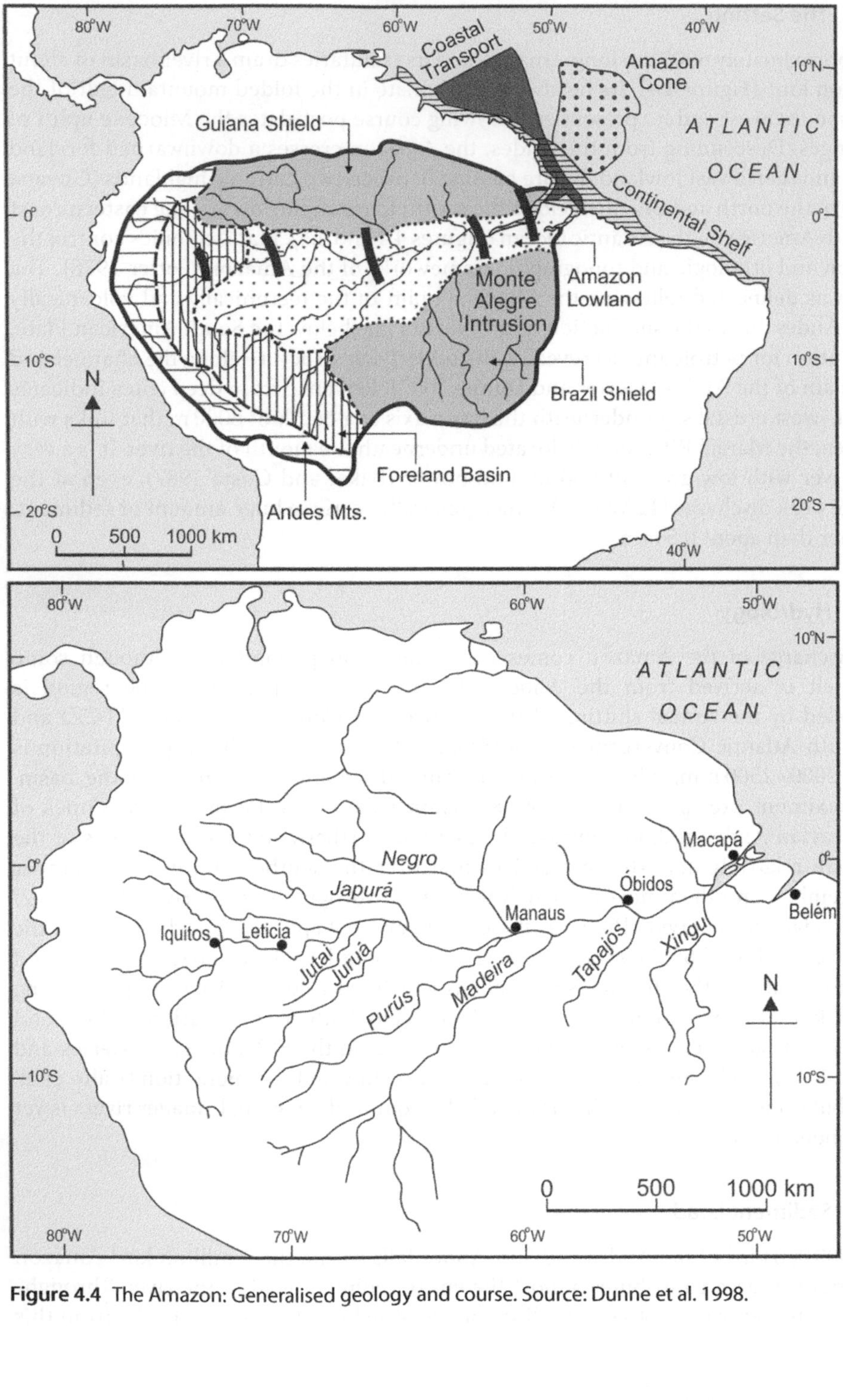

Figure 4.4 The Amazon: Generalised geology and course. Source: Dunne et al. 1998.

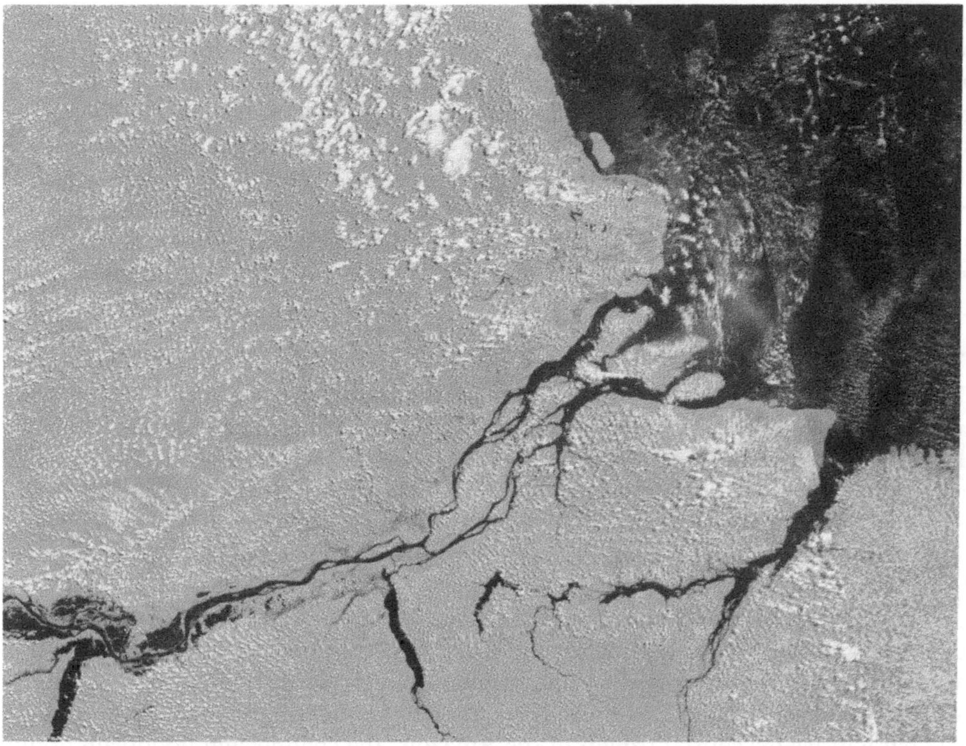

Figure 4.5 The Amazon from satellite imagery. Source: NASA Worldview application (https://worldview.earthdata.nasa.gov), part of the NASA Earth Observing System Data and Information System (EOSDIS).

source, although the foreland basins trapped approximately half of the amount to build fans (Mertes and Dunne 2007, referring to Guyot 1993). The foreland basins sag in response to the rise of the Andes, and its erosion, reducing the river gradients to very low figures. Only fine sediment (<0.5 mm) reaches the lowland Amazon flowing within Brazil. No significant sediment is supplied from the old rocks of the Brazil and Guyana shields, but from there a huge volume of water reaches the river. In brief, the main stem discharge of the Amazon is augmented cumulatively by a number of tributaries draining different parts of the basin, but the sediment supply comes almost exclusively from mountains of high relief at the head of the basin with tectonically fractured rocks and oversteepened slopes as described by Milliman and Syvitski (1992). Almost the entire sediment arrives either from the Peruvian Andes along the Amazon main stem or from the Bolivian Andes via the Madeira (Meade 2007). A number of large tributaries have been described as clearwater streams, rising below the Andes they bring little sediment to the Amazon.

The average annual sediment load of the Amazon measured at Óbidos, approximately at the tidal limit, is about 1200 million tonnes. A bigger load is carried only by the combined flow of the Ganga and Brahmaputra. However, as calculated by Dunne et al. (1998), who studied sediment transport though the 2010 km of the Amazon in lowland Brazil, a higher amount of sediment is transferred laterally between the channel and the floodplain, and then passed downstream (Figure 3.3). The lateral exchange of sediment involves bank erosion, bar deposition in the main channel, settling from overland flow on the floodplain, and sedimentation in channels within the floodplain. Much of the sediment that leaves the channel in suspension during floods and enters the floodplain is deposited there before clearer water returns to the Amazon during the falling stage of the annual hydrograph. Mertes et al. (1996) estimated that in the reach between the confluences of the Jutai and Madeira with the Amazon, the mean recycling time is between 1000 years and 2000 years. This allows progressive enrichment of quartz grains downstream.

4.3.4 Morphology

The axial trough of the central Amazon Basin exhibits a remarkable suite of fluvial landforms. The lowland Amazon has a straight and anastomosing channel within its floodplain. The main channel of the Amazon in Brazil has a sinuosity of 1.0–1.2 for most of the course. Over a measured length of 2000 km, the low water width of the river increases from 2 to 4 km and the corresponding depth from 10 to 20 m (Mertes and Dunne 2007). The width of the flooded Amazon is much bigger and the morphology and ecology of the river in flood is discussed in Chapter 5.

The regional pattern of the channel of the Amazon depends on its changing discharge, sediment, and slope. Steep channels in bedrock and gravel characterise the valleys in the mountains. In the downwarped foreland zone with high sedimentation, rivers begin to build bars and shift smaller channels. Low-gradient meandering channels in fine material characterise the central trough of the Amazon. The channel and floodplain of the Amazon are incised into the central low trough of the basin displaying a complex pattern of channels of various size, scroll bars and levees, and lakes. Towards the east, tributaries are dammed by alluvium of the trunk river, forming characteristic river-mouth lakes. The gradient of the lower river being very gentle, tidal effects extend about 1000 km up the Amazon to Óbidos.

Beyond the Andes foreland, the floodplain of the Holocene Amazon lies below a landscape of low hills under thick forest cover. The forest is interspersed with savanna, and recent deforestation is visible towards the northern and eastern margin of the basin. The channel and the floodplain continue between discontinuous terraces which are about 5–15 m above the general flood surface. The Holocene sediment in the channel and the floodplain consists of medium sand and finer sediment, weathered to clay minerals and enriched quartz grains (Johnsson and Meade 1990).

The river tends to undercut the cohesive terrace above the floodplain at channel bends. Avulsions related to flow switching are common. A dense network of channels cut across the floodplain. Some of these channels are as big as the large tributaries, and floodplain channels about 100 m wide are seen everywhere. The Amazon, its anabranches, and floodplain channels, all shift. As a result, numerous scroll bars and depressions mark the surface of the floodplain. Mertes and Dunne (2007) summarised the Amazon as an entrenched river that is confined to its valley, remains straight, and is relatively immobile over hundreds of kilometres. Large-scale channel migrations or avulsions are common. This huge low-gradient river flows through a 40–50 km wide floodplain decorated with considerable complexity of anabranches, levees, and scroll bars. A complex mosaic of lakes, lake deposits and overbank sedimentation is commonly found on the wide floodplain. The smaller lakes could be due to small low-sediment tributaries being blocked by rapid floodplain alluviation of the Amazon. The lakes increase in size downstream (Figure 3.4).

This general description of the morphology of the Amazon may vary in places. Although the Amazon flows on top of thick sedimentary layers along its central axis, it is still influenced by buried transverse structures (Figure 2.2). Four major structural arches (Iquitos, Jutai, Purús and Gurupa) and the Monte Alegro Intrusion, occur underneath the sedimentary layers and modify the river near their location. The Amazon crosses the structural highs on a straightened course. The water surface steepens slightly, the floodplain narrows, the river tends to hug the foot of the terraces, scroll bars are found only at channel margins, and channel migration becomes less common. Gravity measurements for the Lower Amazon area show a change in the direction of the river and its form with gravity anomalies (Nunn and Aires 1988), indicating that the Amazon crosses its floodplain only in specific places. Structure controls even the biggest river in the world.

Many examples of tectonic control are found In the Amazon and its tributaries. For example, Franzinelli and Igreja (2002) associated the alignment of the lower Negro River (which joins the Amazon at Manaus) with a NW-SE tectonic lineament. Sunken crustal blocks and depressions have been created by intersecting a set of blocks along a half graben that is inundated to the depth of 20 m across a 20 km reach of the river. The Negro carries very little sediment and this small amount is flushed through the fault-lined depression with accumulation of only little sand bodies along river borders. The banks of the Negro are in cohesive cliffs of lacustrine deposits, and bedrock is exposed in river channels because of a shortage of sediment for cover (Franzinelli and Igreja 2002). Straight reaches of the middle Amazon have been related to recent activities along a set of NW-SE and SW-NE fractures (Latrubesse and Franzinelli 2002). These fractures were originally described in 1952 by Sternberg and Russell from river and lake alignments (Mertes and Dunne 2007). Latrubesse and Franzineli (2002) also interpreted the confluences of the Amazon as tectonically controlled, coinciding with a wide V-shaped sunken block (Figure 4.6).

An impressive amount of sediment is deposited on the lowermost Amazonian floodplains, and in the floodplain lakes (Meade 2007). An average annual sediment discharge of 1240 (±130) million tonnes at Óbidos has been computed from a number of measurements by Dunne et al. (1998). The arrival of the turbid sediment in the sea is clearly seen in a satellite image (Figure 3.4). About half of the sediment that passes Óbidos has been estimated to settle on the sea bed of the Amazon mouth (Kuehl et al. 1986) but a considerable amount is carried by the North Brazil Current to the northwest along the coastline and transported longshore around the northern promontory of Cabo Norte to ultimately reach the outer coastline of the delta of the Orinoco. Such sediment has travelled about 1600 km after leaving the mouth of the Amazon (Meade 2007). It is an extraordinary travelogue.

Three factors interact through the late Cenozoic, and particularly the late Quaternary, to determine the form of the Amazon and its floodplain: basin tectonic setting, climate, and sea-level fluctuations (Mertes and Dunne 2007). These factors control erosion, sediment transport and deposition in the Amazon valley and determine the morphology of its channel and floodplain. The final product is a huge river, often nearly straight, influenced by geological structural features underneath, wider between such structural features with characteristic floodplain features, and a levee-bounded main lower channel that carries sediment to the sea, leaving lakes unfilled behind the embankments.

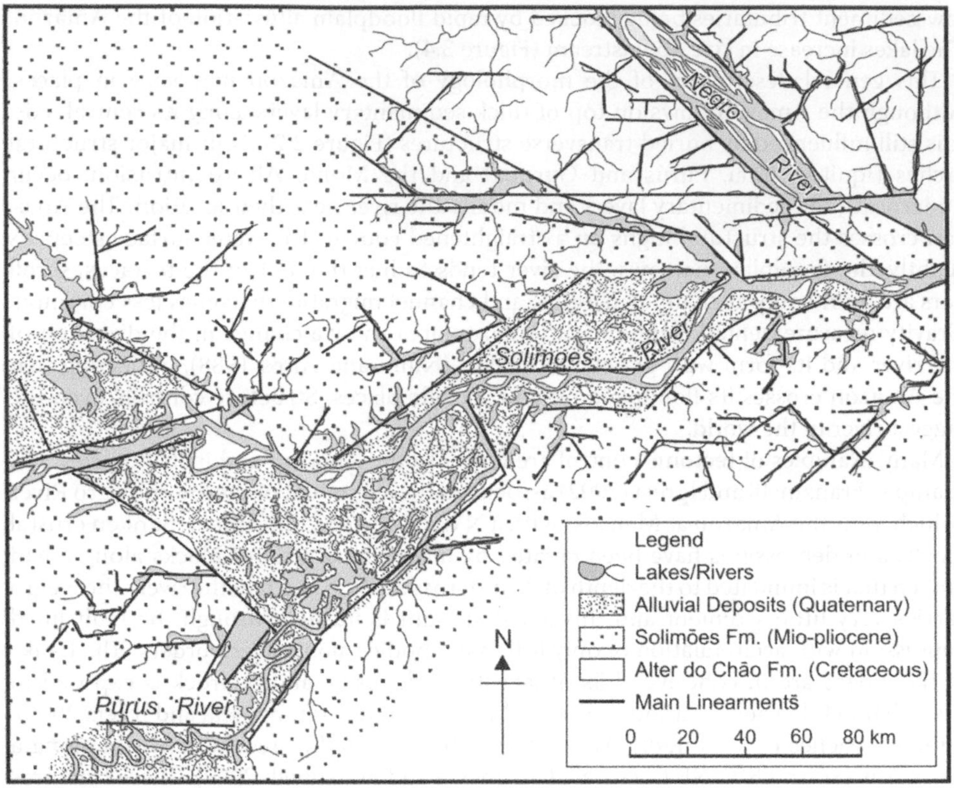

Figure 4.6 Structural control at the Amazon-Negro confluence. In Brazil the Amazon is known as the Solimoes. Source: Adapted from Latrubesse and Franzinelli 2002.

4.4 The Ganga River

4.4.1 The Setting

Rising in the Himalaya, a convergent fold mountain range, the Ganga passes along a foredeep, joins with two other large rivers, the Brahmaputra and Meghna, and flows into the Bay of Bengal through the largest delta on Earth. The three rivers together drain over a million square kilometres, discharging the fourth highest water volume ($970 \times 10\,km^3$) and the second highest annual suspended sediment load (about 1000 million tonnes) to the sea. Such figures are reached because the river drains a high tectonic mountain range with intense monsoon rain falling over the catchment area. Unlike the Amazon, the Ganga has a long history of anthropogenic utilisation of the water of its channel and the land of its basin. The two major diversions of the Ganga happen via the upper Ganga Canal at Haridwar and Lower Ganga Canal at Narora. Part of its water is diverted down the Hugli distributary channel at the Farakka Barrage. The falling quality of the river water, transferring more and more polluted material, is probably the biggest concern in river management.

The Bhagirathi, the main headwater of the Ganga, rises at 3800 m from the Gangotri Glacier. The river is known as the Ganga from Devprayag where it is joined by the

Alaknanda. After another 300 km of river distance, the Ganga descends to 290 MSL at Haridwar. Following the steep descent, the river flows parallel to the mountains along the extensive low-gradient alluvial plains of the Himalayan foreland, a subsiding continental interior setting.

The Ganga turns east and flows along a wide alluvial plain built on the foreland by the main river and its tributaries (Figure 4.7). The river receives both water and sediment from the Himalayas to the north and the cratons of the Peninsular India to the south. All along its course, the river is joined from the north by a number of large Himalayan tributaries at intervals of hundreds of kilometres. The Ganga and these tributaries have built huge alluvial fans (megafans) at the Himalayan highland–lowland contact. Some of these megafans have been studied in detail, e.g. the Kosi (Box 4.1).

The tributaries from the south drain the old rocks of the northern edge of the Indian Peninsula, contributing a smaller amount of water and coarser sediment. Several of these tributary streams flow into the Yamuna, the biggest tributary of the Ganga, instead of directly joining the Ganga. The Yamuna is a Himalayan river and at its confluence with the Ganga at Allahabad, it contributes about 59% of a combined discharge of $130 \times 10\ m^3$ (Das Gupta 1984).

The Ganga turns south at the eastern margin of its large alluvial plain through a gap in the basaltic low hills of Rajmahal, a short distance above the apex of its delta. One of its two main distributaries (the Bhagirathi or Hugli) flows south within India, collecting drainage from the northeastern corner of the Indian Peninsula. The other major distributary, the Padma, carries most of the water into Bangladesh where it meets with the

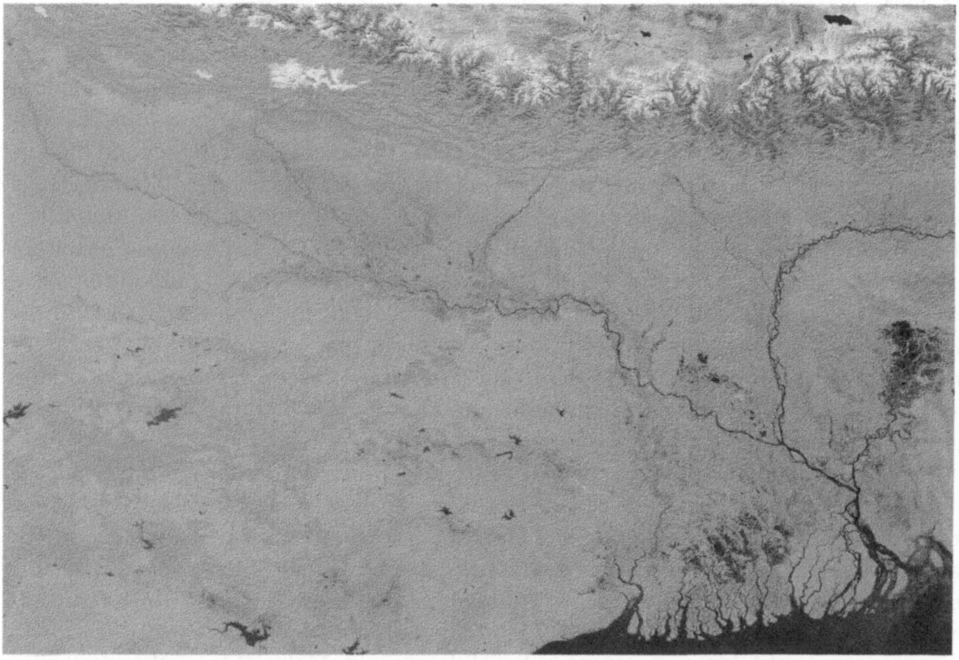

Figure 4.7 The Ganga from satellite imagery. Source: NASA Worldview application (https://worldview.earthdata.nasa.gov), part of the NASA EOSDIS.

Brahmaputra and then the Meghna. These three rivers combine to build and maintain the large Ganga-Brahmaputra Delta.

4.4.2 Hydrology

The Ganga essentially is a rainfed seasonal river sustained by the summer rain of the Indian monsoon. The annual total rainfall decreases from the east to west, from 1600 to 500 mm. The southwestern basin is relatively dry. As expected, rainfall increases on the Himalayan slopes, reaching 1500–2300 mm (Singh 2007). The upper basin also receives some summer snowmelt from the Himalaya. More than 70% of the annual rainfall, rising to 80% in certain locations, arrives between July and early October in the wet monsoon system. The rain often falls intensely and in episodic tropical storms, some reaching cyclonic status, over the lower basin leading to floods.

The flow of the Ganga reflects both seasonality of rainfall and stepwise increment in discharge where the major tributaries, such as the Yamuna, Gomati, Ghaghara, Gandak, Son, and Kosi join the trunk steam. Half of the annual rainfall enters the river as surface runoff, 30% is lost by evaporation, and 20% seeps to the subsurface. During the dry season, part of the subsurface water flows through the high alluvial banks of the Ganga to its channel as baseflow. The mean discharge of the Ganga at Farakka, before it divides into its deltaic distributaries, is 70 547 $m^3 s^{-1}$. About 60% of this arrives from the Himalaya and the northern plains (Das Gupta 1984).

4.4.3 Sediment Load

The sediment load of the Ganga comes mostly from the tectonic Himalaya Mountains. Chemical weathering is not significant in the mountains and the solution load is low, being diluted even further when the discharge is high in the wet monsoon. The suspended and bed load of the Ganga is very high, the suspended load the second highest in the world, superseded only by the Amazon. The annual suspended load of the Ganga has been estimated by Milliman and Syvitski (1992) as 520 million tonnes. About 90% of the sediment travels during the wet monsoon (Singh 2007). The bed load of a large river is difficult to measure but Wasson has estimated that 600–2500 million tonnes of bed load reaches the delta of the river each year. Most of the sediment arrives in the Ganga from the Himalayas along the large tributaries that originate in the mountains and the foothills (Sinha and Friend 1994). The tributaries that come from the south drain the old cratonic rocks of Peninsular India and contribute a high proportion of coarse sediment.

The change in bed material of the Ganga from Haridwar at the foot of the Himalaya Mountains to Ganga Sagar where one of the major distributary channels, the Hugli, enters the Bay of Bengal is plotted in Figure 3.5. Measured from bar samples, it indicates the general downstream fining characteristic of the Ganga, interrupted by periodic coarsening of the bed by contributions from large tributaries (Singh 1996). The bar sediment of the Ganga is essentially sand, the mineralogy of which is primarily quartz with minor amounts of feldspars, micas, and rock fragments. Material from the weathered source-rocks undergoes further alteration when grains are stored as part of floodplain alluvium between their transportation in high flows (Wasson 2003).

4.4.4 Morphology

In the Himalaya Mountains, Ganga and its headwater tributaries flow in narrow, gorge-like valleys flanked by small discontinuous patches of floodplains and terraces. After emerging from the mountains, in the wide alluvial plain between the Himalaya and the cratonic Peninsular India, the rivers are entrenched below the surface of the plain. The channel of the Ganga is confined within a 10–25 km wide elongated lowland, bounded by alluvial cliffs, several metres high. The cliffs are eroded by gullies and small ravines. The lowland bounded by the cliffs comprise: (i) the channel of the Ganga commonly displaying braid bars and meander scars, (ii) the floodplain, (iii) terrace-like features which are higher than the floodplain, and (iv) miscellaneous wetlands. The river channel and the floodplain together extend up to 3 km in width. Large sand bars, kilometres in length, are common in the channel. The Ganga remains confined within this cliff-bounded valley and rarely overtops the cliffs, even in large floods. The floodplain and terrace-like features, however, are inundated periodically, the frequency depending on their height above the channel. The channel pattern of the Ganga changes from place to place (Figure 4.8), and the wide valley between alluvial cliffs is modified by various agencies: the main channel of the Ganga; the smaller channels; the miscellaneous water bodies on the floodplain; and the alteration of the cliff slopes.

Being a seasonal river, the channel of the Ganga is full of bars and multiple channels in the dry season. Huge kilometre-scale bars consisting of braid bars, lateral bars, and point bars are visible at low flow. The low flow effect is enhanced by the large-scale transfer of water into irrigation canals or to meet other demands. In places, the river displays a meandering pattern with point bars and local narrowing of the channel due to extensions of peninsular lineaments under alluvium. The river may briefly change its direction in such locations. The meandering pattern using the entire channel commonly

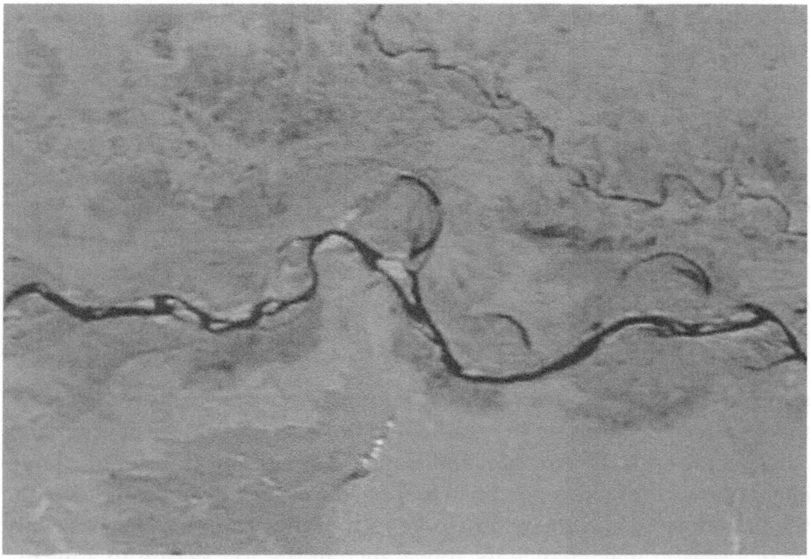

Figure 4.8 The Ganga from satellite imagery in alluvium, bars and bends. Source: NASA Worldview application (https://worldview.earthdata.nasa.gov), part of the NASA EOSDIS.

appears in high flows when nearly the entire channel is under water and mid-channel bars are submerged and removed. During the dry season, however, a braided pattern may re-emerge between the cliffs, as has been described for the Narmada (Gupta et al. 1999), for rivers of monsoon areas.

The bars occur at several levels related to the frequency of inundation. The higher ones are under vegetation and usually farmed. Sediment transfer varies between seasons. During the dry period, it is confined to deeper sub-channels. During the wet monsoon, sediment travels across the entire channel and occasionally even over the floodplain. Several metres of sediment are scoured from temporary storage on top of the floodplain in high flows (Shukla et al. 1999). The general channel pattern remains the same but the location and geometry of the bars vary over time. The river currently tends to shift only several kilometres within the high cliffs.

The huge Ganga-Brahmaputra Delta is discussed in Chapters 6 and 7.

4.5 Morphology of Large Rivers: Commonality and Variations

A review of the morphology and behaviour of the Amazon and Ganga highlights characteristics common to many large rivers. The origin, geographic extension and physical characteristics for many depend primarily on plate tectonics. Certain large rivers have existed for a long time, and most of them reflect repeated changes they have undergone during the Quaternary, concerning the geography of their basins and nature of their course. Regional structural features such as an arching bedrock underneath the channel alluvium or a network of faults modify the general characteristics for the river flowing over such structural features. Large rivers commonly consist of a number of reaches of variable morphology longitudinally assembled to form a big river. Smaller rivers, in contrast, tend to be monotonic in nature (Lewin and Ashworth 2014).

All large rivers are maintained by a large volume of precipitation falling on their basins, at least over a significant part of them. The precipitation can be uniform or seasonal. Given the large size of their basins, large-scale climatic variations such as the ENSO are related to fluctuations in discharge. Such fluctuations may bring in both dry and wet periods. Floods tend to occur in the wet years and from cyclonic disturbances.

Most of the sediment load comes from the high mountains and travels downstream in stages, interrupted by periods of storage in floodplains, on bars, and on bed. As a result, the sediment of a long river becomes progressively enriched in quartz grains and demonstrates textural sorting along the river, the modal class being medium and fine quartz sand. Where tributaries carrying coarse sediment join the main river, local coarsening of the bed material happens for a short distance immediately downstream of the confluence.

Morphologically a large river includes a channel, floodplain, and probably terrace fragments. The channel pattern depends on the gradient of the river and nature of water and sediment as described for both the Amazon and Ganga. Storage and transfer of a large amount of sediment gives a braided appearance to the river, especially in the dry season. The nature of the bars depends on the width and depth of the river. A wide river with braid bars in the dry season may change its appearance in the wet season to a meandering pattern with point bars and lateral bars. During a flood, water and sediment of the river may move laterally, building up the floodplain, as in the Amazon. Sediment in other big rivers may be stored in the same way, and later, after localised erosion of the floodplain and banks of various channels, it is transferred downstream.

Dietrich et al. (1999) described the deposition of the floodplain of the Fly River in the wet equatorial hilly environment of Papua New Guinea. They mentioned three processes of lateral transfer of sediment-laden water into the floodplain: advection of sediment with overland flow when the river is high and the level of water in the floodplain low; lateral diffusion from the sediment-rich water of the river; sediment-laden water travelling upstream into tributaries or small floodplain channels. Sediment is stored for a time in the floodplains of large rivers prior to their erosion and transfer.

On rock, large rivers look and behave differently as their slopes steepen and their courses often follow structural lineations. Their velocity is higher, fewer bars are deposited in the channel, and if they are, they tend to occur near banks or boulder protrusions. Fans occur at tributary mouths, and given the location such fans and bars that evolve from them are built by coarse sediment contributed by tributaries and organised by floods of the main river. Given their length, large rivers may pass through rocky gorges between reaches in alluvium, displaying changing morphology.

The Irrawaddy provides good illustrations. South of Myitkyina, the Irrawaddy is an alluvial river about 1600 km from the Bay of Bengal. Its downstream passage as a low-gradient alluvial river is interrupted by three steep gorges in Palaeozoic rocks. It is a freely meandering river in alluvium south of Myitkyina with a number of abandoned channels and oxbow lakes in the floodplain. The river is nearly 1 km wide when it enters it first gorge which is 56 km long, locally narrows to 50 m, and is tortuous with sharp bends, pools and rapids. The river widens below the gorge and this section in alluvium displays meanders, abandoned channels and bars. Further downstream, below Bhamo, the Irrawaddy turns west to enter the second gorge to cross an upland area before the next alluvial reach where the river returns to its wide meandering course skirting the Gangaw Ranges. This part of the river is flowing in a very wide floodplain with numerous abandoned channels. Numerous bars and islands occur in the main channel, indicating both storage and transport of a large volume of sediment.

Figure 4.9 shows the transition of the Irrawaddy from this alluvial section to the third gorge where the river takes a straight course along the Sagaing Fault. The effect of increased gradient and stream power on the transporting capacity of the river is clearly visible. The gorge is confined in steep forested hillsides, and sand bars occur in the channel next to the banks. Downstream, where the river emerges from the gorge, it changes back to a wide, meandering course along a floodplain of active deposition. Striking alterations occur thus in large rivers as their local environment changes. The Salween flows almost entirely in 1000 m gorges cut into plateaux and mountains and its morphology is different from a large river in alluvium. The morphology and channel material of the Mekong changes several times along its course adjusting to local geology and structure (see Chapter 10). The river is free to move only for the last several hundred kilometres in alluvium.

The basic characteristics of the morphology of large rivers are:

1. A large river simply is a river with impressive dimensions. It may have a channel, even 10–20 km wide at high stage and 30–40 m deep in places. Even deeper scours have been measured, especially at confluences with tributaries. Satellite imagery is an appropriate tool for observing the channel at this scale, for measuring kilometre-scale bends and bars, etc.

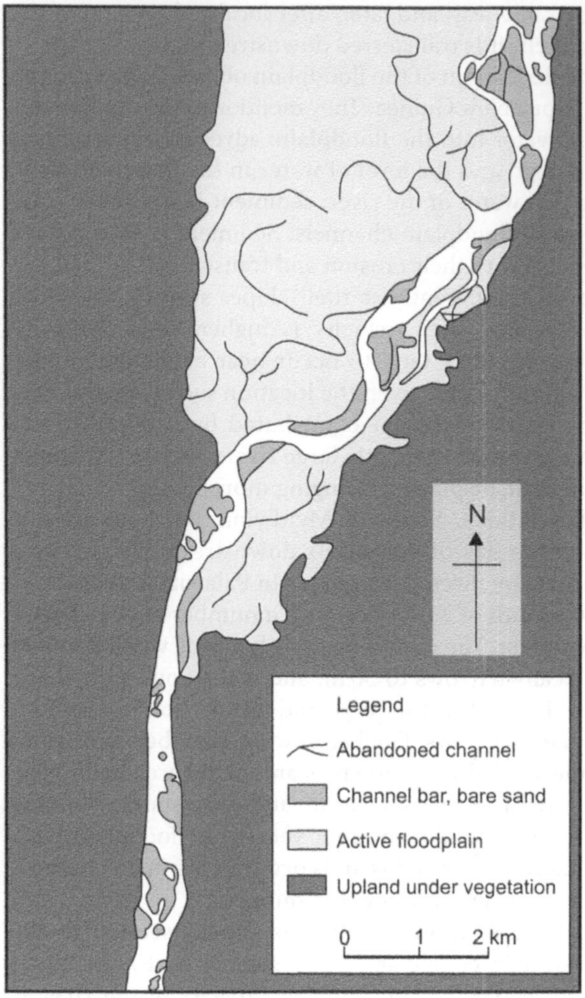

Figure 4.9 Diagrammatic sketch of the Irrawaddy leaving an alluvial segment to enter a rocky gorge along the Sagaing Fault. Scale approximate. Interpreted from SPOT image on the web and topographical maps. Source: Gupta 2005.

Sarkar et al. (2014) derived the following hydraulic relationships for the Jamuna River in Bangladesh in alluvium

$$W = 8.4{Q_f}^{0.55}$$

$$D = 0.278{Q_f}^{0.29}$$

where W = average width (m), D = average depth (m) and Q_f = annual average flood ($m^3\ s^{-1}$). With a rising stage, the Jamuna gets wider rather than deeper with associated morphological changes in the channel.

Large rivers can also be bedrock-constricted or wide when crossing alluvial basins. They may be naturally composite, as reflected in their form and behaviour (see Chapter 7).

2. Large rivers display low-intensity planforms. Traditionally, channel patterns are classified as straight, meandering, braided, and anastomosing. The terms anabranching or wandering have been used instead of anastomosing, but in

general all three terms refer to a river divided into multiple channels by one or several semi-stationary bars or islands. These channels and bars or islands, unlike in a braided river, are not particularly mobile. These planform terms have been further subclassified to describe a particular river but such practices may lead to considerable confusion. Confusion also arises as a river can meander or braid over a short distance but because of its low sinuosity, it appears to be anastomosing over a longer stretch. Furthermore, rivers in seasonal climate may rise considerably in the wet season, and then appear to present a different planform. The Brahmaputra has been reported to rise 8 m (Coleman 1969; Baker and Costa 1987), and the Amazon 12 m (Archer 2005). The channel pattern of large rivers therefore may look different between the wet and dry seasons. The wet season river tends to display less sinuosity.

3. Large-scale avulsions may happen in low-gradient large rivers leading to conspicuous changes in river channels and floodplains.
4. A large river usually does not display the same channel pattern all along its course because of variations in underlying lithology and structure. Segments of a channel reach may also vary in form and behaviour. In contrast, smaller rivers are often monotonic. Ashworth and Lewin (2012) stated that large rivers do not flow through unified valley systems but appear as chains of interlinked domains with varying fluvial functions and scales. They have provided a generalised list. In sequence, these rivers flow through (i) headwater mountain belts providing sediment and runoff, (ii) intermontane and foreland depositional basins and ramps (megafans), (iii) low-sediment yielding cratons, (iv) transverse tectonic controls that shape riverine forms and behaviour over parts of the river, and (v) coastal environment with the history of regional sea-level change.
5. Sedimentary forms of different dimensions tend to occur within large rivers. Mid-channel islands match the depth of the entire stream, large bars are associated with the dimensions of an individual anabranch, and large dunes are dependent on channel depth. These morphological features are often exposed during the dry season, when large dunes are found on top of bars or attached transversely to bars. A series of sedimentary features of different sizes are seen. Many of these sedimentary features remain inherited from the previous wet period.
6. Some of these sedimentary features are scale invariant (Best et al. 2003) but not all. Large rivers thus can be complex phenomena.
7. A large river in alluvium may possess an active wide channel belt. A significant relationship exists between river channels and floodplains, especially at flood stages.
8. Different channels in a large river may function differently, depending on river stage. As a result, the same river may appear and behave different at times. A detailed description of such complexity of large rivers has been reviewed in several papers (Ashworth and Lewin 2012; Lewin and Ashworth 2014; Fielding accepted for publication).
9. Flood pulses and sediment transfer in floods commonly travel both longitudinally and transversely in a large river. Floods commonly move downstream in small rivers. Large rivers tend to have a complex relationship between the main channel and their floodplains as illustrated by the movement of water and sediment in the valley of the Amazon (Mertes and Dunne 2007; and see Chapter 5 of this book).

10. Ashworth and Lewin (2012) have summarised large rivers to be plural systems because of coupling (Harvey 2002). Partial decoupling has been seen between large rivers and their floodplains and between main and subsidiary channels. Coupling is limited in small rivers.
11. Basins of large rivers may include polyzonal sub-basins.

4.6 Conclusion

Large rivers are huge systems that transfer water and sediment from the continents to the oceans. Their nature and behaviour determine the morphology of their drainage basins, which are often subcontinental in size, as the entire drainage network is connected to these massive conduits. Thus large rivers may shape the physiography of the land surface over time.

Their location, morphology and behaviour depend on plate tectonics, regional and local geology, and large-scale climatic systems. Several of these rivers have existed for a long period, and all of them have been affected by the repeated climate and sea level changes in the Quaternary. The majority of large rivers are related to human habitation because of the availability of water, fertile fine-grained sediment, extensive floodplains, and ease of irrigation. Several floodplains and deltas have been anthropologically modified, even impounded, and altered from their natural forms and functions.

The following chapters highlight special cases of the morphology of large rivers, their long association with human civilisation, and their probable adjustments to the changing climate of the future.

Questions

1 What are the principal properties of a large river? What effect do such properties have on the morphology of a large river?

2 What is a megafan? Where are megafans found?

3 Describe the physiography of a megafan.

4 Describe the movement of flood water and sediment in a large river valley.

5 Discuss the morphological complexity of a large river.

6 Large rivers have been described as plural systems by Ashworth and Lewin (2012). Why?

7 Discuss the polyzonal sub-basins which may form part of a large river basin.

References

Aalto, R., Maurice-Bourgoin, L., Dunne, T. et al. (2003). Episodic sediment accumulation on Amazonian floodplain influenced by El Niño/ Southern Oscillation. *Nature* 425: 493–497.

Archer, A.W. (2005). Review of Amazonian depositional systems. In: *Fluvial Sedimentology VII* (eds. M.D. Blum, S.B. Marriott and S. Leclair), 17–39. Blackwell.

Ashworth, P.J. and Lewin, J. (2012). How do big rivers come to be different? *Earth Science Reviews* 114: 84–107.

Baker, V.R. and Costa, J.E. (1987). Flood power. In: *Catastrophic Flooding* (eds. L. Mayer and D. Nash), 1–21. London: Allen and Unwin.

Best, J.L., Ashworth, P.J., Bristow, C.S., and Rodin, J. (2003). Three-dimensional sedimentary architecture of a large mid-channel sand braid bar, Jamuna River, Bangladesh. *Journal of Sedimentary Research* 73: 516–530.

Chakraborty, T., Kar, B., Ghosh, P., and Basu, S. (2011). Kosi megafan: historical records, geomorphology and the recent avulsion of the Kosi River. *Quaternary International* 227: 143–160.

Coleman, J.M. (1969). Brahmaputra River: channel processes and sedimentation. *Sedimentary Geology* 3: 129–239.

Das Gupta, S.P. (1984). *The Ganga Basin, Part 1*. New Delhi: Central Board for the Prevention and Control of Water Pollution.

Dietrich, W.E., Day, G., and Parker, G. (1999). The Fly River, Papua New Guinea: inferences about river dynamics, floodplain sedimentation and fate of sediment. In: *Varieties of Fluvial Form* (eds. A.J. Miller and A. Gupta), 346–376. Wiley: Chichester.

Douglas, I. and Guyot, J.L. (2005). Erosion and sediment yield in the humid tropics. In: *Forests, Water and People in the Humid Tropics* (eds. M. Bonell and L.A. Bruijnzeel), 407–421. Cambridge University Press.

Dunne, T., Mertes, L.A.K., Meade, R.H. et al. (1998). Exchanges of sediment between the flood plain and channel of the Amazon River in Brazil. *Geological Society of America Bulletin* 110: 450–467.

Fielding, C.R. (accepted for publication). Sedimentology and stratigraphy of large river deposits: recognition and preservation potential in the rock record. In: *Large Rivers: Geomorphology and Management*, 2e (ed. A. Gupta). Chichester: Wiley.

Franzineli, E. and Igreja, H. (2002). Modern sedimentation in the Negro River, Amazonas State, Brazil. *Geomorphology* 44: 259–271.

Gole, C.V. and Chitale, S.V. (1966). Inland delta-building activity of the Kosi River. *Journal of the Hydraulics Division, American Society of Civil Engineers* 92 (2): 111–126.

Gupta, A. (2005). Rivers of Southeast Asia. In: *The Physical Geography of Southeast Asia* (ed. A. Gupta), 65–79. Oxford: Oxford University Press.

Gupta, A. (2007). *Tropical Geomorphology*. Cambridge: Cambridge University Press.

Gupta, A., Kale, V.S., and Rajaguru, S.N. (1999). The Narmada River, India, through space and time. In: *Varieties of Fluvial Form* (eds. A.J. Miller and A. Gupta), 113–143. Chichester: Wiley.

Guyot, J.L. (1993). Hydrogéochimie des fleuves de l'Amazonie bolivienne: Colllection Etudes et Thèsis. Paris, Editions de l'ORSTROM, 261p. (as listed in Mertes and Dunne, 2007).

Harvey, A.M. (2002). Effective timescales for coupling within fluvial systems. *Geomorphology* 44: 175–201.

Johnsson, M.J. and Meade, R.H. (1990). Chemical weathering of fluvial sediments during alluvial storage: the Macuapanim Island point bar, Solimões River, Brazil. *Journal of Sedimentary Petrology* 60: 827–842.

Junk, W.J. (ed.) (1997). *The Central Amazon Floodplain: Ecology of a Pulsating System*. Berlin: Springer.

Knox, J.C. (2007). The Mississippi River system. In: *Large Rivers: Geomorphology and Management* (ed. A. Gupta), 145–182. Chichester: Wiley.
Kuehl, S.A., DeMaster, D.J., and Nittrouer, C.A. (1986). Nature of sediment accumulation on the Amazon continental shelf. *Continental Shelf Research* 6: 208–225.
Latrubesse, E. and Franzinelli, E. (2002). The Holocene alluvial plain of the middle Amazon River, Brazil. *Geomorphology* 44: 241–257.
Leier, A.L., DeCelles, P.G., and Pelletier, J.D. (2005). Mountains, monsoons and megafans. *Geology* 33: 289–292.
Lewin, J. and Ashworth, P.J. (2014). Defining large river channel patterns: alluvial exchange and plurality. *Geomorphology* 215: 81–98.
Meade, R.H. (2007). Transcontinental moving and storage: the Orinoco and Amazon Rivers transfer the Andes to the Atlantic. In: *Large Rivers: Geomorphology and Management* (ed. A. Gupta), 45–63. Chichester: Wiley.
Mertes, L.A.K. and Dunne, T. (2007). Effects of tectonism, climate change, and sea-level change on the form and behaviour of the modern Amazon River and its floodplain. In: *Large Rivers: Geomorphology and Management* (ed. A. Gupta), 115–144. Chichester: Wiley.
Mertes, L.A.K., Dunne, T., and Martinelli, L.A. (1996). Channel-floodplain geomorphology along the Solimões-Amazon River, Brazil. *Geological Society of America Bulletin* 108: 1088–1107.
Milliman, J.D. and Syvitski, J.P.M. (1992). Geomorphic/tectonic control of sediment discharge to the ocean: the importance of small mountainous rivers. *Journal of Geology* 100: 525–644.
Nunn, J.A. and Aires, J.B. (1988). Gravity anomalies and flexure of the lithosphere at the middle Amazon basin, Brazil. *Journal of Geophysical Research* 83: 415–428.
Potter, P.E. (1978). Significance and origin of big rivers. *Journal of Geology* 86: 13–33.
Sarkar, M.H., Thorne, C.R., Aktar, M.N., and Ferdous, M.S. (2014). Morpho-dynamics of the Brahmaputra-Jamuna River, Bangladesh. *Geomorphology* 115: 45–59.
Scatena, F.N. and Gupta, A. (2013). Streams of the montane humid tropics. In: *Treatise in Geomorphology*, vol. 9 (eds. J. Schroder and E. Wohl), 595–611. San Diego: Academic Press.
Shukla, U.K., Singh, I.B., Srivastava, P., and Singh, D.S. (1999). Palaeocurrent patterns in braid-bar and point-bar deposits: examples from the Ganga River, India. *Journal of Sedimentary Research* 69: 992–1002.
Singh, I.B. (2007). The Ganga River. In: *Large Rivers: Geomorphology and Management* (ed. A. Gupta), 347–371. Chichester: Wiley.
Singh, M. (1996). *The Ganga River: Fluvial Geomorphology: Sedimentation Processes and Geochemical Studies*, vol. 8. Heidelberg: Heidelberger Beiträge zur Umwelt-Geochemie.
Singh, H., Parkash, B., and Gohain, K. (1993). Facies analysis of the Kosi megafan deposits. *Sedimentary Geology* 85: 87–113.
Sinha, R. and Friend, P.F. (1994). River systems and their sediment flux, Indo-Gangetic plains, northern Bihar, India. *Sedimentology* 41: 825–845.
Wasson, R.J. (2003). A sediment budget for the Ganga-Brahmaputra catchment. *Current Science* 84: 1041–1047.
Wells, N.A. and Dorr, J.A. Jr. (1987). Shifting of the Kosi River, northern India. *Geology* 15: 204–207.

5

Large Rivers and their Floodplains: Structures, Functions, Evolutionary Traits and Management with Special Reference to the Brazilian Rivers

Wolfgang J. Junk[1], Florian Wittmann[2], Jochen Schöngart[3], Maria Teresa F. Piedade[3] and Catia Nunes da Cunha[1]

[1] *National Institute of Science and Technology in Wetlands (INCT-INAU), Federal University of Mato Grosso (UFMT), Cuiabá, MT, Brazil*
[2] *Institute of Wetland Ecology, Karlsruhe Institute of Technology - KIT, Rastatt, Germany*
[3] *National Institute of Amazon Research (INPA), Manaus, Amazonas, Brazil*

5.1 Introduction

Floodplains fringe the majority of lowland rivers worldwide. They usually occupy large areas that become periodically inundated but fall dry afterwards marking the changing patterns of local precipitation and the discharge of the parent rivers. In the past, tropical floodplains played an important role in the development of human societies, as shown by cultures that emerged along the Nile, Euphrates, Tigris, Yangtze, Indus, and Ganga. The relative predictability of the flood pulses allowed synchronisation of farming practices with floods in the fertile floodplains, and in turn, favoured cultural development. In contrast, the floodplains of the temperate rivers with their unpredictable flood patterns, were used less. Thus, in Europe and the USA, priority was given to flood control and reclamation. The costs of these policies have turned out to be very high, they are still rising, and the respective countries have been forced to implement complex legislation for wetland protection and reclamation, e.g. the Water Framework Directive (WFD) in Europe (Ball 2012) and the Clean Water Act in the USA (Mitsch and Hermandez 2013). Such legislation allows restoration of parts of the former floodplains, as much as is reasonably possible, in densely populated countries. However, floodplain destruction has also negatively impacted the adjacent ecosystems as shown by the eutrophication of marine environments adjacent to the Mississippi Delta. There the oxygen concentration in deep water layers was reduced and the fauna including fish were damaged (Mitsch et al. 2001).

The ecology of large river floodplains has been properly examined only during the last few decades. A major step towards a comprehensive ecological approach was the introduction of the Flood Pulse Concept (FPC) (Junk et al. 1989, 2013). Many aspects of our current knowledge about floodplains in the tropics and subtropics are still rudimentary and inadequate, especially for floodplain management and protection. The impact of flood pulse on organisms and floodplain processes, the interaction between terrestrial and aquatic phases, and the feedback between a floodplain and its parent river are

Introducing Large Rivers, First Edition. Avijit Gupta.

poorly understood. Comparative studies among different floodplain systems are scarce because the respective databases are often insufficient.

The requirement of water for modern societies increased as the demand for water expanded from traditional application in fishery, forestry, farming, and animal ranching to land reclamation, water abstraction for irrigation, domestic and industrial use, and hydroelectric power generation. These new demands threatened not only the livelihood of the traditional riverine population but also the ecological integrity of river–floodplain systems (RFSs). Brazil has a surface area of 8.516 million km^2 extending across the tropics and subtropics of South America. Junk (2013) estimated that 20% of Brazil's land surface is occupied by wetlands, most of which experience a strongly oscillating water level. Recent studies have classified these wetlands and their macrohabitats as large river floodplains, internal deltas, and interfluvial wetlands (Junk et al. 2014a; Nunes da Cunha and Junk 2014).

The middle and lower courses of the Amazon River and many of its larger tributaries are accompanied by floodplains fringing the river. In the lower reaches of several sediment-poor blackwater and clearwater tributaries, such as the Negro and Tapajós rivers, these floodplains resemble inland deltas. At the upper course of the Paraguay River lies a large internal delta, the Pantanal of Mato Grosso. A major part of the middle Araguaia River Basin is covered by a huge floodplain that includes the large Bananal Island. The upper and middle Paraná formerly carried extended floodplains. Such floodplains, however, have been modified by a series of large reservoirs for generation of hydropower. Only a small floodplain now exists between the Porto Primaveira Dam and the Itaipu Reservoir. The river beds of the São Francisco and Tocantins are deeply entrenched and do not provide space for large floodplains. The São Francisco has been modified by the construction of reservoirs for hydroelectricity generation.

The floodplains of the Amazon and Negro rivers are amongst the best studied tropical river floodplains as shown by a very large number of publications since the 1980s (Goulding 1980; Sioli 1984; Goulding et al. 1988, 1996; Junk et al. 1997, 2010a; Padoch et al. 1999; Smith 1999; Junk 2000; Wittmann et al. 2010a and others). The Pantanal of Matto Grosso has been the focus of scientific research on Brazil's savanna (*cerrado*) regions (Hockman 1998; Junk et al. 2011a). There are also numerous studies on the surviving floodplains of the upper Paraná River (Thomaz et al. 2004). Research on the floodplain of the Araguaia River is still scarce and the data difficult to access but information on hydromorphology and vegetation cover is now available because of new projects (Latrubesse and Stevaux 2002; Latrubesse et al. 2009; Valente and Latrubesse 2012; Valente et al. 2013; Irion et al. 2016). However, little is known about the floodplains of the large tributaries of the Amazon (the Purus, Juruá, Japurá, Madeira, and Tocantins) and many interfluvial wetlands in Amazonia have not been studied at all because of the difficulty in access (Junk et al. 2011b).

Despite such gaps in knowledge, Brazil is well-suited for a comparative analysis of the structures, functions, and evolutionary traits of different types of tropical and subtropical floodplains and their actual and predicted use. The major Brazilian floodplains have been discussed, basing our analysis on the theoretical framework of the FPC (Junk et al. 1989). New biogeographical scenarios regarding the role of rivers and their floodplains also have been developed. The traditional uses of rivers and their floodplains are also described and future development following the rapidly changing economic and political demands of a rapidly developing tropical country analysed.

The chapter discusses the nature of floodplains of large rivers, their environmental characteristics, and requirements for management. The ecological characteristics of floodplains and their macrohabitats are treated in detail including the ecological responses of organisms to flood-pulsing conditions. Both plants and animal biodiversity of floodplains are presented in detail. The discussion on floodplain management followed such data sets and policies for floodplain preservation have been suggested. Although the examples are mostly from Brazilian wetlands, the discussion is relevant for tropical and subtropical floodplains in general, and complements the geological and geomorphological analysis of floodplains elsewhere in the book.

5.2 Origin and Age of Rivers and Floodplains

Rivers of the temperate zone generally developed after the retreat of the ice of the last glacial age. RFSs in the tropics and subtropics, however, could be several million years old and with a complex history. During the Quaternary, the RFSs passed through a number of wet and dry periods and were also affected by tectonic events. Tropical RFSs therefore could be complex, and at times including older hydro-geomorphologic features of former wetter and drier periods in the form of accumulations of strongly weathered palaeo-sediments along with recent and subrecent sediments.

This is displayed in the Brazilian RFSs. The Amazon used to drain towards the west when South America began to separate from Africa, about 100 million years ago. With the emergence of the Andes, the drainage changed to the North along the Andes. The depression was flooded several times by the sea or covered by freshwater swamps as evidenced by alternate layers of salt and freshwater sediments. Connection to the north was interrupted by the emergence of the Vaupes Arch during the Neogene about 8 million years ago. The northern part of the continent drained through the newly formed Orinoco to the Caribbean Sea. About 5 million years ago, the Amazon eroded its way between the Guyana Shield in the north and the Brazilian Shield to the south to the Atlantic Ocean (Lundberg et al. 1998). Near the city of Óbidos, the Amazon River is squeezed between the two shields and has a width of only about 900 m. The climate was usually hot and humid. A drier and cooler climate during the glacial periods and a retreat of the rain forest to isolated areas (summarised by Haffer and Prance 2001) remains a controversial hypothesis (Colinvaux et al. 2001).

Geological evidence, however, indicates that the lower and middle Amazon rivers reacted to fluctuations in sea level associated with glacial and interglacial periods. During glacial periods, the sea level was lower which led to a higher declivity of the river and the erosion of the older sediment in the Amazon Valley. During interglacial periods, rivers filled the large valleys with sediment. The sea level was about 130 m lower at the Last Glacial Maximum (LGM). Near Manaus, the floodplain surface of the Amazon was 20–25 m lower than today (Muller et al. 1995). With a rising sea level, the waters of the Amazon and its tributaries were dammed back and recent sediments filled the broad valleys. Only a small amount of sediment was transported by the clearwater and blackwater tributaries such as the Negro, Tapajos, and Tefé rivers and many small tributaries as happens today, and the lower valleys remained unfilled giving rise to ria lakes. The present Amazon River floodplain therefore is composed of sediment deposited after the LGM and also sediments of earlier interglacial periods, called paleo-várzeas. The paleo-várzeas sediments are strongly weathered and of lower fertility than the current sediments (Irion et al. 1997, 2010).

Pantanal of Mato Grosso is situated in the depression of the upper Paraguay River, covering 16–20 °S and 55–58 °W. The main period of subsidence is very likely related to the last compressional pulse of the Andes during the upper Pliocene–lower Pleistocene, about 2.5 million years ago. Several rivers drain a surrounding catchment area of about 496 000 km^2, mostly sandstone and limestone with minor granitic outcrops. The rivers have deposited their sediment as alluvial cones in the upper Paraguay River depression, the largest being the internal delta of the Taquari River. The internal deltas are of different age and formed during different paleo-climatic times. Over time, they became deactivated and subject to erosion and deposition internally. The Pantanal is an old complex system, made of internal deltas of varying ages that originated from different rivers. The inundation of large areas occurs mostly after local rainfall and a reduced drainage by high river levels.

The floodplain of the upper Aragauia River provides a different example of the regional complex hydro-geomorphological conditions that formed a huge internal delta, of about 80 000 km^2, in the upper course of the river. Only about 20% of the area consists of recent sediment. The rest of the area is covered by older river sediments (Latrubesse and Stevaux 2002; Valente and Latrubesse 2012; Irion et al. 2016). Mineralogical analyses indicate strong weathering and a change from feldspars and micas to kaolinite, gibbsite, goethite, and aluminium chlorite and a reduction in grain size. This points to weathering over a very long period. These old sediments represent the active paleo-floodplain of the river. They store rainwater and then release it during the dry period, and also participate in the exchange of biota and nutrients between the river and the current floodplain. Figure 5.1 indicates the studied river floodplains and their water qualities.

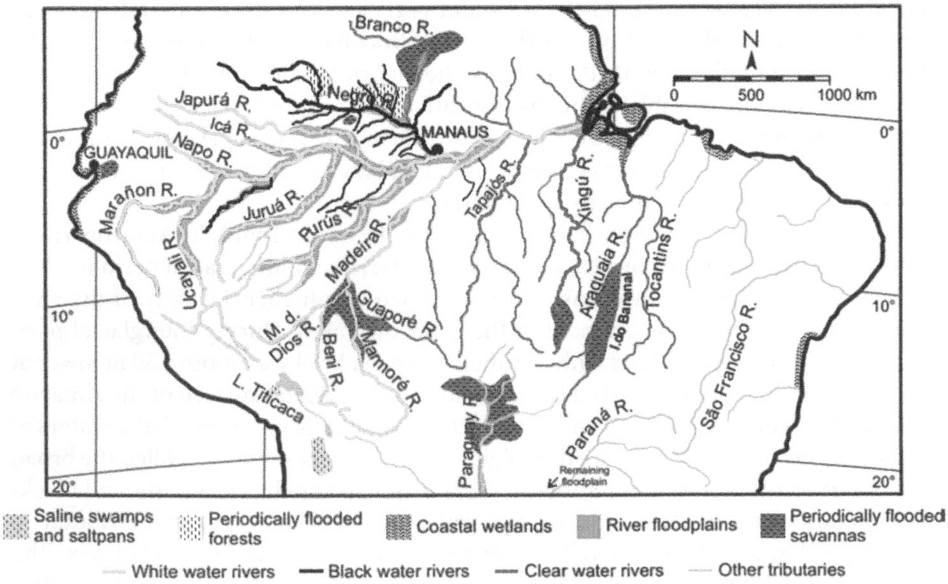

Figure 5.1 Major Brazilian rivers and wetlands.

5.3 Scientific Concepts and their Implications for Rivers and Floodplains

In 1980, Vannote et al. (1980) published the River Continuum Concept (RCC). It describes and explains the dynamic processes that take place along a river, from its low-order headwaters to the high-order channels near its mouth. Shadowing of the stream channel by the riparian forest in the headwaters reduces the incident light reaching the water, limiting primary production by algae and aquatic macrophytes. Since most organic matter derives from allochthonous sources, the production: respiration ratio (P/R) is <1. With increasing river order the channel widens, light reaches the shallow water, autochthonous organic matter production by algae and aquatic macrophytes increases, and the importance of allochthonous organic matter decreases. The P/R ratio thus becomes >1. In bigger rivers, the situation changes again. The water depth increases, autochthonous organic matter production decreases, and the P/R is <1. Invertebrate communities in the headwaters are dominated by shredders which destroy the coarse allochthonous organic matter. In the middle reaches, grazers and collectors become more frequent, while in the lower reaches collectors dominate. They make use of the remaining fine particulate organic material from the upper reaches. Dissolved organic matter reaches its highest value in the upper reaches and diminishes towards the river mouth because of microbial activity. This scenario, described by the RCC, is based mostly on data from temperate rivers, and has not been tested in tropical river systems.

From a hydrological point of view, the river and its floodplain are considered as a unit, the RFS. The ecological implications of the lateral dimension of the RFS were first described by fish ecologists and fishery biologists. Lowe McConnell (1964) showed the relationship between the fish communities of rivers and streams and the periodically flooded savannas of the Rupununi District in Surinam. Holcik and Bastl (1976) described the impact of floods on fish populations in the Danube River floodplain. Welcomme (1979) analysed environmental factors in river floodplains and described the importance of floodplains for fisheries in tropical river systems, identifying three types of river floodplains according to their genesis:

1. Fringing floodplains. These accompany river courses that effectively become part of the major channel during high discharge. These are left dry at low discharge when the river water is confined within deeper channels.
2. Internal deltas. These are large floodplains developed due to geographical accidents, e.g. blockage of the river by geological features and the subsequent deposition of alluvial material.
3. Coastal deltaic floodplains. These develop at the mouth of the river due to the deposition of alluvial material as funnel-shaped deltas extending into the sea, a lake, or a larger river. In marine coastal areas, interference with river flood and tidal activities causes specific hydrologic conditions that often reflect tidal rhythms without the interference of saline water.

Junk and Welcomme (1990) summarised the data on floodplain systems. They identified floodplains as 'areas of low-lying land that are subject to inundation by the lateral overflow of water from river and lakes with which they are associated. The floods further bring about such changes in the physico-chemical environment that biota react by morphological, anatomical, physiological or ethological adaptations or by change in community structure. In some cases flooding may directly result from rainfall or snow and ice melt whereby water periodically accumulates in shallow depressions (sheet flooding). While such area are not true floodplains in that they are not associated with any permanent water body, they do merit consideration as seasonally flooded wetlands' (Junk and Welcomme 1990). Junk et al. (1989) expanded this approach to formulate the FPC, which was later complemented by Junk and Wantzen (2004) and Junk (2005). The concept was extended to lakes by Wantzen et al. (2008). The FPC treats the river channel and its floodplain as one unit: the RFS. The periodic inundation, also called the flood pulse, connects the floodplain and its permanent water bodies to the river channel and separates the two entities when the water recedes. The area that is periodically inundated is called the Aquatic Terrestrial Transition Zone (ATTZ). The border between the dry and wet areas moves according to the changes in water level and is called the moving littoral. During rising discharge, water enters the floodplain along with dissolved nutrients and sediment from the river channel. When the water level drops, the water and a portion of the nutrients and organic matter return to the channel. At the highest water level, parts of the floodplain act as a flow-through system, increasing the discharge capacity of the river channel. The river channel and floodplain exchange organisms with the river channel acting as a highway for the passive transport and active migration of organisms. The floodplain deals with the aquatic phase during high water and the terrestrial phase when the water is low. These stages interact with respect to the production and decomposition of organic material, nutrient cycles, and the life history of the affected organisms.

5.4 Water Chemistry and Hydrology of Major Brazilian Rivers and their Floodplains

The large Brazilian rivers differ from each other with respect to the chemical quality of their water and sediment load (Table 5.1). These differences also affect the connected floodplains. Sioli (1956) classified the Amazonian rivers as (i) whitewater rivers, rich in sediment and nutrients with a near-neutral pH, (ii) acidic blackwater rivers, poor in sediment and nutrients, and (iii) clearwater rivers of intermediate status. This classification has been discussed and modified on the basis of new data by Villamizar (2013). Although Sioli's classification has been shown to be over-simplified, it remains valuable with respect to the characterisation of the major Amazonian large rivers and interfluvial floodplains. Whitewater rivers are accompanied by fertile floodplains, locally known as *várzeas*, and blackwater rivers by nutrient-poor floodplains, locally called *igapós*. The

Table 5.1 Electrical conductivity and pH values of several major Brazilian rivers.

	Amazon near Manaus	Negro near Manaus	Paraguay at Caceres	Araguaia at Bananal Island	Paraná below Porto Primavera Dam
Conductivity (μS cm^{-1})	50–93	7–16	28–101	16–28	30–110
pH	6.5–7.5	4.4–6.2	5.2–8.2	7.6–8.3	6–8

two floodplains differ in plant and animal species that occur on them, and clearwater rivers occupy an intermediate position regarding water and sediment fertility. The physical and chemical parameters of river water vary considerably over the annual cycle.

The quality of the water in the river affects the quality of water in floodplains, but the internal absorption and release processes operating in a floodplain – the influx of seepage water, rainwater dilution, concentration of dissolved substances by evaporation – can strongly modify the water's chemical composition. Water from the floodplain flows back into the river channel as the river level drops and may modify its chemical composition. Isolated lakes in the southern Pantanal of Mato Grosso, locally called *salinas,* show an increased salt concentration by evaporation of 4000–9000 μS cm^{-1}, whereas the electrical conductance of the contributing rivers is only more than 100 μS cm^{-1} (Medina-Junior and Rietzler 2005). The electrical conductance of water in the remaining pools of the várzea can exceed 1000 μS cm^{-1} as a consequence of the decomposition of organic matter and the inflow of seepage water as shown for Camaleão Lake (Furch and Junk 1993; Weber et al. 1996). A large variability in hydrochemical parameters has also been described for the water bodies of the Paraná River floodplain and was related to the level of connectivity with the parent river (Thomaz et al. 1997; Pagioro et al. 1997).

Hydrology is the single most important determinant of the establishment and maintenance of specific types of wetlands and wetland processes (Mitsch and Gosselink 2000). Figure 5.2 gives the mean annual water level fluctuations for major Brazilian rivers. Those of the parent rivers, however, do not always represent the fluctuation of the water level inside the connected floodplains. Major parts of the *várzeas* and *igapós* of the Amazonian whitewater and blackwater rivers are strongly affected by parent rivers and may undergo deep flooding of 10 m and more, depending on the river level. Large parts of the Pantanal of Mato Grosso and the floodplain of the Araguaia River that are distant from the river course are usually only shallowly flooded (≤2 m), mostly from rainwater. In the Pantanal, the flood regime of the main tributaries influences the hydrograph of the respective floodplain areas (Hamilton et al. 1996), and leads to a specific regional flood pattern. The food pulse of the Paraná River has been strongly modified by the release of water from large reservoirs, depending on the daily and seasonal energy demand.

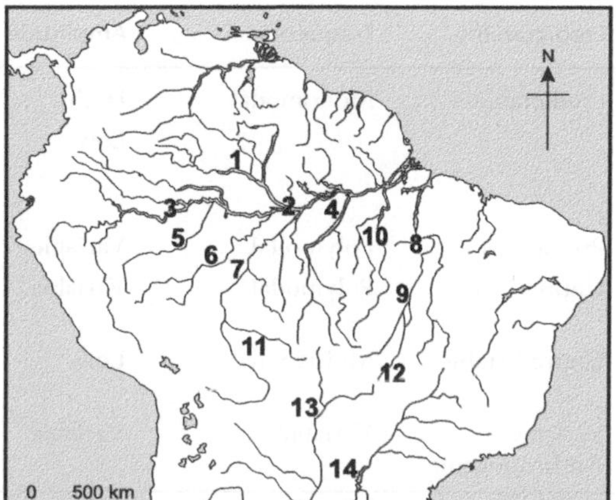

Figure 5.2 Mean annual water level fluctuations of major Brazilian rivers between 1970 and 2010 (in m). 1, Rio Negro, Barcelos: 5.23; 2, Rio Negro/Amazonas, Manaus: 9.47; 3, Rio Solimões, São Paulo de Olivença: 8.09; 4, Rio Tapajós, Santarém: 4.79; 5, Rio Juruá, Gavião: 12.79; 6, Rio Purús, Lábrea: 15.25; 7, Rio Madeira, Porto Velho: 11.38; 8, Rio Tocantins, Itupiranga: 9.03; 9, Rio Araguaia, Conceicão do Araguaia: 5.83; 10, Rio Xingú, Altamira: 4.38; 11, Rio Guaporé, Mato Grosso: 2.75; 12, Rio Araguaia, Barra dos Garças: 3.62; 13, Rio Paraguay, Ladário: 2.90; 14, Rio Paraná, Porto São José: 2.14. Source: Agēncia Nacional de Aguas (ANA).

This has reduced the floods and thus the connectivity between the water bodies in the floodplain and the river channel (Souza Filho 2009).

Water depth is an important criterion as it affects the light in the water. The shallowly flooded *cerrado* floodplains exhibit an exuberant growth of submerged aquatic macrophytes that are rooted in the sediment. These are absent in the deeply flooded Amazonian *várzeas* and *igapós*. In the Paraná River, reservoir construction has reduced the amount of suspended sediment and increased transparency in the river channel from 60 cm to up to 4 m. This has resulted in increased phytoplankton growth and the establishment of submerged aquatic macrophytes. Aquatic chemistry thus tends to change between the channel and floodplains of major rivers.

5.5 Ecological Characterisation of Floodplains and their Macrohabitats

Wetland classification systems generally ascribe little importance to hydrology, it is considered comparable with geomorphology, e.g. the system used in the USA (Cowardin et al. 1979) and for the Ramsar Convention (Scott and Jones 1995). However, whereas in the temperate zones many wetlands, e.g. bogs and marshes, have a rather stable water level, most wetlands in the tropics and subtropics are subject to strong water level fluctuations. In 2014, a hierarchical wetland classification was elaborated on the basis of types of flood pulses by Brazilian wetland ecologists (Junk et al. 2014a) as presented in Table 5.2.

The Brazilian approach, like other classification systems, differentiates at the system level among coastal, inland, and artificial wetlands. Subsystems, orders, suborders, and classes are determined according to hydrological parameters. The large floodplains of Brazil which cover tens of thousands of kilometres are established at the class level as they represent predictably monomodal pulsing wetlands with a high or low flood amplitude. Hydrological, hydrochemical, and vegetation parameters are used in order to differentiate among subclasses and macrohabitats.

Table 5.2 Types of flood pulses and affected wetlands.

Predictability	Frequency	Amplitude	Wetland type
Predictable	Monomodal	High	Wetland along large rivers
		Low	Large interfluvial wetlands, including those in coastal dune fields, (e.g. Lançóis Maranhenses, Brazil)
Predictable	Polymodal	Variable	Tidal wetlands
Unpredictable	Polymodal	Variable	Wetlands along low order rivers, in small depressions, and in coastal lagoons
Unpredictable	Multiannual	Low	Wetlands in the semi-arid Brazilian northeast, flooded only in wet years
Variable Anthropogenic	Variable	Variable	Reservoirs, paddy rice fields

Source: Data from Junk et al. 2014a.

The classification system therefore is organised as:

Systems: coastal, inland, and artificial wetlands.
Subsytems: three coastal subsystems, two inland subsystems.
Orders: three inland orders: monomodal predictable pulsing of long duration; polymodal unpredictable pulsing of short duration; multiannual unpredictable pulsing of short duration.
Suborders: two inland suborders: monomodal predictable pulsing with high or low amplitudes.
Classes: Different classes with varying flood amplitudes.
Subclasses: numerous local variations.
Macrohabitats: basic units, numerous in inland, coastal, and artificial wetlands.

The smallest units are characterised by different associations of higher plants or their absence. The floodplains of the Amazon, Negro, and Paraná rivers are monomodal predictable pulsing wetlands with high flood amplitude whereas the Pantanal of Mato Grosso and most of the floodplain of the Araguaia River belong to the suborder of monomodal predictable flood pulse of low amplitude. The wetlands are very large, more than 100 000 km^2 in area with a mosaic of geomorphic features. Their position on the floodplain gradient extends from permanently aquatic to permanently terrestrial conditions, ranging from floodplain lakes and river channels to paleo-levees a few metres above the highest flood level or terrestrial outcroppings inside the floodplain. They are accompanied by different communities of higher plants and harbour different animal communities.

In the classification of these large wetlands six *functional units* were introduced below the class level, according to its hydrological status: permanent aquatic, predominantly aquatic, predominantly terrestrial, permanent terrestrial, swampy, and anthropogenic (independent from its hydrological status) (Junk et al. 2018). A functional unit is defined as 'a large landscape unit in a floodplain that is subjected to specific hydrological conditions during the annual cycle'. The subclass is defined as 'a subunit of a functional unit with specific hydrological conditions and a characteristic cover of higher vegetation'. A macrohabitat is 'a subunit of a subclass which is subjected to similar hydrological conditions and carries indicator plant species or groups of species'.

The number of macrohabitats of the floodplains of the Amazon and Negro rivers (*várzeas* and *igapós,* respectively) and the Pantanal are compared in Table 5.3. However, the flood pulse changes environmental conditions throughout the hydrological cycle, and the recognition of macrohabitats is challenging.

Table 5.3 Macrohabitat diversity in three large Brazilian floodplains belonging to suborders with predictable monomodal flood pules of low (Pantanal) and high (Amazon River *várzeas* and Negro River *igapós*) amplitudes.

Location	Functional units	Subclasses	Macrohabitats	Source
Pantanal	6	16	57	Nunes da Cunha and Junk 2014
Várzeas	6	13	36	Junk et al. 2014b
Igapós	6	12	25	Junk et al. 2015

The location of the Pantanal in the savanna belt enables the coexistence of different types of semi-deciduous forests in the higher area and flood-adapted forests in low ones. Drought-adapted savanna vegetation of dry areas is found together with wet-adapted marshlands in permanently moist depressions. The shallow flooded old floodplain is hardly affected by erosion and sedimentation processes but multiannual very wet and dry periods with fires cause frequent disturbances of the vegetation and the coexistence of successional stages. In the active recent floodplains along the river courses, river dynamics lead to permanent renewal of macrohabitats.

The Amazon River *várzeas* and the Negro River *igapós* are situated in the Amazonian rainforest biome. All areas inundated for 8 (*várzeas*) and 9 (*igapós*) months are covered by various types of very species-rich floodplain forest. Low-lying areas of the nutrient-rich *várzeas* are covered by different communities of herbaceous plants which occur only in small areas of the *igapós* where sandy beaches prevail. There are no savanna macrohabitats. *Várzeas* are richer in macrohabitats than *igapós* because of stronger sedimentation, erosion, and renewal of macrohabitats there than compared with the rather uniform and stable Negro River floodplains.

Macrohabitats are not randomly distributed in wetlands and vary considerably in size. Several macrohabitats cover only small areas but they are essential, requiring full protection, whereas others extend over large areas and can survive with a certain level of human activities without disturbing the general ecological integrity of the wetland landscape. Important differences among large areas of wetland often become evident because of particular macrohabitats which influence key functions depending on their size. For instance, macrohabitats dominated by herbaceous plants occur in the Amazonian *várzeas* and the Negro River *igapós*. But in the *várzeas*, they cover huge areas and consist of highly productive C4 plants, whereas in the *igapós* such areas are very small and covered by low-productivity C3 grasses and sedges. The consequences for total primary production and the food webs of the two dramatically differentiate one system from the other.

5.6 Ecological Responses of Organisms to Flood-Pulsing Conditions

Flood-pulsing ecosystems require specific adaptations of plants and animals between terrestrial and aquatic environmental conditions. This can be predictable or unpredictable and of varying amplitudes and durations (Table 5.2). Floodplains of large rivers and interfluvial wetlands are subjected to monomodal, long-lasting, predictable flood pulses with high or low flood amplitudes. The specific adaptation depends on the predictability of flood pulse and the stability and age of the respective RFS. The Amazon River runs parallel to the Equator. Its geology and climatic history indicate that floodplains have existed for millions of years. Flood amplitude and river discharge may have changed between glacial and interglacial times, but the flood pulse has probably been monomodal and predictable for a long time. Periodic local extinction of organisms could have been compensated by re-immigration from less-affected areas of the huge catchment. This stability explains the high level of complex adaptations to the flood pulse in plants and animals and the large number of endemic species.

In comparison, the Paraguay River runs north–south, from the tropics to the subtropics. During the Holocene, the Pantanal passed through a set of different climatic episodes still not yet fully understood (Assine and Soares 2004). The following climatic episodes (Iriondo and Garcia 1993; Stevaux 2000) can be identified: cool and dry (40 000–8000 years BP), warm and wet (8000–3500 years BP), warm and dry (3500–1500 years BP), and warm and wet (1500–present). During these periods plants and animals had to cope with flood stress, severe drought, and fire stress. Such changes favoured species with a broad ecological amplitude over wetland specialists and hindered the development of endemic species as shown by wetland trees.

5.6.1 Trees

Drought stress during the terrestrial phase is moderate in the Amazonian rain forest, allowing the growth of a species-rich flood-adapted forest. A large number of morphological, anatomical, physiological, and phenologic adaptations have been described at the level of roots, stems, leaves, and fruit. Furthermore, specific phenological characteristics, with respect to leaf shedding and fruiting, are related to periodic flooding and droughts. The adaptations are summarised in Table 5.4 (Junk et al. 2010c).

Studies on the adaptation of trees on the floodplains of the savanna belt are scarce but observations in the field have shown that many of the adaptations described for

Table 5.4 Morphological, physiological, and phenological adaptations used by trees of central Amazonia to periodic flooding.

Adaptation type to periodic flooding	Description
Leaves	• Xeromorphic with thick cuticle and outer epidermis walls • Periodic leaf shedding at high water • Maintenance of emergent leaves • Maintenance of leaves under water
Diaspores	Commonly hydroichthyochorus
Stems	• Periodic reduction of diameter increment (annual growth rings) • Stem water storage • Pressure ventilation • Internal oxygen transport
Roots and underground area	• Adventitious roots • Lenticels • Stem hypertrophy • Pneumetophores (in various palms in brief-duration floods), • Oxygen diffusion and oxygenation of the rhizosphere • Suberinisation • Anaerobic metabolism
Others	• Excretion of aldehydes and other volatile organic compounds through the canopy • Low-light plants (some with underwater photosynthesis)

Source: Adapted from Junk et al. 2010c.

Amazonian floodplain trees are also exhibited by flood-tolerant species of the Pantanal and the Araguaia River floodplains. But many trees in the floodplains of the savanna belt also show adaptations against wild fires and droughts.

5.6.2 Herbaceous Plants

Herbaceous plants fall into three major groups: (i) annual and perennial species growing during the terrestrial phase, (ii) annual and perennial species growing mostly during the aquatic phase, and (iii) a group of species growing on permanently waterlogged, swampy habitat. In Amazonian wetlands with high flood pulses, swampy habitats are provided mostly by floating islands of organic matter, locally called *matupá*, which accompany the rise and fall of the water level. These islands are called *batumes* in the Pantanal. In the shallow flooded area of Pantanal and in the Araguaia River floodplain, many swampy areas occur with specific, often monodominant herbaceous plant communities.

Many herbaceous species form large banks of flood- or drought-resistant seeds in the sediment which become activated during favourable conditions (Junk and Piedade 1997).The activation is only partial and a large stock of propagules is maintained for additional wet periods if the first attempt is not successful. Experiments with Pantanal sediment displayed a growth of new seedlings of aquatic and terrestrial species after three very dry phases. This is a very well-known behaviour worldwide, specifically in drier wetlands, as in Australia, where flooding occurs only at intervals of several years (Reid and Capon 2011).

High flood pulses in Amazonian river floodplains restrict the growth of submerged herbaceous plants because the suitable habitats are occupied by the floodplain forest which eliminates such plants by shading. Low-lying areas, unoccupied by forests, are deeply inundated, and only few very large aquatic species rooted in the sediment can become established. These habitats are occupied by free-floating aquatic species. Annual herbaceous species with very short lifecycles colonise these areas during the terrestrial phase. In savanna floodplains, competition for light with forests is restricted to relatively small areas and water is sufficiently shallow to allow the growth of all life forms of the aquatic species as demonstrated by a much higher species diversity (Section 5.7). Perennial terrestrial herbaceous species are restricted, because they do not tolerate flooding but instead tend to colonise permanent terrestrial macrohabitats in the floodplain.

5.6.3 Invertebrates

Permanent water bodies in the floodplain are crucial for certain members of the invertebrate population to survive during the low-water period. Alternatively, many aquatic invertebrates overcome the terrestrial phase by entering resting stages and various other techniques. A large body of information about the adaptations of terrestrial invertebrates to periodic flooding has come from the vast research of Adis and collaborators in Amazonian river floodplains (Adis 1981, 1982, 1984, 1992, 1997; Adis and Messner 1997; Adis and Junk 2002; Erwin and Adis 1982; Franklin et al.,1997; Gauer 1997; Martius 1997a, 1997b; Höfer 1997). Many terrestrial invertebrate groups have representatives that are endemic to the *várzeas* and/or *igapó* (Table 5.5).

Table 5.5 Species richness of terrestrial arthropods in Amazonian uplands and floodplains.

	Terra firme	*Várzea*	*Igapó*
Spiders	472 (53)	130 (35)	210 (39)
Termites	90 (3)	12 (2)	11 (2)
Oribatid mites	71–74	13–18	11–47
Springtails	74		65

5.6.4 Fish

Fish respond very well to the monomodal predictable flood pulses in large river floodplains and interfluve wetlands as summarised by Junk et al. (2007). Similar adaptations have been described for species in the Pantanal of Mato Grosso (Resende 2011) and the upper Paraná River floodplain. Most species show a high tolerance for low oxygen concentrations, have a broad food spectrum including terrestrial material in their diet, tolerate changes in physical and chemical parameters of water quality, and are very mobile. All this confers protection against changes in habitat conditions because of the fluctuating water level. Many species accumulate fat during the high-water period when food is abundant; starve during the low-water period when food is scarce and gonadal development occurs; and migrate over long distances to spawn as the water table starts to rise. These features can be observed in floodplains of large rivers throughout the tropics and subtropics. Table 5.6 summarises the information.

Table 5.6 Large Amazonian river flood plains: effect of flood pulse on the fish.

Flood pulse	Changes: Area	Effects: Spawning at rising and high water High mortality at low water	Results: Selective pressure towards r-strategies	Flood pulse induced explicit seasonality in the non-seasonal evergreen rain forest biome
	Changes: Food supply	Effects: (a) Food: wide range (b) Detritus: high amount (c) Terrestrial food items: high amount (d) Physiologically conditioned starvation periods (e) Periodical accumulation of fat	Results: High mobility: (a) Small-scale horizontal and vertical movements (b) Migration for spawning (c) Migration compensating downstream drift of larvae and eggs	
	Changes: Physical and chemical habitat conditions	Effects: (a) Extreme local hypoxia (b) Changes in pH and electrolyte concentration (c) Temperature fluctuation, daily	Results: Many morphological and physiological adaptations to low oxygen concentrations	

Source: Adapted from Junk et al. 1997.

5.6.5 Other Vertebrates

Other vertebrates have synchronised their life cycles with the hydrological cycle. For Amazonia, this was shown for mammals, reptiles and amphibians by Junk and da Silva (1997) and for birds (Petermann 1997), and for the Pantanal by Petermann (2011) (birds), Strüssmann et al. 2011 (amphibians and reptiles), and Tomas et al. (2011) (mammals). Reptiles and aquatic mammals move to the river channel or permanent water bodies to escape the drought and enter their starvation period. Amphibians, reptiles, birds, and mammals have synchronised their reproductive cycles with the flood pulse to optimise reproductive success. Fish-feeding birds reproduce during low or rising water levels when there is ample food available. The large river turtles (*Podocnemis* spp.) and certain shore birds reproduce during the low-water period on sandy beaches when nesting places are protected from flooding. Caimans (*Melanosuchus niger* and *Caiman crocodilus*) reproduce at the end of the low-water period when there is little danger that the habitats of the offspring would dry out. Calvation and lactation of herbivorous manatees (*Trichechus inunguis*) coincide with the high-water period when the food supply for females is the highest. The birth rate of capybaras (*Hydrochoerus hydrochaeris*) peaks with the rising water when the animals have access to terrestrial and aquatic herbaceous plants. The breeding season of river dolphins (*Inia geoffrensis* and *Sotalia fluviatilis*) starts with the receding water level when fish become abundant in the river channel and the food supply for females is the best. Otters (*Pteronura brasiliensis* and *Lutra longicaudis*) breed all year round, with peaks during the low-water period.

5.7 Biodiversity

The RFS is positioned between permanently aquatic deep-water ecosystems, such as lakes and river channels, and the surrounding permanently terrestrial uplands. It results in permanent or periodic invasions of species from both areas in addition to the wetland-specific species. The wetland species have been defined as 'all those plants, animals, and microorganisms that live in a wetland permanently or periodically (including migrants from adjacent or distant habitats) or depend directly or indirectly on the wetland habitat or on other organisms, living in the wetland' (Gopal and Junk 2000). Junk (2000) provided the first synopsis on mechanisms for development and maintenance of biodiversity in neotropical floodplains.

In order to classify river-floodplain organisms, Junk et al. (2006a) differentiated distinct groups according to their location and/or behaviour:

1. Residents of the wetlands proper.
2. Regular migrants from deep-water habitat.
3. Regular migrants from terrestrial uplands.
4. Regular migrants from other wetlands.
5. Occasional visitors.
6. Those dependent on wetland biota (epiphytes, canopy invertebrates, and parasites).

This broad classification allows further subclassification of animal groups according to their adaptation and life-history traits. This approach is essential for biodiversity management from a landscape perspective.

5.7.1 Higher Vegetation

The vegetation cover of RFSs corresponds to a certain extent to the vegetation cover of the surrounding uplands. When precipitation and temperature are suitable for the development of closed forest in the uplands, large river floodplains would also be dominantly covered by a flood-tolerant forest, which is shown also for temperate zones. About 150 flood-tolerant species of trees grow in the Mississippi River floodplain and 33 in the Rhine River floodplain. Large floodplains in savanna areas will be covered by savanna vegetation, with fringes of floodplain forest along river channels and permanent lakes. Estimates of total tree species richness in Amazonian *várzea* floodplain indicate approximately 1000 flood-tolerant tree species with valid species names (Wittmann et al. 2006a). About 600 flood-tolerant tree species occur in Amazonian *igapó* floodplains (Wittmann et al. 2010b). These numbers are likely to increase with more sampling in the remote areas of the Amazon Basin.

Locally, however, similarity of tree species between *várzea* and *igapó* floodplains is only about 20%, mainly due to the differences in nutrient concentrations in flood waters and alluvial soils (Wittmann et al. 2010a,b). As Amazonian whitewaters are rich in suspension load, an additional factor explaining the comparatively low floristic similarity between whitewater and blackwater floodplains is alluvial dynamics which enable many pioneer species to colonise freshly deposited substrates in *várzea* floodplains (Wittmann et al. 2004; Peixoto et al. 2009) but not low dynamic blackwater and clearwater floodplains (Junk et al. 2015).

Amazonian floodplains are characterised by a strong tree species turnover along the flood-level gradient, with few tree species continuing along the entire amplitude (Ferreira 1997; Wittmann et al. 2004). Constant erosion and depositional processes allow for the establishment of several forest types and successional stages in *várzea* floodplains. The tree species in these forests differ with respect to establishment strategies, structure (diameters, height, etc.), ages, growth rates, and related functional traits (Wittmann et al. 2004, 2006b; Schöngart et al. 2010). Because environmental site conditions and the functional traits, composition, and richness of tree species are strongly tied to the flood-level gradient, the establishment of different forest types is predictable to a certain degree when considering the replacement of a few tree species and ecological niches along latitudinal and longitudinal gradients (Wittmann et al. 2004, 2006a). As alluvial processes in blackwater and clearwater floodplains are less active than those of the whitewater, the establishment of different forest types and successional stages are less pronounced. The compositional differences in *igapó* forest instead are mainly triggered by local differences in physical and chemical soil characteristics which may vary according to the physical characteristics of the rivers. This is shown by the *igapó* forests along the Negro River (Montero et al. 2014).

Amazonian floodplain forests contain a relatively large number of ecologically and geographically restricted (endemic) tree species. In the Amazonian *várzea*, about 10% of the 658 most important species of trees have been classified as endemic (Wittmann et al. 2013). Comparable data are still unavailable for the Amazonian *igapó*, but Prance (1979), Kubitzki (1989), and Mori (2001) postulated similar high level of endemism for tree species in the *igapó* along the Negro River.

In Brazilian savanna floodplains, however, the richness of tree species along episodically flooded riparian habitats and seasonal floodplains is much lower than in the perhumid Amazon. In the Pantanal, there are 756 woody species (Pott and Pott 1996, 1997, 2000), none endemic. Many flooded communities are composed of a few dominant species (Pott and Pott 1994; Nunes da Cunha and Junk 2001). An analysis of habitat preferences of 85 tree species showed that only 18 species were restricted to habitats with long inundation periods. Forty-five species preferred terrestrial habitats or habitats with short inundation periods and 22 species were found along the entire flood gradient (Nunes da Cunha and Junk 1999). Extrapolating these numbers to the 756 woody species occurring in the Pantanal, about 160 would show a preference for habitats with long inundation periods, and about 195 would occur along the entire flood gradient.

Similarly, tree vegetation of the riparian forests of the *cerrado* is a subset of the woody *cerrado* flora, and no endemic tree species exist. However, species richness in episodically flooded riparian habitats is generally higher than in adjacent terrestrial habitats (Oliveira-Filho et al. 1994; Felfili 1995; Damasceno-Junior et al. 2005). This is explained by groundwater availability and deficiency of soil water over most of the savanna landscape. Seasonal drought and fire frequency is comparatively low in riparian forests (Kellman and Meave 1997). Single species dominated tree communities occur in shallow flooded plains of the Pantanal but are scarce along river channels where high abundances are generally shared by several co-dominant tree species. The comparatively high richness in riparian forests is thus probably attributable to the high habitat diversity. Riparian zones consist of a gradient ranging from seasonally long-term floods to occasional short-term floods (Wittmann 2012).

Floristic inventories along *cerrado* streams have determined a high compositional variability. For example, Rodrigues and Nave (2000) analysed 43 inventories in riparian forests of the *cerrado* and found a generally low floristic similarity among different rivers and forest types. Similar results were obtained by Souza et al. (2004) in their comparison of nine floristic inventories in riparian forests of the upper Paraná River where the majority of all inventoried woody species were exclusively situated at one location. In their study of rare species, Metzger et al. (1997) reported that 42% occurred in one of the four investigated riparian forests. The comparatively high species richness in riparian forests possibly reflects also the high diversity of the *cerrado* landscape, where water and alluvial soils differ in sediment load and nutrients (Wittmann 2012). An exceptionally high diversity was also reported in the biodiversity hot spots along the lower Doce River in the southeastern Atlantic rainforest where species-rich assemblages of episodically flooded riparian forest were observed (Rolim et al. 2006; Resende et al. 2010).

The number of herbaceous plants is smaller in forested floodplains than in savanna floodplains. In the Amazonian *várzeas*, Junk and Piedade (1993a,b) list 388 herbaceous plant species, 47 of which are considered as aquatic or palustric. Of these, 28 are free-floating and 19 are rooted in the sediment. In the Pantanal, 247 of the 1147 registered herbaceous plants are aquatic or palustric. The difference between the Amazonian *várzea* and the Pantanal is due to the fact that the Amazonian habitats suitable for higher plants are covered by forests, restricting herbaceous plants by shading. Furthermore, the large flood amplitude is inappropriate for the growth of submersed plants. In the Pantanal, shallow flooding and lack of a dense forest cover provide suitable growth conditions for all types of aquatic macrophytes.

From the Paraná River floodplain, 955 species of higher plants have been recorded (Souza et al. 1997, 2009, Thomaz et al. 2004). Of these, 440 species (44.22%) are herbaceous, 62 aquatic macrophytes, 47 are epiphytes, 239 (24.02%) are woody plants, 149 scrubs, 145 lianas, 6 palms, and 16 are unclassified.

5.7.2 Animal Biodiversity

Data on animal diversity are scattered among locations and vary between taxa. Lists of the aquatic and terrestrial invertebrates are not yet complete and restricted to a few locations only. Vertebrates are the best studied but the available information varies between taxa and floodplain systems, and the numbers of recorded species are still undetermined.

Only a few lists of species have been recorded for the Amazonian floodplains. A total of 2500–3000 species of fish have been suggested for the Amazon River Basin. It has been estimated that about half of the total number occur in the large rivers and their floodplains, and the rest in the smaller tributaries. Most of the former species are distributed over a large area whereas the rest tend to have a restricted distribution and therefore are more vulnerable to extinction. Goulding et al. (1988) collected more than 450 fish species in a 1200 km long reach of the lower Negro and estimated that it contained over 700 species. Bayley (1983) reported more than 226 species in Camaleão Lake, a floodplain lake on an island in the Amazon near Manaus with a surface area of about 2 km^2 during periods of high water.

A comparison of the bird fauna of major South American wetlands indicates a clear difference between temperate wetlands with many swimming species as in the Andes (Fjeldsa and Krabbe 1990) and lowland tropical wetlands with many wading species and kingfishers (Reichholf 1975). The majority of lowland aquatic bird species occur in all tropical South American wetlands (e.g. the Pantanal, Amazonia, and the Llanos de Orinoco), but their abundance varies considerably. High flood pulses in forested floodplains of the Amazon hinder the occurrence of wading species such as the Jabiru (*Jabiru mycteria*) and Wood stork (*Mycteria americana*) which are frequent in the savanna, periodically flooded by low flood pulses. A census of 390 confirmed bird species in the Pantanal included 64 aquatic birds, 40 terrestrial but wetland dependent, and 286 terrestrial without any wetland dependency (Junk et al. 2006a).

In the Amazonian lowlands, 1042 bird species have been recorded and in the floodplains 729 species. Of the latter, 83 species are considered aquatic and 132 as occurring exclusively in floodplain macrohabitats, from sand bars to different types of floodplain forests. The remaining 514 species have been observed in the floodplains, but occur mostly in the uplands (P. Petermann, personal communication, 2011). With its 132 species the number of wetland-dependent terrestrial species is much higher in Amazonia than in the Pantanal. Many of these species have also been found in the Llanos, but few in the Pantanal (Remsen and Parker 1983). Petermann (1997) has recorded 204 bird species from Marchantaria Island in the Amazon River floodplain near Manaus.

For aquatic herpetofauna, habitat diversity in combination with extreme longevity and environmental stability explains the explosive radiation of these animals in wetlands (McCoy 1984). In the Pantanal, however, the lack of strictly endemic amphibians and reptiles is consistent with its environmental instability.

5.8 The Role of Rivers and their Floodplains for Speciation and Species Distribution of Trees

The composition of tree species in RFSs only partially matches the vegetation cover of the surrounding uplands because of alteration of local sites in flood (Sabo et al. 2005).

The extent of environmental alterations relative to the adjacent uplands is influenced by the length, depth, and periodicity of flooding, the chemical and physical energy of floodwaters, and the underlying geology and substrate (Naiman et al. 1993). Alluvial soils may have different substrate whose fertility contrasts with that of the uplands, leading to environmental filters that select tree species by their life history and functional traits that are necessary to colonise flooded habitats. This is found in the Amazonian *várzea* which contains many tree species whose biogeographical origins are in the Andes, the *cerrado*, and even temperate regions (Wittmann et al. 2013).

As *várzea* substrates are much more nutrient rich than soils of adjacent uplands, colonising tree species can dispose of traits linked to rapid establishment and growth rates and to elevated productivity (Wittmann et al. 2013). As the Amazon Basin is predominantly covered by mature forest, the comparatively high geomorphic dynamism in sediment-rich *várzea* is another environmental alteration that constantly creates new areas for tree colonisation and thus attracts light-demanding pioneer species which otherwise are rare in the Amazonian uplands. Riparian forests of savanna landscape such as the Brazilian *cerrado* add a shaded and moist ecological niche to the landscape, and thus attract drought-sensitive tree species from rainforest biomes as the Amazon and Atlantic domain (Meave et al. 1991; Oliveira-Filho and Ratter 1995).

Climatic variability, including that related to altitudinal changes, also contributes to riparian plant diversity, as streams flow from high to low altitudes across entire biomes (Naiman et al. 1993). Large rivers provide important vectors of range expansion for many semi-aquatic plant species and flood-tolerant trees by propagule dispersion by water and aquatic organisms, and, perhaps more importantly, by mitigating environmental difference, such as temperature and soil factors, in the aquatic and semi-aquatic environment against adjacent terrestrial systems (Sculthorpe 1985). This highlights the importance of large rivers and wetlands for biogeography, especially during periods of climate change, when river and wetland habitats serve as refugia for terrestrial and semi-aquatic plant species. These invasive lineages may significantly increase the local diversity in river wetlands (Sabo et al. 2005).

At evolutionary time scale, species invasion from surrounding uplands and from other biomes and climatic zones significantly contributes to wetland species richness, once the formerly terrestrial species adapted to moderate and brief and/or seasonal waterlogging as shown for the Amazonian large river floodplains (Wittmann et al. 2006a,b, 2013). The presence of endemic tree species in Amazonian seasonal floodplains indicates a relative geotectonic and climatic stability of rivers and floodplain systems in the Amazon Basin, probably since the evolution of the modern neotropical flora in the Upper Palaeocene (Burnham and Johnson 2004; Hoorn et al. 2010). This evolutionary period allows for the continuous or repeated addition of tree species in the floodplains which required the development of specific adaptations that allowed species to cope with transient oxygen deficiencies and the associated constraints (Parolin et al. 2004; Parolin 2009). The development of traits necessary to colonise flooded sites (reduction of metabolic rates and rapid growth and regeneration rates) suggests a trade-off between these traits and others associated with persistence (e.g. long-term storage of carbon, defence mechanism against herbivory, etc.) (Mitch and Gosselink 2000; Hultine et al.2013). Tree species adapted to prolonged periods of inundation hardly complete their life cycles when they become established outside flooded areas increasing the chance of their becoming endemic ("Tree Species Colonisation Concept' for *várzea* trees sensu Witmann et al. 2010b). In fact, most endemic tree species of the central Amazonian *várzea* floodplains colonise mean flood levels of more than 3 m, and up to 7 m, corresponding to an inundation period of 50–270 days a year.

In non-Amazonian river wetlands, endemic tree species are rare or even absent (Montes and San José 1995; Vencklaas et al. 2005; Junk et al. 2006a). This can be traced to climate changes during the past, in which there were intermittent periods of large-scale flooding or drought that interrupted the evolution of flood-adapted tree communities in the Brazilian savannas, Atlantic rainforests, and South Brazilian *campos*. Instead, most river wetlands of these biomes are either colonised by flood-tolerant tree species from adjacent terrestrial habitats or share widely distributed flood-tolerant tree species (Wittmann 2012). The occurrence of an increased number of endemic terrestrial invertebrates adapted to Amazonian floodplain conditions suggests a similar pattern in other groups of organisms.

5.9 Biogeochemical Cycles in Floodplains

Biogeochemical cycles of the floodplain are driven not only by flood pulses but also by other factors, mainly rainfall and the influx of water from the surrounding uplands. Water from the river enters the floodplain as the water rises, carrying nutrients and sediments of different quantities depending on the physical environment. Large amounts may come from Amazonian whitewater rivers; very small amounts from blackwater rivers; and intermediate amounts from clearwater rivers. The distribution of the material in the ATTZ is not homogenous owing to the complex relief of the floodplain and its plant cover. Normally, coarse sediments are deposited near the river channel, whereas small amounts of very fine material reach remote floodplain areas. Dissolved material is diluted by rainfall and water from the uplands. Nutrients such as nitrogen and phosphorous are quickly taken up by the algae and aquatic macrophytes.

5.9.1 Biomass and Net Primary Production

Algae (phytoplankton and periphyton), herbaceous plants (terrestrial and aquatic species), and woody plants (shrubs and trees) contribute to primary production in the floodplain. Production by algae and aquatic macrophytes is limited to the aquatic phase and that of terrestrial herbaceous plants to the terrestrial phase. For woody plants, primary production occurs during the terrestrial and aquatic phases with a reduction in wood production during the flood period. Biomass and production of different plant communities of the *várzea* are summarised in Table 5.7.

5.9.1.1 Algae

Algae (phytoplankton and periphyton) have a short life cycle. Schmidt (1973) calculated a turnover time of 1.7 days for phytoplankton. In Lago Castanho, net primary production (NPP) was estimated at 3 MgC ha^{-1} year^{-1}. Values of the same order of magnitude were reported in other studies (Fisher 1979; Melack and Fisher 1990 and others). NPP and algal density vary considerably during the annual cycle.

Phytoplankton studies in the lakes of the Pantanal show a pattern different from that in the Amazon River floodplain as summarised by Loverde-Oliveira et al. (2011). Most of

Table 5.7 Biomass and net primary production per hectare and year of different plant communities in the floodplain of the Amazon River.

Plant Community	Biomass	Net primary production
Phytoplankton	10–40 kg	6 t
Periphyton	68–? kg	7.6 t
Annual terrestrial plants	Up to 10 t	Up to 20 t
Perennial grasses	30–80 t	Up to 100 t
Emergent aquatic macrophytes	3–20 t	Up to 60 t
Várzea forest	300–600 t	Up to 33 t

Source: Data from Junk 1997.

the ATTZ is shallow flooded with light penetrating to the bottom. It favours the growth of rooted and free-flowing aquatic macrophytes which compete successfully with algae for nutrients and light. As the water level rises, the water in the macrophyte stands stay clear. As it recedes and macrophytes decline and decompose, algae start to grow. The most intense and largest blooms of cyanophytes occur during the low-water period in the remaining water bodies.

Periphyton grows on surfaces such as leaves, branches, and stems of the floodplain forest, at the bottom of shallow water bodies, and in the root bunches of aquatic macrophytes. However, field studies on periphyton are difficult to carry out.

5.9.1.2 Herbaceous Plants

Flood pulses divide herbaceous plants into two major groups that reflect their growth during the terrestrial and aquatic phases. The same area can be occupied during the terrestrial phase by one plant community and by another during the aquatic one, thereby increasing total primary production per unit area.

Critical factors for the growth of herbaceous plants are light availability, nutrient content in soil and water, and the availability of water during the dry season. Light penetration in *várzea* and *igapó* lakes varies between tens of centimetres to about 4 m depth. The high flood pulse, however, hinders the growth of submerged plants, which are thus easily subjected to unsuitable light conditions. Because a considerable percentage of the light is absorbed by the canopy, the growth of terrestrial and aquatic plants is hindered in the flooded forest.

In the *várzea*, large-scale erosion and deposition processes lead to the destruction of the floodplain forest and provide favourable light conditions for emergent and free-floating herbaceous plants. The nutrient status in soil and water promote high productivity. A lack of soil moisture for a few months of the dry period may negatively affect the growth of terrestrial herbaceous plants.

Forests cover most of the ATTZ in blackwater river floodplains because of limited erosion and deposition processes. The low nutrient status in water and soils limits the growth of herbaceous plants both on land and in water. Drought stress on sandy beaches is an additional stress factor for plant growth.

In flooded savannas, as in the Pantanal, light conditions for the growth of herbaceous plants are much better than in Amazonia because large areas are not shaded by trees. The depth of flooding reaches only up to 3 m, and light may penetrate to the bottom. Despite the relatively low nutrient status of the water and soils, the growth of emergent aquatic macrophytes is exuberant. During the terrestrial phase, the growth of herbaceous plants is hindered by drought stress and frequent wild fires.

Plants rooted in the ground generally grow vertically. The productivity can be expressed as weight increase per unit time and unit area. In the *várzea*, the biomass of annual herbaceous plant communities growing during the terrestrial phase and shortly before flooding was 3.2–8.7 Mg ha^{-1}.The biomass of the perennial C4 grass *Paspalum fasciculatum* during 230 days was 60 Mg ha^{-1}, of which about 10% was dead material (Junk and Piedade 1997). In the aquatic phase the following biomass values (in Mg ha^{-1}) were recorded for well-developed stands: *Paspalum repens* (12.7–22.1), *Oryza perennis* (12.6–19.1), *Hymenachne amplexicaulis* (17.2–22.7), and *Luziola spruceana* (5.5–6.4) (Junk and Piedade 1997). The highest recorded biomass (approximately 80 Mg ha^{-1}) and productivity (approximately 100 Mg ha^{-1} year^{-1}) values were measured for the tall perennial C4 grass *Echinochloa polystachya* (Piedade et al. 1991). The successful occupation of the same area by three different plant communities was observed by Junk and Piedade (1993a, 1993b). Different annual species start during the terrestrial phase, the tall grass *Hymenachne amplexicaulis* develops during the beginning of the aquatic phase, and later free-floating plants take over at high water. Accumulative biomass reaches about 30 Mg ha^{-1} and NPP about 60 per year assuming a monthly rate of loss of 25%.

Free-floating plants expand their population horizontally. A better indicator of productivity is the relative growth rate (RGR) which is the increase in grams of biomass per gram of plant material per day. This unit is used to determine the comprehensive doubling time, indicating the time that the plants need to double their biomass or the occupied area (Mitchell 1974). According to Junk and Williams (1984), the quickest growth are those of *Pistia stratiotes, Eichnornia crassipes*, and *Salvinia auriculata* with a doubling time of 7.2–6.9 days. Exponential growth of these species results during rising water level in the occupation of large areas by a few specimens, which survived the dry phase. Low nutrient status and low pH values inhibit the growth of free-floating macrophytes in blackwater floodplains. Very few studies have been carried out on the biomass of herbaceous plants in the Pantanal. The values achieved are much smaller than those determined for the Amazonian *várzea* and reflect the low nutrient status of the water and soils.

5.9.1.3 Trees of the Flooded Forest

Above-ground wood biomass (AGWB) and NPP in the Amazonian floodplains show a wide range, depending on edaphic conditions and the successional stage as defined by ages and canopy heights of varying stands, mean wood density, and diameter growth rates (Worbes et al. 1992; Piedade et al. 2001; Schöngart et al. 2010; Wittman et al. 2010b). AGWB increases rapidly during the first 50 years of successional development in the highly dynamic *várzea* forests of the central Amazon, from 3 to 18 Mg ha^{-1} (stand ages of 2–7 years) to 98–117 Mg ha^{-1} (stand ages of 12–20 years) and may reach the maximum values of 258–261 Mg ha^{-1} (stand ages corresponding to 44–50 years) (Worbes 1997; Schöngart et al. 2010). However, the estimated AGWB of mature *várzea* forests with stand ages of 80–240 years does not increase farther. Estimates for the eastern Amazon regions indicate AGWBs of 193 ± 18 Mg ha^{-1} for the tidal *várzeas* in the estuary of the Amazon (Almeida et al. 2004), and 226 ± 87 Mg ha^{-1} for the secondary várzea in the Santarém region (Lucas et al. 2014), while the western Amazon Basin has higher stocks of 270S ± 40 Mg ha^{-1} (Malhi et al. 2006).

Few estimates are available for the blackwater *igapós*. Stadtler (2007) estimated an AGWB of 227–304 Mg ha^{-1} for the *igapó* of the Amanã Lake in the central Amazon Basin, while the *igapó* of the Uatumã River had much lower AGWB of 126–173 Mg ha^{-1} (Taghetta 2012). Both studies indicate an increase in biomass stock with decreasing flood heights. Several estimates of the AGWB are available from stands in other Brazilian wetlands. Wittmann et al. (2008) estimated an AGBW of 125 Mg ha^{-1} in an inundated riparian forest at the Miranda River, southern Pantanal. In the wetland forests in northern Pantanal, dominated by *Vochysia divergens*, AGWB stocks (37–218 Mg ha^{-1}) increase with rising stand age of 64–124 years (Schöngart et al. 2011).

The few estimates of NPP in floodplain forests include that of Schöngart et al. (2010 a,b) who estimated the above-ground NPP for three successional stages in the central Amazon *várzea*, considering fine litterfall (6.4–13.7 Mg ha^{-1} $year^{-1}$) and AGWB increment (6.8–16.7 Mg ha^{-1} $year^{-1}$) as well as estimates for volatile organic compounds and above-ground losses through herbivory and leaching (Clark et al. 2001). Young successional stages in the *várzea* are among the most productive tropical forests worldwide, with NPP values even higher than those of tropical mahogany plantations or pine of comparable stand age. For the nutrient-poor *igapó* floodplain forests no estimate for above-ground NPP is available. Considering the low production of AGWB and the reduced fine litterfall, a much lower NPP may be expected.

5.9.2 Decomposition

An accumulation of dead organic matter in floodplains is expected because of the large amount of primary production by herbaceous plants and the floodplain forest. But organic material under the tropical climates decomposes very rapidly, and the change between terrestrial and aquatic phases favours decomposition.

Decomposition experiments in water and on land indicate different dynamics. In experiments with different species of herbaceous plants in litter bags, in the floating mats of *Paspalum repens* in a *várzea* lake, the weight loss after a 6-month exposure was 80–90%, and after 338 days 3–9% of the exposed material remained. The most important changes occurred at the beginning of the decomposition. After 14 days, about 50% of the exposed material but 50–95% of the K, Na, Mg, and P contents were lost as summarised in Furch and Junk (1997).

The release of dissolved substances into water was confirmed by Furch and Junk (1992) in container experiments with *Echinochloa polystachya* and *Paspalum fasciculatum*. An abrupt decrease in oxygen content and an increase in dissolved organic carbon in the water indicated high decomposition rates. The concentrations of K, Ca, Mg, P (PO_4), and N ($NO_3 + NO_2 + NH_4$) increased abruptly. The ion-poor acid groundwater used for the experiment was converted very quickly to ion-rich, well-buffered water, with K, NH_4 and HCO_3 ions as the dominant causes.

On land, the speed of decomposition of herbaceous plants depends on humidity. No weight loss was recorded in experiments with sun-dried aquatic macrophytes maintained under a roof in dry conditions. But after being exposed to sixteen 30-min periods in water during intervals of 2, 4, and 8 days, reductions in dry weight and in bioelements (sum of K, Ca, Mg, N, and P) of 30% and 55–70%, respectively, were found. The leaching of soluble organic and inorganic compounds is therefore of major importance at the beginning of the decomposition process.

With about 6.4–13.8 Mg ha^{-1} in várzea and 6.7–6.8 Mg ha^{-1} in *igapó* floodplain forests (Schöngart et al. 2010), leaf litter contributes a considerable amount of organic material to the organic matter budget. The decomposition of leaf litter in water varies among species. After 120 days *Cecropia latiloba* was observed to lose 86%, *Salix humboldtiana* 82.2%, and *Pseudobombax munguba* 58.5% of the original biomass and 89% of their bioelements. However, in species with very sclerophyllic leaves, decomposition is retarded.

A similar study of leaf litter exposed on land showed a weight loss of about 20% in the *várzea* and 10% in the *igapó* after 100 days, despite 500 mm of rainfall during the experimental period (Irmler and Furch 1980). The authors calculated that 6.3 years in the *igapó* and 2.8 years in the *várzea* were needed for 95% decomposition. Experiments with dried fresh leaves of *igapó* tree species *Eschweilera coriacea* and *Buchenavia ochroprumna* showed their quicker decomposition rates of 1.0–1.7 years for 95% decomposition. Meyer (1990) calculated 1.9–2.3 years in the *igapó* and 0.9–1.4 years in the *várzea* for 95% decomposition of leaf litter.

Wood decay is one of the least studied major processes in ecosystems in general, and in river floodplains in particular. According to Martius (1997b), the annual input of wood litter to the forest floor of the *várzea* is about 6 t ha^{-1}. During the terrestrial phase, fragmentation occurs mainly by termites and to a much lesser extent by wood-feeding beetles. During the aquatic phase the larvae of the mayfly *Asthenopus curtus* build their tunnels even in hardwood. Fragmentation by terrestrial and aquatic animals, greatly enhances microbial attack and thus accelerates wood mineralisation.

5.9.3 The Nitrogen Cycle

The different sources of nutrients, the quick uptake and turnover by algae, the floodplain forest and herbaceous plants in combination with the water-level fluctuations make nutrient balances difficult. In the case of nitrogen compounds, biogenic nitrogen fixation and denitrification makes the situation even more complex as shown by Kern and Darwich (1997) and Kern et al. (2010) for Camaleão Lake (Tables 5.8 and 5.9).

The balance indicates that the floodplain forest acts as an important source of nitrogen in the *várzea*. Whereas non-symbiotic nitrogen fixation is restricted to

Table 5.8 Nitrogen flux in the Camaleão Lake over an area of 650 ha from June to March 1992–1993.

	Input		Output		Balance
	t	%	t	%	t
Water flow	6.8	70.1	3.9	43.8	2.9
Rain	1.3	13.4			1.3
Nitrogen fixation	1.6	16.5			1.6
Denitrification			5.0	56.2	-5.0
Total nitrogen flux	**9.7**	**100**	**8.9**	**100**	**0.8**

Source: Data from Kern and Darwich 1997.

Table 5.9 Nitrogen input and output fluxes from *várzea* forest at Camaleão Lake.

	$kgN\ ha^{-1}\ year^{-1}$	Percentage
Nitrogen input		
Symbiotic N_2 fixation	12.9–16.1	48–54
Non-symbiotic N_2 fixation	4.1	12–17
Dry and wet deposition	2.6	8–11
Exchange with the river	4.4–10.5	18–32
Total N input	**24.0–33.3**	
Nitrogen output		
Denitrification	12.5	56–63
Leaching	1.2–3.8	6–17
Exchange with the river	6.0	27–30
Total N output	**19.7–22.3**	

Source: Data from Kern et al. 2010.

the terrestrial phase, there is a whole year of nitrogen fixation in the rhizosphere of nodulated legume trees. Nitrogen output by nitrogen fixation within the forest was calculated to be 17–20 kg ha^{-1} for one hydrological cycle (one year).

5.9.4 Nutrient Transfer Between the Terrestrial and Aquatic Phases

The terrestrial and aquatic phases of the floodplain interact with each other. During the low-water period, the dry parts of the floodplain are colonised by terrestrial plants which take up nutrients from the sediments. With rising water level, these plants are flooded, decomposed, and also transfer their organic matter and nutrients into the water, increasing the productive capacity of the aquatic phase. When the floodplain becomes dry, the organic matter of the decomposed aquatic vegetation acts as a fertiliser for the sediments carrying terrestrial plants. Therefore, floodplains with a large ATTZ and an abundant terrestrial and aquatic herbaceous vegetation have a larger productivity than expected only based on the nutrient content of the incoming river water. In floodplains of the blackwater rivers, this exchange is restricted without a major cover of terrestrial herbaceous plants. The floodplain forest reduces nutrient losses by a thick layer of roots at the soil surface that quickly take up nutrients from the decomposing leaf litter.

The water in the remaining shallow *várzea* lakes can become enriched with electrolytes, reaching conductivity values that are about 20 times higher than in the main river channel, as shown for the Camaleão Lake by Furch and Junk (1993).

A budget of bioelements introduced by the Amazon River into the ATTZ of a hypothetical *várzea* lake which is about 4 m deep and store and release bioelements from leaf litter, wood litter, and annual and perennial herbaceous plants was calculated by Furch and Junk (1997). They showed that up to 80% of K and more than 90% of N and P were derived from decomposing terrestrial vegetation, with the remainder introduced in dissolved form by the river. Internal nitrogen cycles have been discussed in Section 5.9.3. In contrast, the main source of sodium, calcium, and magnesium is the river water. Sediments, unlike other sources, are a long-term nutrient store in the floodplain but the readily available portion is relatively small. They, however, are used by rooted herbaceous and woody plants.

Decomposing organic matter consumes oxygen. Therefore, most *várzea* lakes are strongly hypoxic, and at high water level, anoxic near the bottom. A simple model of the oxygen budget of a floodplain lake indicates that 647 g of oxygen per square metre is available during a 240-day flood period. Of this 89% is derived from diffusion, 4.6% from the river, and 6.4% from phytoplankton production. This amount is sufficient to oxidise about 605 g of dry organic matter per square metre. Biomass produced by the annual terrestrial plants in the ATTZ varies between 320 and 720 g dry organic matter per square metre, of which about 90% decomposes in water. The biomass of perennial plants reaches 1560–5700 g of dry organic matter per square metre, of which up to 50% decomposes in water. Leaf litter and woody litter produced by the *várzea* forest reach values of 780–1360, and 600 g of dry organic matter per square metre, respectively, a considerable part of which decomposes during the aquatic phase. These amounts are large enough to use up all oxygen available in the lake and explain the strong hypoxia in most *várzea* lakes (Furch and Junk 1997). With a reduction of the ATTZ against the permanent area of the lake, nutrient recycling from decomposing organic material is reduced but availability of oxygen improves. It has been postulated that floodplain lakes with a stable water level behave like the classical model of a lake.

5.9.5 Food Webs

Food webs can be divided into terrestrial and aquatic parts, but the two interact with each other. Aquatic food webs rely strongly on algae with a high protein content, as shown by stable isotope studies on certain fish species (Forsberg et al. 1993; Oliveira et al. 2006). Wood is used mostly during the terrestrial phase by termites and some beetles and their larvae. They facilitate the attack of fungi and bacteria during the aquatic phase. The aquatic larvae of the mayfly *Asthenopus curtus* are important wood destroyers in the aquatic phase and build their tunnels in the wood. The large number of detrivorous fish species may make greater use of the organic matter than the detritus itself.

Aquatic and palustric herbaceous plants contribute little to the diet of the fish. Forsberg et al. (1993) and Melack and Forsberg (2001) have concluded that because of their low food values, C4 plants are selectively avoided by fish and other herbivores of the *várzea*. In fact, the average nutrient value of C4 grasses does not differ considerably from that of terrestrial C3 grasses in the *várzea* (Ohly and Hund 2000). The practical experience of cattle and buffalo ranchers demonstrated that C4 grasses are highly valuable food items for domestic animals in the *várzea*. We suggest that their low contribution in Amazonian food webs is related to the dramatic reductions in post-Columbian times of the populations of large native herbivores rather than to a low food value. Pre-Columbian populations of capybara, turtle, and manatee that colonised the Amazonian river floodplains and that of marsh deer and capybaras in the Pantanal were very large. Today, however, these populations are nearly extinct and their place in food webs has been taken over by domestic animals. In the other tropical and subtropical floodplains, herbaceous plants are essential for terrestrial ungulates as also shown for the floodplains of Africa, e.g. the Okavango Delta (Ramberg et al. 2006).

The use of flowers, fruits, and seeds from the floodplain forest and the consumption of terrestrial invertebrates by fishes and other aquatic animals well demonstrate the importance of the terrestrial phase to the aquatic one. Fishing birds make use of the fish stocks, ducks use detritus, algae, aquatic macrophytes, and invertebrates, and some passeriforme birds and bats feed on the adults of aquatic insects. The co-evolution between fruit-feeding fishes and trees for at least 13.5 million years has been postulated. We postulate that this behaviour occurs in all floodplains where fish has access to fruits and seeds during the aquatic phase.

5.10 Management of Amazonian River Floodplains

5.10.1 Amazonian River Floodplains

The natural vegetation cover and the nutrient status of the river floodplains determine their management options. For the forested, highly productive floodplains such as the Amazonian *várzeas*, there are multiple options but all of them are linked to the flood pulse. This has been clearly demonstrated by the management activities of the traditional population in the Sustainable Management Reserve of Mamirauá near Tefé. Timber harvesting concentrates on the period of rising and high water, which allows felling under dry conditions and subsequent extraction of the logs by boat during peak flood (Table 5.10).

Agriculture is restricted to higher areas near the river channel which remain dry long enough to allow planting and harvesting of staple crops and their transport to consumers. The production of vegetables is highly profitable but labour intensive and restricted to areas near urban centres. Only a few fruit trees (e.g. mango and breadfruit) are flood-resistant enough to grow on elevated sites in the floodplain.

Extensive animal ranching is spreading all over the *várzea*, but land productivity is low and requires large-scale destruction of the floodplain forest, with negative impacts

Table 5.10 Annual cycle of economic activities of the riparian human population of the várzea related to the seasonal variation of the water level.

Season	Month	Economic activities of riparian human population
Dry	January	Cattle ranching, agriculture, fishery
	February	Cattle ranching, forestry, agriculture, fishery
Floods	March	Cattle ranching, forestry, agriculture, fishery
	April	Cattle ranching (first half of April), forestry, agriculture, fishery
	May	Forestry
	June	Forestry
	July	Forestry
Dry	August	Cattle ranching starts, agriculture, fishery
	September	Cattle ranching, agriculture, fishery
	October	Cattle ranching, agriculture, fishery
	November	Cattle ranching, agriculture, fishery
	December	Cattle ranching, agriculture, fishery

Discussed in detail in Junk et al. 2010c.

on timber production and fisheries. High floods place animal stocks at risk and require the transfer to upland pastures which is costly. Conversely, during periods of extreme low water, access to water becomes problematic for the animals.

Timber extraction has been a profitable but unsustainable activity for several decades but it is regulated mainly by demand for timber with specific properties, with no consideration for the need for a floodplain-adapted management strategy (Schöngart and Queiroz 2010). Species and site-specific difference in felling cycles and minimum logging diameter (MLD) were not considered for many years in forest legislation (e.g. Normative Instruction No. 5, 11 December 2006 [IBAMA]), thus placing the management of timber resources derived from many commercial species at risk which meanwhile disappeared from local and regional markets. Depending on wood density, various species of floodplain trees show different diameter and volume growth rates as well as highly varying felling cycles and MLD (Rosa 2008; Schöngart 2010).

Based on tree growth models, the concept of Growth-Oriented Logging (GOL) was developed for the central Amazonian *várzea* (Schöngart 2008). GOL relies on species specific felling cycles and MLDs as well as management systems that also take into account information on the population structure and regeneration of tree species in different successional stages. Basic statements and information related to GOL were recently included in innovative forest legislation established for the *várzea* of the State of Amazonas by the Secretary of Environment and Sustainable Development (Normative Instruction No. 9, 12 November 2010). In addressing felling cycles, the legislation considered species-specific MLDs, distinguishing between commercial tree species with wood densities less than $0.60\,g\,cm^{-3}$ and more than $0.60\,g\,cm^{-3}$.

Slow growing high-density woods in late secondary stages can be used by polycyclic, selective systems with a felling cycle of 30 years and species-specific MLDs. The sustainable use of fast growing low-density woods in early and late secondary stages is achieved by monocyclic systems with a rotation period of 20–60 years according to species-specific MLDs (Junk et al. 2010a,b,c).

Fishery is a very important activity along all large Amazonian rivers. Its potential has been estimated at about 900 000 t annually (Bayley and Petrere 1989). However, uncontrolled use of the stocks has greatly reduced the numbers of several large highly valuable species such as *Arapaima gigas (Pirarucu)*, *Colossoma macropomum (Tambaqui)*, and large catfishes *Brachyplatystoma vailantii (piramutaba)* and *B. flavicans (dourado)*. Furthermore, the construction of reservoirs for hydroelectric power generation on some tributaries, e.g. the Madeira River, is interrupting the migration of some species to their spawning grounds as is the case for large catfish. By contrast, controlled fishery of *Arapaima gigas* in the Sustainable Management Reserve of Mamirauá has led to the recovery of the stocks and increasing yields.

The economic potential of the extremely nutrient-poor floodplains along the Negro River and its tributaries is very low. Trees grow very slowly and do not allow sustainable timber extraction (Fonseca et al. 2009; Schöngart 2010; Scabin et al. 2012). The infertile soils are suitable only for subsistence agriculture and animal ranching by the local population. Professional fishery concentrates on a few species (e.g. *Semaprochilodus* spp. and *Brycon melanopterus*) during spawning migration at the beginning of the rising water level, and on large catfish. In the tributaries of the middle Negro River, ornamental fish are of great regional importance, with the neon tetra *Paracheirodon axelrodii* accounting for 65–80% of the export. Junk et al. (2007) consider this fishery as beneficial for the region, because the previous high losses during the catch and transport of these fish have been reduced and do not place the stocks at risk. The riverine human population involved in these activities is not involved in environmentally determined timber extraction and animal ranching. Because of their greater variability and high species diversity, the river floodplains of the blackwater rivers should be protected and used mostly for sports fishing and ecotourism.

5.10.2 Savanna Floodplains

The most natural use of shallowly inundated savanna floodplains is low-density cattle ranching as practiced in the Pantanal of Mato Grosso for about three centuries. During this period, slowly ranchers have increased the area under pasture, modifying the landscape but maintaining the natural habitat and species diversity. Maintenance of the natural flood pulse has been very important. Other economic activities are less important in this location. Ecotourism is concentrated in areas of easy access, e.g. along the Transpantaneira Park Road, and in several ranches along rivers. There is a local demand for timber and fish.

Changes recently have become apparent. The ranch owners have come under strong economic pressure to increase productivity which required an increase in pastureland by replacing natural vegetation with exotic grasses (*Brachyaria* spp.) (Junk and Nunes da Cunha 2012a). Frequently, anthropogenic fire negatively affects the species of trees, many of which cannot tolerate fire stress. The development of fishery as an uncontrolled sport has reduced fish population with negative socio-economic impact on the local riverine human population.

Modern agro-industrial developments have provided the technical means to transform floodplains with a low flood amplitude into areas of conventional agriculture and intensive animal ranching in polders. These polders exclude periodic floods and thus permit cattle ranching on artificial pastures for the entire year. Polders with a large internal channel system that excludes floods but distributes water are also used, pumping water from the river to irrigate melons and staple crops throughout the year. Both systems are in use, covering large areas of the Bananal Island floodplain in the middle Araguaia River basin (W.J. Junk et al., unpublished).

This type of management not only transforms the floodplain concerned by modifying its natural hydrologic regime but also affects the connected rivers. Intensive agriculture involves the application of large amounts of fertilisers and pesticides, part of which travels back to the river with surplus water from the polders, but the related statistics and the environmental impact of the agrochemicals are not available.

Although the Pantanal of Mato Grosso is protected by law it is already being invaded by agro-industries for soybean and sugar cane plantations (Junk and Nunes da Cunha 2012b). Channelization of the Paraguay River (*hidrovia* project) has long been a case of heated political discussion, despite the many scientific studies indicating multiple negative impacts on the entire system (Ponce 1995; Hamilton 1999 and many others).

5.11 Policies in Brazilian Wetlands

The requirement of modern societies for the use of RFSs are changing worldwide. Increasing importance is given to land reclamation, use of water for domestic and industrial purposes, irrigation agriculture in the surrounding uplands, production of hydroelectric energy, flood control, and navigation. Such requirements tend to mobilise a number of outside stakeholders who exert strong pressure on politicians and planners to the detriment of the local traditional floodplain population (Junk et al. 2014a,b).

Water availability has become a political problem in Brazil. In the densely populated regions of southern Brazil, destruction of riparian wetlands has led to a shortage of water during the dry season and catastrophic flooding during the rains (Piedade et al. 2012; Junk and Nunes da Cunha 2012b, 2014c; Bleich et al. 2014). An extended dry period since 2014 has amplified this problem. Scientists predict an increase in extreme wet and dry events because of changes in global climate. The government intends to ameliorate the increasing periodic shortage of water in southern Brazil with technical solutions such as wastewater purification, reduction of water loss in a distribution system, improvement of industrial water use, and water transfer among river stems, even from Amazonia to São Paulo. The ecologically intelligent approach to protect and restore wetlands and riparian forest along the headwaters of streams and rivers, has not been taken into consideration, even though it is an indispensable step that must accompany the technological measures.

In 2012, despite the warnings of the ecologists, the Brazilian government enacted the new Forest Code (Law No. 12.652, 25 May 2012) although neither this version nor the old one specifically mentions wetlands. In the old code (Law No. 4771, 15 September 1965) the protection of forests along streams and rivers was determined by the width and maximum water level of the rivers, which at least allowed for the protection of wetland areas. In the new Forest Code, protection is established using the 'regular' water level. This definition is weak and opens up opportunities for the destruction of the fringing wetland areas (Souza et al. 2011). The new code enables agroindustry to transform floodplain areas into dry croplands.

Understanding the new Brazilian Forest Code was seriously hampered by the lack of definitions for water resources and wetlands, and their delimitations. According to Junk et al. (2014a) 'Wetlands are ecosystems at the interface between aquatic and terrestrial environments; they may be continental or coastal, natural or artificial, permanently or periodically inundated by shallow water or consist of waterlogged soil. Their water may be fresh, or highly or mildly saline. Wetlands are home to specific plant and animal communities adapted to their hydrological dynamics'. This definition is similar to those derived by Junk and Welcomme (1990) and Junk et al. (1989).

Delimitation of large wetlands now includes internal, permanently terrestrial areas as parts of the wetland complex, as they are indispensable for the maintenance of wetland biodiversity. 'The extent of a wetland can be determined by the border of the permanently flooded or waterlogged area or in the case of fluctuating water levels, by the limit of the area influenced during the mean maximum flood. The outer borders of wetlands are indicated by the absence of hydromorphic soils and/or hydrophytes and/or specific woody species that are able to grow in periodically or prominently flooded or waterlogged soils. The definition of a wetland area should include, if present, internal permanently dry areas as these habitats are of fundamental importance to the maintenance of the functional integrity and biodiversity of the respective wetland' (Junk et al. 2014a).

These definitions can be used by the government to establish modern environmentally friendly legislation, which in the long run would guarantee water for a growing population while avoiding both damage to the environment and economic and social harm to local, low-income populations that occupy riparian areas at increased risk of floods.

A major problem of Brazilian legislation in dealing with water resources is the lack of a general definition of the term. The following definition was proposed in 2013 for the Ministry of the Environment and later published (Junk et al. 2014b): 'Water resources include rain water and all water bodies, natural and artificial, superficial and subterraneous, continental, coastal, and marine, of freshwater, saline or salt water, standing (lakes and reservoirs) and flowing (rivers – intermittent, ephemeral, or perennial – and its tributaries, waterways and artificial channels) and all types of permanent and temporary wetlands'. This definition confers upon wetlands the legal status of a water resource and thus specific protection. The change in global climate will affect the many types of wetlands found all over Brazil, requiring immediate steps for their protection, to maintain water in the landscape. Preliminary indications of the changes to come can already be seen even in the discharge pattern of the huge Amazon River which during the last few decades has been influenced by increases in extreme droughts and floods (Piedade et al. 2013).

The appointment of responsibilities for the protection of the environment between the central government and the state governments has led to a complex judicial situation that can hardly be understood by non-specialists in the many-faced issues. Additionally, as law enforcement occurs in Brazil's state and municipalities, the lack of a precise guiding policy of the federal government has hampered the establishment of laws easily transferable to practical levels. The result, complicated by the dimensions of Brazil, and the multiplicity of biomes, is that many private and governmental activities may take place without proper legal control.

5.12 Discussion and Conclusion

The periodic flooding of terrestrial environments and periodic droughts affecting aquatic ones are common. The scale of the affected area is frequently underestimated. Traditional sciences, dealing with ecology and different types of land use tend to classify ecosystems of the continents in three categories: aquatic, wet, and terrestrial. Until recently, even permanently wet ecosystems were not properly recognised by national, state, and municipal planners. As a result, enormous wetland areas have been transformed into permanently terrestrial or, to a minor extent, permanently aquatic ecosystems. In the USA, about 53% of the wetlands were lost between the 1780s and 1980s (Mitsch and Hermandez 2013). In Russia, wetlands are still not recognised as separate or distinct ecosystems (Robarts et al. 2013).

The periodic change between terrestrial and aquatic conditions establishes flood-pulsing ecosystems. Their study, at the interface between ecology and limnology, has included specific scientific theories, and their management requires new approaches based on the acquired knowledge. Despite the elaboration of a theoretical framework and the numerous publications on river floodplains over the last few decades, many countries are still not dealing adequately with flood-pulsing ecosystems.

Considerable differences exist in the environmental conditions among large flood-pulsing wetlands in different climatic zones. In the floodplains of the humid tropics, the flood pulse is the principal driver of ecological processes. In high latitudes, the impact of the light-temperature pulse (summer–winter pulse) partially overrides the impact of the flood pulse. In savanna wetlands, fire, and drought become additional stress factors besides floods.

Biogeographic peculiarities determine the composition of plants and animal communities and their role in floodplains. For example, the large number and diversity of ungulates has a strong impact on African floodplains, whereas in South America these animals are less abundant and so less important. A comparison among the different Brazilian floodplains provides an understanding of the functioning of these complex ecosystems and facilitates the development of strategies for their sustainable management and protection. Furthermore, the following generalisations to a certain extent can be extrapolated to other tropical and subtropical floodplains. These generalisations are highlighted to indicate their importance.

The area recognised as wetland is heavily underestimated in many countries, as inventories only consider permanently wet area which corresponds to the area at the low-water level in flood-pulsing systems. For example, Eva's estimate for South America includes 270 500 km^2 of flooded tropical forests, 320 900 km^2 of flooded savanna, 13 000 km^2 of flooded shrublands, 106 900 km^2 of moorlands and heath, 9400 km^2 of salt pans, and 220 200 km^2 of natural and artificial water bodies (Eva et al. 2004) but the estimates of Junk (2013) are about three times higher and correspond to about 20% of the land area of South America. In Brazil about 20% of the land surface is periodically inundated. Kandus et al. (2008) estimated a similar figure for Argentina.

The major driving force in wetlands is hydrology (Mitsch and Gosselink 2000). In flood-pulsing wetlands the extent, amplitude, frequency, predictability, and timing of flood and drought events determine species occurrence, distribution, and behaviour as well as biotic and abiotic processes in the ATTZ. A classification of macrohabitats according to hydrology and vegetation cover is necessary for comparative scientific studies, environmental impact analysis, and management purposes. The large Brazilian rivers and their floodplains are characterised by a monomodal, predictable flood pulse induced by clear annual rainy and wet seasons. Near the Equator this seasonality does not impede the growth of a rainforest, but with increasing distance from the Equator, savanna (*cerrado*) vegetation replaces the rainforest. Climatic stability decreases with increasing distance from the Equator, with savanna wetlands being subjected to multiannual periods of heavy floods and droughts, the latter with increased fire stress. In large Amazonian rivers, the flood pulse is of high amplitude. The pulse amplitude is low in various savanna wetlands (the Pantanal, the Araguaia River floodplain, the Llanos dos Moxos in Bolivia, the savanna of Roraima and the Llanos baixos of Venezuela).

The fertility of floodplains depends on the quality and quantity of sediments deposited inside the ATTZ and is related to the physical environment of the catchment of the parent river, as shown by the Amazonian *várzeas* which receive fertile sediment from the Andes. The *igapós* of the Amazonian blackwater rivers receive a limited amount of infertile sediment. The Everglades (USA) and the Okavango Delta (Botswana) also receive nutrient-poor water and little sediment and are oligotrophic and mesotrophic, respectively, whereas the floodplains of the Mekong River (Southeast Asia) receive nutrient-rich water and sediment from the parent river and are eutrophic (Junk et al. 2006b).

The age and genesis of floodplains are important. Many tropical and subtropical floodplains can be separated into an active recent and a much older but still active paleo-floodplain. About 80% of the Araguaia floodplain can be recognised as an old active palaeo-floodplain. Large areas of the Pantanal also belong to active palaeo-floodplains. These areas are mostly inundated by rainwater, delivered slowly to the parent rivers when the floods recede but do not receive new sediment.

Processes internal to floodplains can considerably modify water quality and influence productivity as for the *várzea* where nutrients are transferred internally by aquatic and terrestrial plants, leading to an increase of nutrients in the floodplain. The degree of connectivity with the parent river becomes important for the hydrochemistry of floodplain lakes. In savanna floodplains, a strong increase in salinity can be observed in isolated water bodies when evaporation surpasses precipitation.

The vegetation of floodplains corresponds to the vegetation cover of the surrounding uplands. Floodplains in the rainforest biome carry forests; those in the savanna belt carry savannas with forested areas along river channels and around floodplain lakes. The species composition differs between upland and floodplain forests because of the flood stress. There is a strong immigration pressure of upland species in the floodplains but their success depends on various factors, such as the species richness in the adjacent uplands and the hydrological stability of the floodplain over geological time scale which influence adaptations to flooding by respective species. Additional stress factors inside the floodplain, such as drought and fire stress, reduce the adaptation to flooding. In the *várzeas* and *igapós* of the Amazonian rainforest belt where the annual precipitation is more than 2000 mm, about 1000 and 600 flood-tolerant tree species occur, respectively, and many of them are endemic. In the Pantanal, where the annual precipitation is between 1100 mm and 1400 mm, there are 355 flood-tolerant species, each with a wide distribution range. In the Okavango Delta with a mean annual precipitation of 460–490 mm, only 10 of the 191 tree species tolerate periodic flooding (Ramberg et al. 2006).

The dynamics of the vegetation cover are also wide-ranging. Multiannual periods of high floods lead to the setback of poorly food-adapted invasive species. Fire stress is low in Amazonian flood basins, but it has a huge impact on the vegetation of savanna floodplains. During multiannual dry periods, large wild fires hinder the expansion of population of flood-adapted species as in the Pantanal (Nunes da Cunha and Junk 2004). For many plant and animal species in Amazonian floodplains, the shift between exclusively terrestrial and aquatic phases requires morphological, anatomical, and physiological adaptations and specific life-history traits. Similar adaptations can be expected in all large tropical and subtropical floodplains because of the adjustment of plants and animals with the specific environment. The frequency of occurrence of these adaptations and their variability differ because their development occurs gradually and depends on the geological and palaeoclimatic stability of the floodplains.

The presence of endemic species is a good indicator of long-term stability. Amazonian floodplain forests show a high number of endemic tree species. However, only one endemic copepod (*Argyrodiaptomus nhumirim*) and a few endemic herbaceous plants have been described from the Pantanal which is situated in the circumglobal belt of climatic instability. The Everglades provide an exception, and 65 taxa of higher plants are found concentrated in the rocky pinelands to the east and with tropical affinity. This happens as a result of isolation of the area of on a peninsula that hinders gene exchange with the Caribbean tropics (Junk et al. 2006b).

In spite of the drastic change in environmental conditions between terrestrial and aquatic phases, floodplain biodiversity is high because in addition to the characteristic wetland species, explicitly aquatic immigrants from the parent rivers and terrestrial immigrants from the surrounding uplands are present. Higher plants and vertebrates are the best studied, but every year additional species are described. In forested wetlands, tree species dominate the vegetation; in savanna wetlands, terrestrial, amphibious, and aquatic herbaceous species are more common as shown for the Everglades, Okavango Delta, and the Pantanal (Junk et al. 2006b).

All large floodplains offer migratory areas for new species. Migration is a natural process, common in Amazonia and the Pantanal by the network of rivers and adjacent riparian forests and accelerated by people. Frequent perturbation by sedimentation and erosion results in establishment of herbaceous ruderal plants which had been introduced by farmers as seeds in Amazonian *várzeas*. These plants also spread along travel routes (Seidenschwarz 1986). The Asian golden mussel (*Limnoperna fortunei*) was introduced into the Rio de la Plata system in 1993 by ballast water from ships, spreading over the entire Paraná-Paraguay River system with the exception of their headwaters in 20 years (Mansur et al. 2012). In the Pantanal, ecologists are concerned about the establishment of the Amazonian predatory peacock bass (*Cichla ocellaris*) and the spread of the introduced exotic grasses. The Everglades, located near trade routes for ornamental plants and pet animals that escape or are released, have become an extensively affected large subtropical floodplain. Junk et al. (2006b) identified 221 introduced plant species, 32 fish species, about 30 amphibian and reptile species, and 10 mammals. Primary production, however, as discussed earlier, depends on the nutrient status of water and soils in the respective floodplains which in turn depends on the geology and geomorphology of the river catchments. In Amazonia this is clearly seen between the whitewater and the blackwater rivers. Interfluvial wetlands and large parts of savanna floodplains are strongly influenced by rainwater and their nutrient values are low except near river channels.

Primary production in the floodplains is the sum of the production during both the terrestrial and aquatic phases. Algae (phytoplankton and periphyton) are major producers in the aquatic phase and herbaceous plants and the floodplain forest in both phases. In shallow flooded savannas, algal production may be reduced by the presence of aquatic macrophytes. Production by terrestrial, palustric, and aquatic herbaceous plants may be very high in nutrient-rich floodplains. In areas covered by a dense floodplain forest, the total primary production nearly equals the forest production.

In spite of high flood stress, wood production reached by the late successional *várzea* forest is much higher than in the adjacent upland and in the nutrient-poor *igapó*. Trees in savanna floodplains suffer more from drought and fire stress, and also from low soil fertility. Primary production is only about half that of the late successional forest stages in the *várzea*. In the moist tropics, the decomposition of organic material in land and water is very fast but depends on the availability of oxygen. The change between the aquatic and terrestrial phases accelerate decomposition processes by altering decomposed communities and the oxygenation of organic layers on land. Despite a very high primary production, organic matter accumulates only in permanently waterlogged areas. In contrast to wetlands with a rather stable water level (e.g. peat bogs in temperate regions), papyrus swamps at the edge of African lakes or Asian swamp forests, flood-pulsing wetlands are not major long-term sinks for organic carbon.

Nutrient cycles in flood-pulsing wetlands are very complex as they include nutrient exchange between the parent river and the ATTZ. The floodplain forest is also an important net nitrogen producer of the *várzea*. The end result is the nutrient enrichment of the floodplain. A considerable amount of nutrient transfer between the terrestrial and aquatic phases occurs via food webs. In African floodplains, herbaceous plants play a major role for terrestrial and aquatic grazers (Ramberg et al. 2006).

All major tropical and subtropical floodplains have been successfully managed by local populations, often for millennia. Calculations of Costanza et al. (1997) and the Millennium Ecosystem Assessment (MEA 2005) have shown the large economic benefits of intact RFSs. The Action Plan for the Environment Initiative (NEPAD 2003) states that the African countries and their people have healthy and productive wetlands and watersheds and can support fundamental human needs: clean water, appropriate sanitation, food security, and economic development. Carrying capacity for human population varies depending on floodplain fertility. In Brazil, human populations are concentrated on whitewater river floodplains and the Pantanal of Mato Grosso. The infertile Negro River *igapó* has always had and continues to have only a small number of people.

In many tropical and subtropical floodplains, the optimised use of multiple floodplain resources faces competition with the maximisation of the use of a single resource as in the Amazonian *várzeas*. Cattle ranching is highly detrimental to the floodplain. It has low land productivity and also causes large-scale destruction of floodplain forests, thereby damaging fisheries, forestry, floodplain stability, biodiversity, etc. However, cattle ranchers, being an economically and politically strong group, have been able to impose their interests on management strategies (Piedade et al. 2012).

In shallow-inundated savanna floodplains of the Araguaia River and in the Pantanal of Mato Grosso, agroindustry has started to modify large areas by construction of polders for irrigation agriculture and animal ranching. This process stated in the Everglades more than 100 years ago (Brown et al. 2006). Currently about half of the Everglades has been transformed and the rest has been affected by hydrological changes, mercury pollution, and eutrophication from storm water release and nutrient-laden runoff from farmland. The Everglades restoration project is one of the largest in the USA, with several billion dollars in support from both the Federal Government and that of the State of Florida (Mitsch and Hermandez 2013). Recovery of decayed wetlands could be a lot more costly than the maintenance of normal ones.

Wetlands in general and floodplains in particular tend to involve special interest groups from the outside as shown for strong impact wetland policies, such as for the Pantanal of Mato Grosso (Ioris 2012; Saffort 2012). The new Brazilian Forest Code may facilitate wetland destruction, neglecting their multiple functions for the environment and people. The negative effects of the many reservoirs of the Paraná River on the remaining floodplain area and its fish fauna below the Porto Primavera Dam have been demonstrated (Agostinho et al. 2008; Souza Filho 2009 and others). Many new reservoirs in other parts of the world, however, are being constructed over many rivers without considering their impact on regional wetlands (Cochrane et al. 2014).

The densely settled southern part of Brazil, with the megacities of São Paulo and Rio de Janeiro, faces serious water shortage. The common technical solutions to the problem have been discussed but they should be complemented by the recovery of destroyed wetlands and riparian forests which formerly provided water to the cities, as should be done worldwide. The impact of human activities on floodplains can easily be detected and quantified at the macrohabitat level. A Brazil-specific definition of hydrological resources, the classification of wetlands and their macrohabitats, and their delimitation has been elaborated (Piedade et al. 2012; Junk et al. 2014a). Information of this kind could be used by various governments to form specific wetland policies, which are urgently needed.

Acknowledgements

This chapter was prepared under the leadership of the National Institute for Science and Technology in Wetlands (INCT-INAU), Cuiabá, Mato Grosso, Brazil. We acknowledge the support of the Brazilian Research Council (CNPq), Brasilia, University of Mato Grosso (UFMT), Cuiabá, State University of Amazonas (UEA), Manaus, and National Amazon Research Institute (INPA), Manaus, through the working group Ecology, Monitoring and Sustainable Use of Wetlands (MAUA), the PELD (Program of Long-tern Ecological Research), and the cooperation programme between INPA and the Max-Planck-Institute of Chemistry, Mainz, Germany.

References

Adis, J. (1981). Comparative ecological studies on the terrestrial arthropod fauna in Central Amazonian inundation forests. *Amazoniana* 7 (2): 87–173.

Adis, J. (1982). Zur Beseidlung zentralamazonischer Überschwemmungswälder (Várzea-Gebiet) durch Carabiden (Coleoptera). *Archiv für Hydrobiologie* 95 (1/4): 3–15.

Adis, J. (1984). "Seasonal Igapó" – forests of Central Amazonian blackwater rivers and their terrestrial arthpod fauna. In: *The Amazon. Limnology and Landscape Ecology of a Mighty Tropical River and its Basin* (ed. H. Sioli), 245–268. Dordrecht: Junk.

Adis, J. (1992). Überlebensstrategien terrestrischer Invertebraten in Überschwemmungswäldern Zentralamazoniens. *Verhandlungen des Naturwissenschaftlichen Vereins in Hamburg* 33 (NF): 21–114.

Adis, J. (1997). Terrestrial invertebrates: Survival strategies, group spectrum, dominance and activity patterns. In: *The Central Amazonian Floodplain: Ecology of a Pulsing System* (ed. W.J. Junk), 299–318. Berlin: Springer.

Adis, J. and Junk, W.J. (2002). Terrestrial invertebrates inhabiting lowland river floodplains of Central Amazonia and Central Europe: a review. *Freshwater Biology* 47: 711–731.

Adis, J. and Messner, B. (1997). Adaptations to life under water: tiger beetles and millipedes. In: *The Central Amazonian Floodplain: Ecology of a Pulsing System* (ed. W.J. Junk), 319–330. Berlin: Springer.

Agostinho, A.A., Pelicice, F.M., and Gomes, L.C. (2008). Dams and the fish fauna of the Neotropical region: impacts and management related to diversity and fisheries. *Brazilian Journal of Biology* 68 (4): 1119–1132.

Almeida, S.S., Amaral, D.D., and Silva, A.S.L. (2004). Análise floristica e estrutura de florestas de várzea no estuário amazônica. *Acta Amazonica* 34 (4): 513–524.

Assine, M.L. and Soares, P.C. (2004). Quaternary of the Pantanal, west-central Brazil. *Quaternary International* 114: 23–34.

Ball, T. (2012). Wetlands and the water environment in Europe in the first decade of the Water Framework Directive: Are expectations matched by delivery? In: *Tropical Wetland Management: The South American Pantanal and the International Experience* (ed. A.A.R. Ioris), 255–274. Burlington: Ashgate Publishing Co.

Bayley, P.B. (1983) Central Amazon fish populations: biomass, production and some dynamic characteristics. PhD thesis. Dalhousie University, Halifax, Nova Scotia, Canada.

Bayley, P.B. and Petrere, M. Jr. (1989). Amazon fisheries: Assessment methods, current status, and management options. *Canadian Special Publication of Fisheries and Aquatic Services* 106: 385–398.

Bleich, M.E., Mortati, A.F., André, T., and Piedade, M.T.F. (2014). Riparian deforestation affects the structural dynamics of headwater streams from southern Brazilian Amazona. *Tropical Conservation Science* 7 (4): 657–676.

Brown, M.T., Cohen, M.J., Bardi, E., and Ingwersen, W.W. (2006). Species diversity in the Florida Everglades USA: A system approach to calculating biodiversity. *Aquatic Sciences* 68: 254–277.

Burnham, R.J. and Johnson, K.R. (2004). South American palaeobotany and the origin of neotropical rainforests. *Philosophical Transactions of the Royal Society B* 359: 1595–1610.

Clark, D.A., Brown, S., Kicklighter, D.W. et al. (2001). Net primary production in tropical forests: an evaluation and synthesis of existing field data. *Ecological Applications* 11: 371–384.

Cochrane, T.A., Arias, M.E., and Piman, T. (2014). Historical impact of water infrastructure on water levels of the Mekong River and the Tonle Sap system. *Hydrology and Earth System Sciences* 18: 4529–4541.

Colinvaux, P.A., Irion, G., Räsänen, M.E. et al. (2001). A paradigm to be discarded: geological and palaeological data falsify the Haffer and Prance refuge hypothesis of Amazonian speciation. *Amazoniana* 16: 609–646.

Costanza, R., d'Arge, R., de Groot, R. et al. (1997). The value of the world's ecosystem services and natural capital. *Nature* 387: 253–260.

Cowardin, L.M., Carter, V., Golet, F.C., and LaRoe, E.T. (1979). *Classification of Wetlands and Deepwater Habitats of the United States*. Washington, DC: United States Fish and Wildlife Service.

Damasceno, G.A. Jr., Semir, J., Santos, F.A.M., and Leitão Filho, H.F. (2005). Structure, distribution of species, and inundation in a riparian forest of of Rio Paraguai, Pantanal, Brazil. *Flora* 200: 119–135.

Erwin, T.L. and Adis, J. (1982). Amazonian inundation forests. Their role as short-term refuges and generators of species richness and taxon pulses. In: *Biological Diversification in the Tropics: Proceedings of the Fifth International Symposium of the Association for Tropical Biology* (ed. G.T. Prance), 358–371. New York: Columbia University Press.

Eva, H.D., Belward, A.S., de Miranda, E. et al. (2004). A land cover map of South America. *Global Change Biology* 10: 731–744.

Felfili, J.M. (1995). Diversity, structure and dynamics of a gallery forest in Central Brazil. *Vegetatio* 117: 1–15.

Ferreira, I.V. (1997). Effects of the duration of flooding on species richness and floristic composition in three hectares in the Jaú National Park in floodplain forests in central Amazonia. *Biodiversity and Conservation* 6: 1353–1363.

Fisher, T.R. (1979). Plankton and primary production in aquatic systems of the central Amazon basin. *Comparative Biochemistry and Physiology* 62A: 31–38.

Fjeldsa, J. and Krabbe, N. (1990). *Birds of the High Andes*. Svendborg: Apollo Books.

Fonseca, S.F. Jr., da, Piedade, M.T.F., and Schöngart, J. (2009). Wood growth of *Tabebuia barbata* (E. Mey.) Sandwith (Bignoniaceae) and *Vatairea guianensis* Aubl. (Fabaceae) in Central Amazonian black-water (igapó) and white-water (várzea) floodplain forests. *Trees – Structure and Function* 23 (1): 127–134.

Forsberg, B.R., Araujo-Lima, C.A.R.M., Martinelli, L.A. et al. (1993). Autotrophic carbon sources for fish of the central America. *Ecology* 74 (3): 643–652.

Franklin, E., Adis, J., and Woas, S. (1997). The oribatid mites. In: *The Central Amazon Floodplain: Ecology of a Pulsing System* (ed. W.J. Junk), 331–349. Berlin: Springer.

Furch, K. and Junk, W.J. (1992). Nutrient dynamics of submersed decomposing Amazonian herbaceous plant species *Paspalum fasciculatum* and *Echinochloa polystachya*. *Revue d'hydrobiologie. tropicale* 25 (2): 75–85.

Furch, K. and Junk, W.J. (1993). Seasonal nutrient dynamics in an Amazonian floodplain lake. *Archiv für Hydrobiologie* 128 (3): 277–285.

Furch, K. and Junk, W.J. (1997). Physiochemical conditions in floodplains. In: *The Central Amazon Floodplain: Ecology of a Pulsing System* (ed. W.J. Junk), 69–108. Berlin: Springer.

Gauer, U. (1997). The Collembola. In: *The Central Amazon Floodplain: Ecology of a Pulsing System* (ed. W.J. Junk), 351–359. Berlin: Springer.

Gopal, B. and Junk, W.J. (2000). Biodiversity in wetlands: an introduction. In: *Biodiversity in Wetlands: Assessment, Function and Conservation*, vol. 1 (eds. B. Gopal, W.J. Junk and J.A. Davis), 1–10. Leiden: Backhuys Publishers.

Goulding, M. (1980). *The Fish and the Forest: Explorations in Amazonia Natural History*. Berkeley: California University Press.

Goulding, M., Carvalho, M.L., and Ferreira, E.G. (1988). *Rio Negro: Rich Life in Poor Water*. The Hague: SPB Academic Publishing.

Goulding, M., Smith, N.J.H., and Mahar, D.J. (1996). *Floods of Fortune: Ecology and Economy Along the Amazon*. New York: Columbia University Press.

Haffer, J. and Prance, G.T. (2001). Climatic forcing of evolution in Amazonia during the Cenozoic: on the refuge theory of biotic differentiation. *Amazoniana* 16: 579–607.

Hamilton, S.K. (1999). Potential effects of a major navigation project (Paraguay-Paraná-Hidrovia) on inundation in the Pantanal floodplains. *Regulated Rivers: Research and Management* 15: 289–299.

Hamilton, S.K., Sippel, S.J., and Melack, J.M. (1996). Inundation patterns in the Pantanal wetland of South America determined from passive microwave remote sensing. *Archiv fur Hydrobiologie* 137: 1–23.

Hockman, C.W. (1998). *The Pantanal of Poconé*. The Hague: Kluwer.

Hofer, H. (1997). The spider communities. In: *The Central Amazon Floodplain: Ecology of a Pulsing System* (ed. W.J. Junk), 373–383. Berlin: Springer.

Holcik, J. and Bastl, I. (1976). Ecological effects of water level fluctuation upon the fish population in the Danube River floodplain in Czechoslovakia. *Acta Scientiarim Naturalium Academic Scientiarum Bojemoslov Brno* 10: 1–46.

Hoorn, C., Wesselingh, F.P., and Ter Steege, H. (2010). Amazonia through time: Andean uplift, climate change, landscape evolution and biodiversity. *Science* 330: 927–931.

Hultine, K.R., Dudely, T.J., and Leavitt, S.T. (2013). Herbivory-induced mortality increases with radial growth in an invasive riparian phreatophyte. *Annals of Botany* 111: 1197–1206.

Ioris, A.A.R. (2012). Reassessing development: Pantanal's history, dilemmas and prospects. In: *Tropical Wetland Management: The South-American Pantanal and the International Experience. Asgate Studies in Environmental Policy and Practice* (ed. A.A.R. Ioris), 199–222. Leeds: University of Leeds.

Irion, G., Junk, W.J., and de Mello, J.A.S.N. (1997). The large central Amazonian river floodplains near Manaus: geological, climatological, hydrological, and geomorphological aspects. In: *The central Amazon Floodplain: Ecology of a Pulsing System* (ed. W.J. Junk), 23–46. Berlin: Springer.

Irion, G., de Mello, J.A.S.N., Morais, J. et al. (2010). Development of the Amazon valley during the Middle to Late Quaternary: sedimentological and climatological observations. In: *Ecology and Management of Amazonian Floodplain Forests* (eds. W.J. Junk, M.T.F. Piedade, F. Wittmann, et al.), 27–42. Berlin: Springer.

Irion, G., Nunes, G.M., Nunes da Cunha, C. et al. (2016). Araguaia river floodplain: size, age, and mineral composition of a large tropical savanna wetland. *Wetlands* 36 (5): 945–956.

Iriondo, M.H. and Garcia, N.O. (1993). Climatic variations in the Argentine plains during the last 18,000 years. *Palaeogeography, Palaeoclimatology, Palaeoecology* 101: 209–220.

Irmler, U. and Furch, K. (1980). Weight, energy and nutrient changes during the decomposition of leaves in the emersion phase of Central-Amazonian inundation forests. *Pedobiologia* 20: 118–130.

Junk, W.J. (1997). Structure and function of the large Central-Amazonian river-floodplains: Synthesis and discussion. In: *The Central Amazon Floodplain: Ecology of a Pulsing System* (ed. W.J. Junk), 455–472. Berlin: Springer.

Junk, W.J. (2000). Mechanics for maintenance and development of biodiversity in neotropical floodplains. In: *Biodiversity in Wetlands: Assessment, Function and Conservation*, vol. 1 (eds. B. Gopal, W.J. Junk and J.A. Davis), 119–139. Leiden: Backhuys Publishers.

Junk, W.J. (2005). Flood pulsing and the linkages between terrestrial, aquatic, and wetland systems. *Proceedings of the International Association of Theoretical and Applied Limnology* 29 (1): 111–138.

Junk, W.J. (2013). Current state of knowledge regarding South America wetlands and their future under global climate change. *Aquatic Sciences* 75 (1): 113–131.

Junk, W.J. and da Silva, V.M.F. (1997). Mammals, reptiles and amphibians. In: *The Central Amazon Floodplain: Ecology of a Pulsing System* (ed. W.J. Junk), 409–417. Berlin: Springer.

Junk, W.J. and Howard-Williams, C. (1984). Ecology of aquatic macrophytes in Amazonia. In: *The Amazon -- Limnology and Landscape Ecology of a Mighty Tropical River and its Basin* (ed. H. Sioli), 269–293. Dordrecht: Junk.

Junk, W.J. and Nunes da Cunha, C. (2012a). Pasture clearing from invasive woody plants in the Pantanal: a tool for sustainable management or environmental destruction? *Wetlands Ecology and Management* 20 (2): 111–122.

Junk, W.J. and Nunes da Cunha, C. (2012b). Wetland management challenges in the South American Pantanal and the international experience. In: *Tropical Wetland Management: The South American Pantanal and the International Experience, Asgate Studies in Environmental Policy and Practice* (ed. A.A.R. Ioris), 315–331. Leeds: University of Leeds.

Junk, W.J. and Piedade, M.T.F. (1993a). Herbaceous plants of the Amazon floodplain near Manaus: species diversity and adaptations to the flood pulse. *Amazoniana* 12 (3–4): 467–484.

Junk, W.J. and Piedade, M.T.F. (1993b). Biomass and primary-production of herbaceous plant communities in the Amazon floodplain. *Hydrobiologia* 263: 155–162.

Junk, W.J. and Piedade, M.T.F. (1997). Plant life in the floodplain with special reference to herbaceous plants. In: *The Central Amazon Floodplain: Ecology of a Pulsing System* (ed. W.J. Junk), 147. Berlin, 186: Springer.

Junk, W.J. and Wantzen, K.M. (2004). The flood pulse concept: New aspects, approaches, and applications – an update. In: *Proceedings of the 2nd International Symposium on the Management of Large Rivers for Fisheries*, vol. 2 (eds. R.L. Welcome and T. Petr), 117–149. Bangkok: Food and Agriculture Organization and Mekong River Commission, FAO Regional Office for Asia and the Pacific.

Junk, W.J. and Welcomme, R.L. (1990). Floodplains. In: *Wetlands and Shallow Continental Water Bodies*, vol. 1 (eds. B.C. Patten, S.E. Joergensen and H. Dumont), 491–524. The Hague: SPB Academic Publishing bv.

Junk, W.J., Bayley, P.B., and Sparks, R.E. (1989). The flood pulse concept in river-floodplain-systems. *Canadian Special Publication for Fisheries and Aquatic Sciences* 106: 110–127.

Junk, W.J., Soars, M.G., and Saint-Paul, U. (1997). The fish. In: *The Central Amazon Floodplain: Ecology of a Pulsing System* (ed. W.J. Junk), 385–408. Berlin: Springer.

Junk, W.J., Nunes da Cuha, C., Wantzen, K.M. et al. (2006a). Biodiversity and its conservation in the Pantanal of Mato Grosso, Brazil. *Aquatic Sciences* 68 (3): 278–368.

Junk, W.J., Brown, M., Campbell, I.C. et al. (2006b). The comparative biodiversity of seven globally important wetlands: a synthesis. *Aquatic Sciences* 68 (3): 400–414.

Junk, W.J., Soars, M.G., and Bayley, P.G. (2007). Freshwater fishes of the Amazon River basin: their biodiversity, fisheries and habitats. *Aquatic Ecosystem Health and Management* 10 (2): 153–J73.

Junk, W.J., Piedade, M.T.F., Wittmann, F. et al. (2010a). *Amazonian Floodplain Forests: Ecophysiology, Biodiversity and Sustainable Management*. Berlin: Springer.

Junk, W.J., Piedade, M.T.F., Wittmann, F., and Schöngart, J. (2010b). The role of floodplain forests in an integrated sustainable management concept of the natural resources of the Central Amazonian várzea. In: *Amazonian Floodplain Forests: Ecophysiology, Biodiversity and Sustainable Management* (eds. W.J. Junk, M.T.F. Piedade, F. Wittman, et al.), 485–509. Berlin: Springer.

Junk, W.J., Piedade, M.T.F., Parolin, P. et al. (2010c). Ecophysiology, biodiversity and sustainable management of central Amazonian floodplain forests: a synthesis. In: *Amazonian Floodplain Forests: Ecophysiology, Biodiversity annd Sustainable Management* (eds. W.J. Junk, M.T.F. Piedade, F. Wittman, et al.), 511–540. Berlin: Springer.

Junk, W.J., da Silva, C.J., Nunes da Cunha, C., and Wantzen, K.M. (2011a). *The Pantanal: Ecology, Biodiversity, and Sustainable Management of a Large Neotropical Seasonal Wetland*, 870. Sofia-Moscow: Pensoft.

Junk, W.J., Piedade, M.T.F., Schöngart, J. et al. (2011b). A classification of major naturally-occurring Amazonian lowland wetlands. *Wetlands* 31: 623–640.

Junk, W.J., An, S., Cizcová, H. et al. (2013). Current state of knowledge regarding the world's wetlands and their future under global climate change: a synthesis. *Aquatic Sciences* 75 (1): 151–167.

Junk, W.J., Piedade, M.T.F., Lourival, R. et al. (2014a). Brazilian wetlands: Definition, delineation and classification for research, sustainable management and protection. *Aquatic Conservation* 24: 5–22.

Junk, W.J., Piedade, M.T.F., Lourival, R. et al. (2014b). Definição e classificação das Areas Úmidas (AUs) Brasileiras: Base Cientifica para uma Nova Politica de PROTEÇÃO e Manejo Sustentável. In: *Classificação e Delineamento das Áreas Úmidas Brasileiras e de seus Macrohabitats* (eds. C.N. da Cunha, M.T.F. Piedade and W.J. Junk), 13–76. Cuiabá: Universidade Federal de Mato Grosso (UFMT).

Junk, W.J., Piedade, M.T.F., Schöngart, J., and Wittmann, F. (2014c). A classificação dos macrohabitats das várzeas Amazônicus. In: *Classificação e Delineamento das Áreas Úmidas Brasileiras e de seus Macrohabitats* (eds. C.N. da Cunha, W.J. Junk and M.T.F. Piedade), 124–153. Cuiabá: Universidade Federal de Mato Grosso (UFMT).

Junk, W.J., Wittmann, F., Schöngart, J., and Piedade, M.T.F. (2015). A classification of the major habitats of Amazonian black-nwater river floodplains and a comparison with their white-water counterparts. *Wetlands Ecology and Management* 23 (4): 677–693.

Junk, W.J., Piedade, M.T.F., Nunes da Cunha, C. et al. (2018). Macrohabitat studies in large Brazilian floodplains to support sustainable development in the face of climate change. *Ecohydrology & Hydrobiology* 18: 334–344.

Kandus, P., Minotti, P., and Malvárez, A.I. (2008). Distribution of wetlands in Argentina estimated from soil charts. *Acta Scientiarum Biological Sciences* 30 (4): 403–409.

Kellman, M. and Meave, J. (1997). Fire in the tropical gallery forests of Belize. *Journal of Biogeography* 24: 23–24.

Kern, J. and Darwich, A. (1997). Nitrogen turnover in the várzea. In: *The Central Amazon Floodplain: Ecology of a Pulsing System* (ed. W.J. Junk), 147–186. Berlin: Springer.

Kern, J., Kreibich, H., Koschorreck, M., and Darwich, A. (2010). Nitrogen balance of a floodpalin forest of the Amazon River: the role of nitrogen fixation. In: *Amazonian Floodplain Forests: Ecophysiology, Biodiversity and Sustainable Management* (eds. W.J. Junk, M.T.F. Piedade, F. Wittman, et al.), 281–299. Springer.

Kubitzki, K. (1989). The ecogeographical differentiation of Amazonian inundation forests. *Plant Systematics and Evolution* 163: 285–304.

Latrubesse, E.M. and Stevaux, J.C. (2002). Geomorphology and environmental aspects of the Araguaia fluvial basin, Brazil. *Zeitschrift für Geomorphologie* 129: 109–127.

Latrubesse, E.M., Amsler, M.L., Morais, R.P., and Aquino, S. (2009). The geomorphologic response of a large pristine alluvial river to tremendous deforestation in the south American tropics: the case of the Araguaia River. *Geomorphology* 113: 239–252.

Loverde-Oliveira, S.M., Adler, M., and Pinto-Silva, V. (2011). Phytoplankton, periphyton and metaphyton. In: *The Pantanal:Ecology, Biodiversity and Sustainable Management of a Large Neotropical Seasonal Wetland* (eds. W.J. Junk, C.J. da Silva, C.N. da Cunha and K.M. Wantzen), 235–256. Sofia-Moscow: Pensoft.

Lowe-McConnell, R.H. (1964). The fishes of the Rupununi savanna district of British Guiana, South America. Part 1. Ecological groupings of fish species and effects of the seasonal cycle on the fish. *Zoological Journal of the Linnean Society* 45 (304): 103–144.

Lucas, C.M., Schöngart, J., Sheikh, P. et al. (2014). Effects of land-use and hydroperiod on aboveground biomass and productivity of secondary Amazonian floodplain forests. *Forest Ecology and Management* 319: 116–127.

Lundberg, J.G., Marshall, L.G., Guerrero, J. et al. (1998). The stage for neotopical fish diversification. In: *Phylogeny and Classification of Neotropical Fishes* (eds. L.R. Malabarba, R.E. Reis, R.P. Vari, et al.), 13–48. Porto Alegre, Brazil: EDIPUCRS.

Malhi, Y., Wood, D., Baker, T.R. et al. (2006). The regional variation of aboveground live biomass in old-growth Amazonian forests. *Global Change Biology* 12: 1107–1138.

Mansur, M.C.D., dos Santos, C.P., Pereira, D. et al. (2012). *Moluscos limnicos, invasores do Brasil:. biologia, prevenção e controle.* Porto Alegre, Brazil: Redes Editora Ltda.

Martius, C. (1997a). The termites. In: *The Central Amazon Floodplain: Ecology of a Pulsing System* (ed. W.J. Junk), 361–371. Berlin: Springer.

Martius, C. (1997b). Decomposition of wood. In: *The Central Amazon Floodplain: Ecology of a Pulsing System* (ed. W.J. Junk), 267–276. Berlin: Springer.

McCoy, C.J. (1984). Ecological and zoogeographic relationships of amphibians and reptiles of the Cuatro Cienegas basin. *Journal of the Arizona-Nevada Academy of Science* 19: 49–60.

MEA (2005). *Ecosystems and Human Well-Being: Wetlands and Water Synthesis.* Washington, DC: World Resources Institute.

Meave, J., Kellman, J., MacDougall, D., and Rosales, J. (1991). Riparian habitats as tropical forest refugia. *Global Ecology and Biogeography Letters* 1: 69–76.

Medina-Junior, P.B. and Rietzler, A.C. (2005). Limnological study of a Pantanal saline lake. *Brazilian Journal of Biology* 65 (4): 651–659.

Melack, J.M. and Fisher, T.R. (1990). Comparative limnology of tropical floodplain lakes with an emphasis of the central Amazon. *Acta Limnologica Brasiliensia* 3: 1–48.

Melack, J.M. and Forsberg, B.R. (2001). Biogeochemistry of Amazonian floodplain lakes and associated wetlands. In: *The Biogeochemistry of the Amazon Basin and its Role in a Changing World* (eds. M.E. McClain, R.I. Victoria and R.E. Richey), 235–274. Oxford: Oxford University Press.

Metzger, J.P., Bernacci, L.C., and Goldenberg, R. (1997). Pattern of tree species diversity in riparian forest fragments of different widths (SE Brazil). *Plant Ecology* 133: 135–152.

Meyer, U. (1990) Feinwurzelsysteme und Mycorrhizatypenals Anpassungsmechanismen in zentralamazonishchen Überschwemmungswäldern – Igapó und Várzea. PhD thesis. University of Hohenheim.

Mitchell, D.S. (1974). The development of excessive populations of aquatic plants. In: *Aquatic Vegetation and its Use and Control* (ed. D.S. Mitchell), 38–49. Paris: UNESCO.

Mitsch, W.J. and Gosselink, J.G. (2000). *Wetlands*, 3e. New York: Wiley.

Mitsch, W.J. and Hermandez, M.E. (2013). Landscape and climate threats to wetlands in North and Central America. *Aquatic Sciences* 75: 133–149.

Mitsch, W.J., Day, J.W. Jr., Gilliam, J.W. et al. (2001). Reducing nitrogen loading to the Gulf of Mexico from the Mississippi River Basin: strategies to counter a persistant ecological problem. *Bioscience* 51: 373–388.

Montero, J.C., Piedade, M.T.F., and Wittmann, F. (2014). Floristic variations across 600 km of inundation forest (Igapó) along the Negro River, central Amazonia. *Hydrobiologia* 729: 229–246.

Montes, R. and San José, J.J. (1995). Vegetation and soil analysis of topo-sequences in the Orinoco llanos. *Flora* 190: 1–33.

Mori, S. (2001). A Familia da Castanha-do-Pará: Simbolo do Rio Negro. In: *Florestas do Rio Negro* (eds. A.A. Oliveira and D.C. Daly), 119–142. São Paulo: UNIP, NYBG e Companhia das Latras.

Muller, J., Irion, G., de Melo, J.N., and Junk, W.J. (1995). Hydrological changes of the Amazon during the last glacier-interglacier cycle in Central-Amazonia (Brazil). *Naturwissenschaften* 82: 232–235.

Naiman, R.J., Décamps, H., and Polllock, M. (1993). The role of riparian corridors In maintaining regional biodiversity. *Ecological Applications* 3: 209–212.

NEPAD *(*2003*)* Action Plan of the Environmental Initiative of the New Partnership for Africa's Development (NEPAD).

Nunes da Cunha, C. and Junk, W.J. (1999). Composição floristica de capões e cordilheiras: localização das espécies lenhosas quanto ao gradiente de inundação no Pantanal de Poconé, MT-Brasil. In: *Anais do II Simpósio sobre Recursos Naturais e Sócio-economicos do Pantanal Manejo e Conservação* (eds. M. Dantas, J.B. Catto and E.K. de Resende), 17–28. Corumbá, Brazil: EMBRAPA.

Nunes da Cunha, C. and Junk, W.J. (2001). Distribution of woody plant communities along the flood gradient in the Pantanal of Poconé, Mato Grosso, Brazil. *International Journal of Ecology and Environmental Sciences* 27: 63–70.

Nunes da Cunha, C. and Junk, W.J. (2004). Year-to-year changes in water level drive the invasion of Vochysia divergens in Pantanal grasslands. *Applied Vegetation Science* 7: 103–110.

Nunes da Cunha, C. and Junk, W.J. (2014). A Classificação dos Macrohabitats do Pantanal Mato Grossense. In: *Classificação e Delineamento das Áreas Úmidas Brasileiras e de seus Macrohabitats* (eds. C.N. da Cunha, M.T.F. Piedade and W.J. Junk), 77–122. Cuiabá: Universidade Federal de Mato Grosso.

Ohly, J.J. and Hund, M. (2000). Floodplain animal husbandry in central Amazonia. In: *The Central Amazon Floodplain: Actual Use and Options for a Sustainable Management* (eds. W.J. Junk, J.J. Ohly, M.T.F. Piedade and M.G.M. Soares), 313–343. Leiden: Backhuys Publishers.

Oliveira, A.C.B., Soares, M.G.M., Martinelli, L.A., and Moreira, M. (2006). Carbon sources of fish in an Amazonian floodplain lake. *Aquatic Sciences* 68: 229–238.

Oliveira-Filho, A.T. and Ratter, J.A. (1995). A study of the origin of Central Brazilian forests by the analysis of plant species ditribution patterns. *Edinburgh Journal of Botany* 52: 141–194.

Oliveira-Filho, A.T., Almeida, R.J., de Mello, J.M., and Gavilanes, M.L. (1994). Estrutura fitosociológica e variáveis ambientais em um trecho de mata ciliar dos córregos das Vilas Boas, Reserva Biológica do Poço Bonito, Lavras (MG). *Revista Brasileira de Botánica* 17: 67–85.

Padoch, C., Ayres, J.M., Pinedo-Vasquez, M., and Henderson, A. (1999). *Várzea: Diversity, Development and Conservation of Amazonia's Whitewater Floodplains*. New York: The New York Botanical Garden Press.

Pagioro, T.A., Roberto, M.C., and Lansac-Toha, F.A. (1997). Comparative limnological analysis of two lagoons on the floodplain of the Upper Paraná River. *International Journal of Ecology and Environmental Sciences* 23: 229–239.

Parolin, P. (2009). Submerged in darkness: Adaptations to prolonged submergence by woody species of the Amazonian floodplains. *Annals of Botany* 103: 359–376.

Parolin, P., De Simone, O., and Haase, K. (2004). Central Amazonian floodplain forests: tree adaptations in a pulsating system. *Botanical Review* 70: 357–380.

Peixoto, J.M.A., Nelson, B.W., and Wittmann, F. (2009). Spatial and temporal dynamics of alluvial geomorphology and vegetation in central Amazonian white-water floodplains by remote-sensing techniques. *Remote Sensing of Environment* 113: 2258–2266.

Petermann, P. (1997). The birds. In: *The Central Amazon Floodplain: Ecology of a Pulsing System* (ed. W.J. Junk), 419–452. Berlin: Springer.

Petermann, P. (2011). The birds of the Pantanal. In: *The Pantanal: Ecology, Biodiversity and Sustainable Management of Large Neotropical Seasonal Wetland* (eds. W.J. Junk, C.J. da Silva, C.N. da Cunha and K.M. Wantzen), 523–564. Sofia-Moscow: Pensoft.

Piedade, M.T.F., Junk, W.J., and Long, S.P. (1991). The productivity of the C_4 grass *Echinochloa polystachya* on the Amazon floodplain. *Ecology* 72 (4): 1456–1463.

Piedade, M.T.F., Worbes, M., and Junk, W.J. (2001). Geo-ecological controls on elemental fluxes in communities of higher plants in Amazonian floodplains. In: *The Biogeochemistry of the Amazon Basin* (eds. M.E. McClain, R.I. Victoria and J.E. Richey), 209–234. New York: Oxford University Press.

Piedade, M.T.F., Junk, W.J., de Sousa, P.T. Jr. et al. (2012). As áreas úmidas no àmbito do Código Florestal brasileiro. In: *Código Florestal e a ciencia o que nossos legisladores ainda precisam saber. Sumários executivos de estudos cientificos sobre impactos do projecto de Código Florestal* (ed. Comité Brasil em Defesa das Florestas e do Desenvolvimento Sustentável), 9–17. Brasilia: Comité Brasil.

Piedade, M.T.F., Schöngart, J., Wittmann, F. et al. (2013). Impactos ecológicos de inundação e seca a vegetação das áreas alagáveis amazônicus. In: *Eventos climáticos extremos na Amazônia: causas e conseqüências* (eds. C.A. Nobre and L.S. Borma), 268–304. São Paulo: Officina de Textos.

Ponce, V.M. (1995). *Hydrological and Environmental Impact of the Paraná-Paraguay waterway on the Pantanal of Mato Grosso, Brazil.* San Diego: San Diego State University Press.

Pott, A. and Pott, V.J. (1994). *Plantas do Pantanal.* Brasilia: EMBRAPA.

Pott, A. and Pott, V.J. (1996). Flora do Pantanal – Listagem atual de Fanerógamas. In: *Anais II Simpósio sobre Recursos Naturais e Sócio-econômicos do Pantanal. Manejo e Conservação* (ed. EMBRAPA – Empressa Brasileira de Pesquisa Agropecuária), 297–325. Corumbá: EMBRAPA.

Pott, V.I. and Pott, A. (1997). Checklist das macrófi tas aquáticas do Pantanal, Brasil. *Acta Botânica Brasílica* II: 215–227.

Pott, V.I. and Pott, A. (2000). *Plantas Aquáticas de Pantanal*, 404. Brasilia: EMBRAPA.

Prance, G.T. (1979). Notes on the vegetation of Amazonia III. The terminology of Amazonian forest types subject to inundation. *Brittonia* 3: 26–38.

Ramberg, L., Hancock, P., Lindholm, M. et al. (2006). Species diversity of the Okavango Delta, Botswana. *Aquatic Sciences* 69 (3): 310–337.

Reichholf, J. (1975). Biogeographic und Ökologie der Wasservögel im subtropisch-tropischen Südamerika. *Anzeiger der Ornithologischen Gesellschaft in Bayern eV* 14: 1–69.

Reid, M. and Capon, S. (2011). Role of the soil seed bank in vegetation responses to environmental flows on a drought-affected floodplain. *River System* 19 (3): 249–259.

Remsen, J.V. Jr. and Parker, T.A. III (1983). Contribution of river-created habitats to bird species richness in Amazonia. *Biotropica* 15: 223–231.

Resende, E.K. (2011). Ecology of Pantanal fish. In: *The Pantanal: Ecology, Biodiversity and Sustainable Management of Large Neotropical Seasonal Wetland* (eds. W.J. Junk, C.J. da Silva, C.N. da Cunha and K.M. Wantzen), 469–496. Sofia-Moscow: Pensoft.

Resende, H.C., Yotoko, K.S.C., Delabie, J.H.C. et al. (2010). Pliocene and Pleistocene events shaping the genetic diversity within the central corridor of the Brazilian Atlantic forest. *Biological Journal of the Linnean Society* 101: 949–960.

Robarts, R.D., Zhulidov, A.V., and Pavlov, D.F. (2013). The state of knowledge about wetlands and their future under aspects of global climate change: The situation in Russia. *Aquatic Sciences* 75: 27–38.

Rodrigues, R.R. and Nave, A.G. (2000). Heterogeneidade floristicca da matas ciliares. In: *Matas Ciliares: conservação e recuperação* (eds. R.R. Rodrigues and H.F. Leitão-Filho), 45–71. São Paulo: EDUSP/FAPESP.

Rolim, S.G., Ivanauskas, N.M., Rodrigues, R.R. et al. (2006). Composição floristica do estrato arbóreo de floresta estacional semidecidual na planicie alluvial do Rio Doce, Linhares, ES, Brasil. *Acta Botãnica Brasilica* 20: 549–561.

Rosa, S.A. (2008) Modelos de Crescimento de Quatro Espécies Madeireiras de floresta de Várzea Alta da Amazônia Central por meio de Métodos Dendrocronológicos. MSc thesis. Instituto Nacional de Pesquisas da Amazônia (INPA), Manaus, Brazil.

Sabo, J.L., Sponseller, R., and Dixon, M. (2005). Riparian zones increase regional species richness by harboring different, not more, species. *Ecology* 86: 56–62.

Saffort, T.G. (2012). Organizational complexity and stakeholder engagement in the management of the Pantanal wetland. In: *Tropical Wetland Management: The South American Pantanal and the International Experience, Agate Studies in Environment Policy and Practice* (ed. A.A.R. Ioris), 173–198. Leeds: University of Leeds.

Scabin, A., Costa, F., and Schöngart, J. (2012). The spatial distribution of illegal logging in the Anavilhanas Archipelago (Central Amazonia) and logging impacts on the primary timber species. *Environmental Conservation* 39 (1): 111–121.

Schmidt, G.W. (1973). Primary production of phytoplankton in the three types of Amazonian waters III. Primary production of phytoplankton in a tropical floodplain lake of Central Amazonia, Lago do Castanho, Amazon, Brazil. *Amazoniana* 4: 379–404.

Schöngart, J. (2008). Growth-Oriented Logging (GOL): A new concept towards sustainable forest management in Central Amazonian várzea floodplains. *Forest Ecology and Management* 256: 46–58.

Schöngart, J. (2010). Growth-Oriented Logging (GOL): The use of species-specific growth information for forest management in Central Amazonian floodplains. In: *Amazonian Floodplain Forests: Ecophysiology, Biodiversity and Sustainable Management* (eds. W.J. Junk, M.T.F. Piedade, F. Wittmann, et al.), 437–462. Berlin: Springer.

Schöngart, J. and Queiroz, H.L. (2010). Timber extraction in the Cengral Amazonian floodplains. In: *Amazonian Floodplain Forests: Ecophysiology, Biodiversity and Sustainable Management* (eds. W.J. Junk, M.T.F. Piedade, F. Wittmann, et al.), 419–436. Berlin: Springer.

Schöngart, J., Witmann, F., and Worbes, M. (2010). Biomass and NPP of Central Amazonian floodplain forests. In: *Amazonian Floodplain Forests: Ecophysiology, Biodiversity and Sustainable Management* (eds. W.J. Junk, M.T.F. Piedade, F. Wittmann, et al.), 347–388. Berlin: Springer.

Schöngart, J., Arieira, J., Fortes, C.F. et al. (2011). Age-related and stand-wise estimates of carbon stocks and sequestration in the aboveground coarse wood biomass of wetland forests in the northern Pantanal, Brazil. *Biogeosciences* 8: 3407–3421.

Scott, D.A. and Jones, T.A. (1995). Classification and inventory of wetlands: A global overview. *Vegetatio* 118: 3–16.

Sculthorpe, C.D. (1985). *The Biology of Aquatic Vascular Plants*. Königstein, Germany: Koeltz Scientific Books.

Seidenschwarz, F. (1986). *Pioniervegetation im Amazonasgebiet Perus: Ein pflanzensoziologischer Vergleich von vorandinem Flussufer und Kulturland*, vol. 3. Langen: J. Margraf. Triops Verlag.

Sioli, H. (1956). Über Natur und Mensch in brasilianischen Amazonasgebiet. *Erdkunde* 10 (2): 89–109.

Sioli, H. (ed.) (1984). *The Amazon – Limnology and Landscape Ecology of a Mighty Tropical River and its Basin. Monographiae Biologicae*. Dordrecht: Dr W. Junk Publishers.

Smith, N.J.H. (1999). *The Amazon River Forest: a Natural History of Plants, Animals, and People*. New York: Oxford University Press.

Sousa, P.T. Jr., Piedade, M.T.F., and Candotti, E. (2011). Brasil's forest code puts wetlands at risk. *Nature* 478: 458.

Souza Filho, E.E. (2009). Evaluation of the Upper Paraná River discharge controlled by reservoirs. *Brazilian Journal of Biology* 69 (2): 707–716.

Souza, M.C., Cislinski, J., and Romagnolo, M.B. (1997). Levantamento floristico. In: *A planicle de inundação do alto rio Paraná: aspectos fisicos, biológicos e socioeconômicos* (eds. A.E.A. de Vazzoler, A.A. Agostinho and N.S. Hahn), 343–368. Marangá: EDUEM.

Souza, M.C., Romagnolo, M.B., and Kita, K.K. (2004). Riparian vegetation: ecotones and plant communities. In: *The Upper Paraná River and its Floodplain: Physical Aspects, Ecology and Conservation* (eds. S.M. Thomaz, A.A. Agostinho and N.S. Hahn), 353–367. Leiden: Bachhuys Publishers.

Souza, M.C., Kawakita, K., Slusarski, S.R., and Pareira, G.F. (2009). Vascular flora of the Upper Paraná River floodplain. *Brazilian Journal of Biology* 69 (2): 735–745.

Stadtler, E.W.C. (2007) Estimaivas de Biomassa Lenhosa, Estoque e Seqüestro de Carbono acima do solo ao longo do Gradiente de Inundação em uma Floresta de Igapó Alagada por Agua Preta na Amazònia Central. MSc thesis. Instituto Nacional de Pesquisas da Amazònia (INPA), Manaus, Brazil.

Stevaux, J.C. (2000). Climatic events during the late Pleistocene and Holocene in the upper Parana Rover: Correlation with NE Argentina and South-Central Brazil. *Quaternary International* 72: 73–85.

Strussmann, C., Prado, C.P.A., Ferreira, V.L., and Ribeiro, R.A.K. (2011). Diversity, ecology, management and conservation of amphibians and reptiles of the Brazilian Pantanal: a review. In: *The Pantanal: Ecology, Biodiversity and Sustainable Management of Large Neotropical Seasonal Wetland* (eds. W.J. Junk, C.J. da Silva, C.N. da Cunha and K.M. Wantzen), 497–521. Sofia-Moscow: Pensoft.

Targhetta N. (2012) *Comparação* floristica e estructural entre florestas de igapó e campinarana ao longo de gradients hidro-edáficos na Reserva de Desenvolvimento sustentável do Uatumã, Amazònia Central. Msc thesis. Instituto Nacional de Pesquisas da Amazònia (INPA), Manaus, Brazil.

Thomas, W.M., Cáceres, N.C., Nunes, A.P. et al. (2011). Mammals in the Pantanal wetland, Brazil. In: *The Pantanal: Ecology, Biodiversity and Sustainable Management of Large*

Neotropical Seasonal Wetland (eds. W.J. Junk, C.J. da Silva, C.N. da Cunha and K.M. Wantzen), 565–597. Sofia-Moscow: Pensoft.

Thomaz, S.M., Roberto, M.C., and Bini, L.M. (1997). Caracterização limnológica dos ambientes aquáticos e influência dos niveis fluviométricos. In: *A Planície de inundação do Alto Rio Paraná: aspectos fisicoe e biológicos e socioeconômicos* (eds. A.E.A.M. Vazzoler, A.A. Agostinho and N.S. Hahn), 73–102. Marangá: EDUEM.

Thomaz, S.M., Bini, L.M., Pagioro, T.A. et al. (2004). Aquatic macrophytes: diversity, biomass and decomposition. In: *The Upper Paraná River and its Floodplain: Physical Aspects, Ecology and Conservation* (eds. S.M. Thomaz, A.A. Agostinho and N.S. Hahn), 331–350. Leiden: Bachhuys Publishers.

Valente, C.R. and Latrubesse, E.M. (2012). Fluvial archive of peculiar avulsive fluvial patterns in the largest Quaternary intercratonic basin of topical South America: The Pantanal Basin, Central Brazil. *Palaeogeography, Palaeoclimatology, Palaeoecology* 356–357: 62–74.

Valente, C.R., Latrubesse, E.M., and Ferriera, L.G. (2013). Relationships among vegetation, geomorphology and hydrology in the Bananal Island tropical wetlands, Araguaia River basin, Central Brazil. *Journal of South American Earth Sciences* 46: 150–160.

Vannote, R.I., Minshall, G.L., Cummins, K.W. et al. (1980). The river continuum concept. *Canadian Journal of Fisheries and Aquatic Sciences* 37: 130–137.

Veneklaas, E.J., Fajardo, A., Obregon, S., and Lozano, J. (2005). Gallery forest types and their environmental correlates in a Colombian Savanna landscape. *Ecography* 28: 236–252.

Villamizar, E.A.R. (2013) Química da água para a classificaçáo dos rios e igarapês da bacia amazônica. PhD thesis. Instituto Nacional de Pesquisas da Amazônia (INPA) and Universidade Estadual do Amazônas (UEA), Manaus, Brazil.

Wantzen, K.M., Junk, W.J., and Rothhaupt, K.-O. (2008). An extension of the flood pulse concept (FPC) for lakes. *Hydrobiologia* 613: 151–170.

Weber, G.E., Furch, K., and Junk, W.J. (1996). A simple modelling approach towards hydrochemical seasonality of major cations in a Central Amazonian floodplain lake. *Ecological Modelling* 91: 39–56.

Welcomme, R.L. (1979). *Fisheries Ecology of Floodplain Rivers*. London: Longman.

Wittmann, F. (2012). Tree species composition and diversity in Brazilian freshwater floodplains. In: *Mycorrhiza: Occurrence in Natural and Restored Environments* (ed. M.C. Pagano), 223–263. New York: Nova Science Publishers.

Wittmann, F., Junk, W.F., and Piedade, M.T.F. (2004). The várzea forests in Amazonia: flooding and the highly dynamic geomorphology interact with natural forest succession. *Forest Ecology and Management* 196: 199–212.

Wittmann, F., Schöngart, J., Montero, J.C. et al. (2006a). Tree species composition and diversity gradients in white-water forests across the Amazon basin. *Journal of Biogeography* 33: 1334–1347.

Wittmann, F., Schöngart, J., Parolin, P. et al. (2006b). Wood specific gravity of trees in Amazonian white-water forests in relation to flooding. *IAWA Journal* 27: 255–266.

Wittmann, F., Zorzi, B.T., Tizianel, F.A.T. et al. (2008). Tree species composition, structure, and aboveground wood biomass of a riparian forest of the lower Miranda River, Southern Pantanal, Brazil. *Folia Geobotanica* 43: 397–411.

Wittmann, F., Schöngart, J., De Brito, J.M. et al. (2010a). *Manual de árvores de várzea da Amazonia Central. Taxonomia, ecologia e uso*. Manaus: Editora INPA.

Wittmann, F., Schöngart, J., and Junk, W.J. (2010b). Phytogeography, species diversity, community structure and dynamics of central Amazonian floodplain forests. In: *Amazonian Floodplain Forests: Ecophysiology, Biodiversity and Sustainable Management* (eds. W.J. Junk, M.T.F. Piedade, F. Wittmann, et al.), 61–102. Berlin: Springer.

Wittmann, F., Householder, E., and Piedade, M.T.F. (2013). Habitat specifity, endemism, and the neotropical distribution of Amazonian white-water floodplain trees. *Ecography* 36: 690–707.

Worbes, M. (1997). The forest ecosystem of the floodplains. In: *The Central Amazon Floodplain: Ecology of a Pulsating System* (ed. W.J. Junk), 223–265. Berlin: Springer.

Worbes, M., Klinge, H., Revilla, J.D., and Martius, C. (1992). On the dynamics, floristic subdivision and geographical distribution of várzea forests in Central Amazonia. *Journal of Vegetation Science* 3: 553–564.

6

Large River Deltas

6.1 Introduction

Located at the end of a river its delta indicates the nature of water and sediment discharge, basin geology, and history of changes in the base level. Deltas are essential for studying the nature and behaviour of a large river. They are part subaerial and part subaqueous accumulation of sediment at the mouth of a river draining into a sea or a lake. The sediment is river-borne, although in certain cases it is reorganised by tides, waves, and coastal currents. These processes also may add to or remove sediment from the accumulation. Deltas therefore vary in their geometry, morphology, operating processes, and nature of the constituting sediment. A delta develops well when a river brings in a large amount of sediment to deposit on a wide, shallow, tectonically inactive continental shelf. Worldwide, rivers annually bring about 36 000 km^3 of freshwater and more than 20 billion tonnes of solid and dissolved material to the oceans (Milliman and Farnsworth 2011). Deltas therefore are common coastal features, particularly at the mouth of rivers, transporting large volumes of sediment to the sea.

A delta is divided primarily into a subaerial and a subaqueous part (Figure 6.1). The subaerial delta may be further divided into a tidal and a non-tidal part, separated from each other by the tidal limit. Deltas begin where the main river first splits into two distributary channels. These channels then divide repeatedly to form a drainage network. Some of the channels may rejoin each other. At a point in time only one or two main channels function efficiently, and these, with a set of local distributaries, carry most of the water and sediment to the sea. The part of the delta with such active channels is known as the active delta. The rest of the subaerial delta, with channels that are relatively inactive, is the inactive delta or moribund delta or abandoned delta. It is an old part of the delta where channels have become dormant. The final shape of deltas is usually triangular like the Nile Delta, but occasionally lobate, like the Mississippi Delta.

A large river usually carries fine material towards the end of its course: fine sand, silt, and clay (Figure 3.5). A delta is commonly made of such fine-grained constituents, although exceptions may occur. Usually, the surface of the delta is nearly at sea level; the gradient of the river is miniscule; and only fine material is transferred from the river to the sea. The high parts of a delta are levees which are natural linear embankments that border the channels. A low alluvial basin occurs in between the levees of two neighbouring channels, and almost the entire surface of the delta, except the levees, may become inundated during a prolonged rainy season or after a single large storm (Figure 6.2).

Introducing Large Rivers, First Edition. Avijit Gupta.

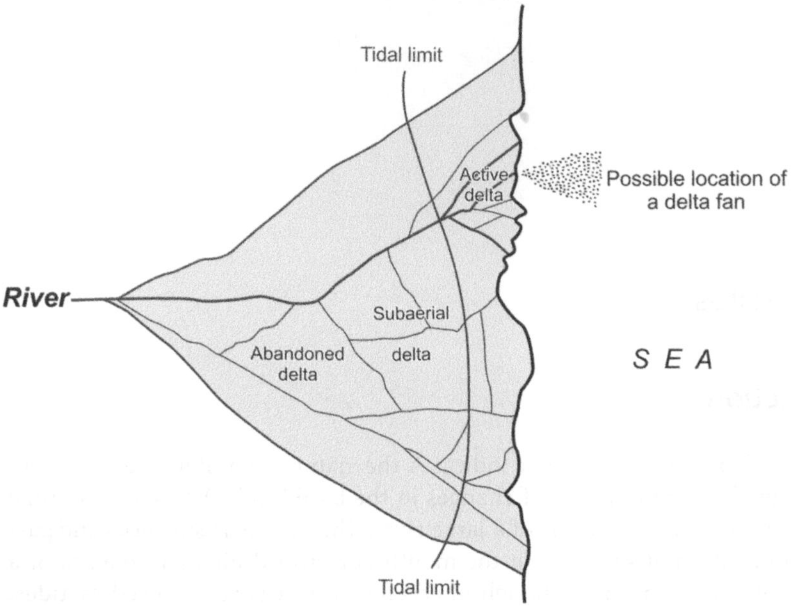

Figure 6.1 Delta landforms. Adapted from Gupta 2011.

Deltas tend to be highly populated and intensely farmed because of their flat gradient, rich aquatic resources, wide biodiversity, availability of water, and easy navigability along the channels. They are also, however, extremely vulnerable to soil acidity, floods, large storms, and channel avulsions. About a quarter of the global population lives on deltas or wetlands exposed to such hazards (Syvitski et al. 2005). Deltas, especially deltas of large rivers, therefore needed to be investigated for both resource utilisation and hazard management.

6.2 Large River Deltas: The Distribution

In general, the size of a delta relates to the scale of the river. Large rivers tend to build major deltas, provided the river carries high discharges of both water and sediment. All except four of the 25 largest rivers have built major deltas, and they deliver 31% of the total fluvial sediment that reaches the sea (Milliman and Meade 1983; Meade 1996). Several large deltas (Table 6.1) have been studied in detail (Figure 6.3).

6.3 Formation of Deltas

Where a river meets the sea, sedimentation occurs for two reasons: (i) deceleration of the river, and (ii) flocculation following mixing of the waters of the river and saline sea. The river enters the sea as a jet of sediment-laden water, sand, and silt being deposited first, followed further on by clay. This clay forms the advanced part of the delta, called the prodelta. The lowest part in front of the subaqueous delta therefore is clay, and as

Figure 6.2 The Effect of at the tropical cyclone Nargis on the Irrawaddy delta. (a) Part of the Irrawaddy Delta in normal conditions with a village on high ground and paddy fields occupying a low alluvial basin. (b) The same area after the arrival of the cyclone. Source: IKONOS satellite image © Centre for Remote Imaging, Sensing and Processing, National University of Singapore (2009), reproduced with permission.

the delta advances into the sea, sand is deposited on top of the clay. Delta sediment characteristically shows a coarsening-upward sequence.

The jet of river water commonly floats above the saline seawater unless it is so heavily laden with sediment that it sinks. This happens rarely but it used to be the case for the heavily turbid Huanghe before dams and reservoirs drastically reduced its flow. The velocity of the outflow is high near the middle of the jet or plume and slow at the edges. The mixing between the two sets of water at the edges of an intruding river jet leads to the deposition of sediment which builds ridge-like forms underwater. More sediment is deposited over time, bars are formed, and these ridges become higher and exposed above water as levees. Thus, the seaward face of the delta becomes a mosaic

Table 6.1 Selected major deltas in the world.

River	Drainage area (10^3 km^2)	Annual water discharge (km^3)	Annual total suspended load (10^6 t)	Delta area (10^3 km^2)
Amazon	6000	6300	1200	467.0
Ganga-Brahmaputra-Meghna	980 (G) +670 (B) +80 (M)	490 (G) +630 (B) +150 (M)	520 (G) +540 (B)	105.6
Mekong	800	550	110	93.8
Changjiang	1800	900	470	66.7
Lena	2500	520	20	43.6
Huanghe	750	15 (43)	150 (1100)	36.3
Indus	980	5 (90)	10 (250)	29.5
Mississippi	3300	490	210 (400)	28.6
Volga				27.2
Orinoco	1100	1100	210	20.6
Irrawaddy	430	380	325	20.6
Niger	2200	160 (190)	40	19.1
Shatt-al-Arab	420	46 (77)	(100)	18.5
Grijalva	50	23 (16)	13 (24)	17.0
Po	74	46	10 (15)	13.4
Nile	2900	30 (80)	0.2 (120)	12.5
Sõng Hóng	160	120	50 (110)	11.9
Chao Phraya	160	30	3	11.3

Note: Approximate measurements of rivers with large deltas. Delta areas are from Coleman (1982). Rest of the figures from Milliman and Farnsworth (2011) and references therein. Figures within parentheses are estimated pre-diversion figures. Construction of dams and reservoirs on several major rivers has drastically reduced the upstream supply of water and discharge (Chapter 8). The figures are better taken as approximate, and treated as relative numbers rather than absolute measures.

of river channels bounded by levees, with waterlogged depressions such as bays and backswamps in between. Breaks in the levees produce distributary sediment channels that run from the main river to the bays and backswamps, depositing sandy splays through levee crevasses and fine-grained sediment in the depressions. The bays and swamps are filled over time and become subaerial deltaic plains but at a lower elevation than levees. Distributary channels are backfilled and abandoned across the delta plain.

Most of the sedimentation happens in episodic large-magnitude floods (Giosan and Bhattacharya 2005). Avulsions and switching of the main river to a lower part of the delta shift the delta-building processes to a different area, enlarging the size of the delta and leaving behind an abandoned sector. Rapid sedimentation is often related to avulsions. Deltas dominated by river action grow in this fashion. The Mississippi is a common example, although recent work indicates that part of the Mississippi delta plain has been

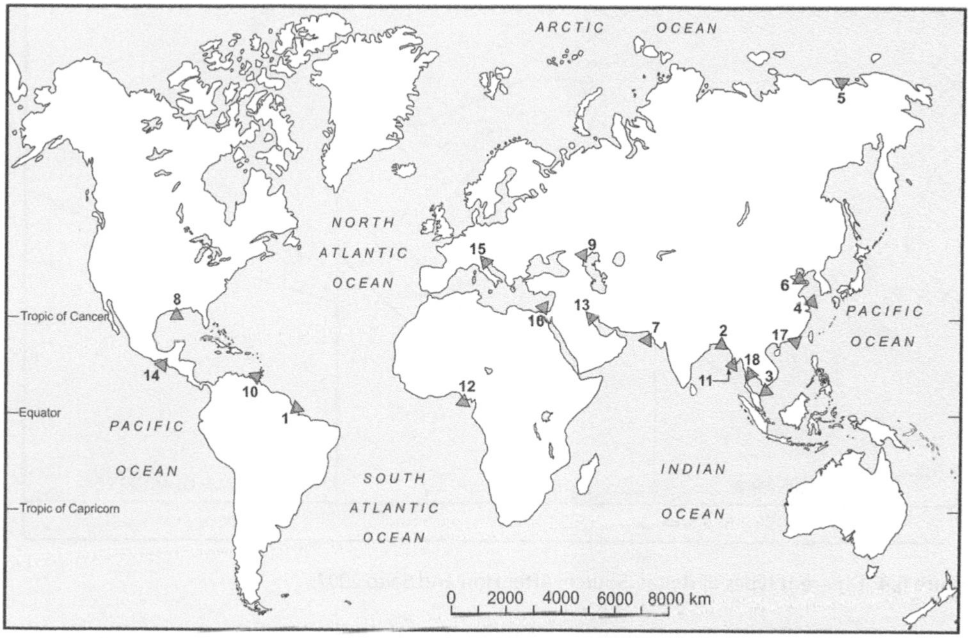

Figure 6.3 Location of major deltas in the world: 1, Amazon; 2, Ganga-Brahmaputra-Meghna; 3, Mekong; 4, Changjiang; 5, Lena; 6, Huanghe; 7, Indus; 8, Mississippi; 9, Volga; 10, Orinoco; 11, Irrawaddy; 12, Niger; 13, Shatt-al-Arab; 14, Grijalva; 15, Po; 16, Nile; 17, Sŏng Hóng; 18, Chao Phraya.

developed also by waves and longshore transport. Headlands in the southcentral part of the Mississippi delta complex carry regressive beach ridges which indicate the important role of waves in delta formation (Kulp et al. 2005). Marine action contributes sands from lobe headlands formed earlier, and such sand is carried in the downcoast direction by young distributaries building wave-influenced lobes. The abandoned lobes are thus disintegrated.

In a river-dominated delta, an underwater bar forms near the mouth of the river from deposition of sand and mud offshore where the waters of river and sea efficiently mix. The mouth bar builds up, progrades into the sea, grows higher and wider, and restricts river discharge. This results in transfer of the flow by avulsion to a different area of the delta face. The old bar is then modified by continuing slow deposition of bioturbated mud or transversely flowing currents. Large river floods episodically deposit erosion-based sand beds in the mouth bar which is subsequently topped by fines during the waning stage of the flood and then lines of mud mark the period between floods (Figures 6.4 and 6.5). Subsidence also plays an important role in the formation of large deltas like that of the Mississippi.

Tides, waves, and currents may reorganise the sediment across the face of the delta to produce its final form and sedimentary sequence. Tide-influenced deltas are formed on coasts with high tidal ranges but limited wave power. River-borne sediment moves up and down the tidal reach, and flocculation is spread over a length of the river. Distributary mouths tend to be choked by sandy shoals where flows of flood tides slow down landward. As a result, tidal deltas are distinguished by several features: funnel-shaped river mouths, bars, and islands elongated in the direction of tidal movement in the

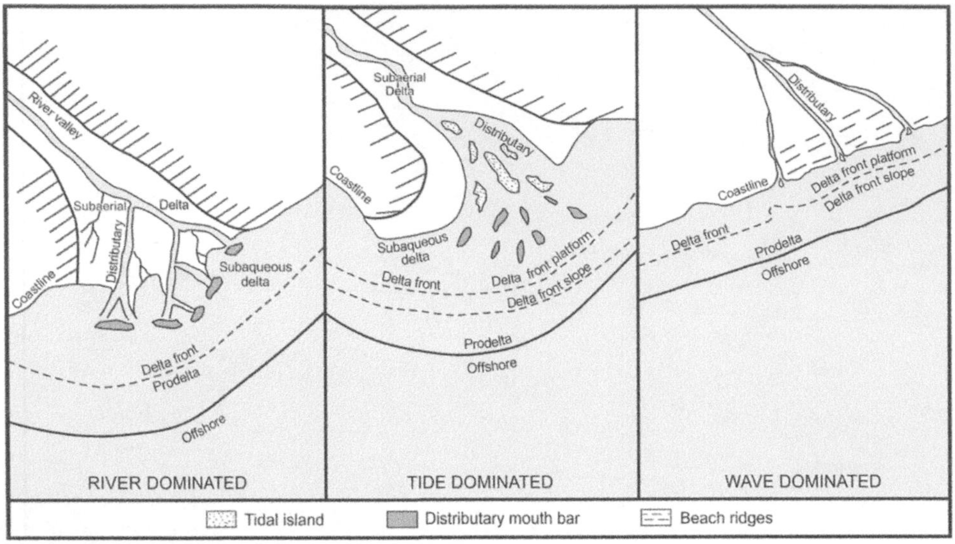

Figure 6.4 Different types of deltas. Source: After Hori and Saito 2007.

Figure 6.5 Photograph of a tidal creek through mangrove forest, Sundarbans, Ganga-Brahmaputra Delta; steep bank in silt-clay. Source: A. Gupta.

channel, a number of small tidal creeks separating the islands. Salt-resistant vegetation, such as mangroves in the humid tropics, anchors the islands. The Ganga-Brahmaputra Delta displays an excellent example.

Where waves are strong, river-borne sediment may be pushed back to straighten the delta edge unlike the common case of deltas protruding into the sea. These deltas often display wide beaches and, if the onshore wind is strong, lines of beach ridges and dunes characterise such coasts. The deltas of Senegal, Nile, and Sâo Francisco are good examples. The lower Mekong Delta displays beach ridges. The upper part of the Mekong Delta was primarily built by fluvial processes in a sheltered bay, a landscape of multiple levee-bordered channels separated by backswamps. As the delta grew beyond the shelter of the headland bordering the bay, it became exposed to waves, and the lower delta became both tide- and wave-dominated, displaying landscape of divergent and bifurcating beach ridges separated by low inter-tide depressions (Ta et al. 2005). Beach ridges, also known as chenier ridges, found to the west of the mouth of the Mississippi were formed by the sedimentation carried by a westerly current across the large delta. Deltas, especially large ones, thus can reflect a range of morphology and multiple operating processes.

A strong longitudinal current flowing past the delta face tends to erode and transfer material away from the delta, limiting its size and probably straightening the coastal edge. The Amazon Delta is a striking example. A considerable part of the sediment that is released by the Amazon into the South Atlantic is carried northwest parallel to the coast by the Guyana Current which transfers this sediment for hundreds of kilometres before it comes ashore to build the mudflats and beach ridges of north Brazil, French Guyana and Surinam. Part of the Orinoco Delta is built with the Amazon sediment. This sediment was originally eroded from the Andes, but it reaches the Orinoco Delta by a very long and circuitous route, unlike Orinoco's own sediment. Sediment carried by the Guyana Current has built mudcapes, a characteristic feature of the Orinoco Delta. Mudcapes are linear, fine-grained, round-ended promontories, several kilometres wide and up to 100 km long (Warne et al. 2002).

Deltas are river-dominated constructive forms, if the sediment is essentially carried by a river to a depositional basin. However, if the sediment is reworked and redeposited by tides or waves, the delta is identified as a tide- or wave-influenced delta. These two types can be both constructive and destructive (Hori and Saito 2007). Deltas tend to be classified as river, tidal or wave dominated, but many deltas of large rivers are complex in nature. The Godavari Delta on the Bay of Bengal coast of India progrades into a microtidal basin and low to moderate wave environment. The water discharge of the Godavari is monsoon-driven and seasonal. Large storms occasionally lead to huge floods. The delta has grown to more than 30–35 km across the continental shelf after the Holocene maximum transgression which occurred regionally around 5500–6000 years ago. The delta has two active lobes which have been morphologically changed by two strong longshore drifts flowing in the opposite directions. The different lobes of the Godavari Delta are distinguished by variable morphology, as a complex interaction happens among river discharge, nature of the waves, and location of distributary mouths. A powerful longshore sediment transport, induced by waves, builds spits and barrier islands and gives rise to a variety of depositional environments and variable efficacy in sediment trapping on the delta plain (Rao et al. 2005). Fielding et al. (2005) discussed the formation of the Holocene delta of the Burdekin River with at least 13 lobes formed over the last 8000–10 000 years. It has been constructed primarily by large river floods that deposited river mouth bars, forming the major part of the deltaic sediment. A minor volume was built and organised by beach ridges and spits which indicates that waves and longshore drifts are partly responsible. The Po Delta system is an assemblage of multiple lobes advancing into a shallow shelf sea, with a significant part of it being subaqueous. The present delta has been identified as an example of a wave-dominated delta being modified by anthropogenic activities up the river (Correggiari et al. 2005). Large deltas can be formed in multiple ways.

The current deltas are geologically very young. They were formed in the Early Holocene when the rise of the sea slowed down after the end of the last glaciation. This provided accommodation space for sediment deposition for building deltas, first higher, and then forward, as described for the Changjiang Delta. Based on radiocarbon dates of materials collected from boreholes sunk into deltas, Stanley and Warne (1994) dated the beginning of modern deltas to 8500–6500 years BP. The dates, however, may vary as arrival of sediment could be crucial. The base of the current Ganga-Brahmaputra Delta has been dated to be earlier, 10 000–11 000 years BP. The large deltas of South, Southeast, and East Asia were formed in the Holocene when (i) the sea level became stable or fell slightly, (ii) there was supply of large volumes of sediment from the Himalaya and the Tibetan Plateau, and (iii) the monsoon system was stronger. The highest Holocene sea level was probably several metres above the present level and occurred about 6000 years ago, but both age and height varied in different locations. In several instances, such as the Ganga-Brahmaputra and the Nile, the present sea level is the highest, and fluvial sediments are actively accumulating on the delta surface now (Hori and Saito 2007). Changes in hydrologic and sediment transfer pattern in large rivers affect sedimentation in their deltas, overriding regional eustatic and regional environments, and thus the formation dates may vary. A second set of older, submerged deltas at the head of a linking channel also has been found for a number of large rivers (Meade and Moody 2010).

6.4 Delta Morphology and Sediment

Morphologically deltas can be divided into three parts: delta plain, delta front, and prodelta. A delta plain is a low, flat plain which forms the subaerial part of the delta. This flat area is cut through by a number of active and abandoned distributary channels that leave the main river at a high angle. The channels are bounded by levees, and the rest of the plain is an assemblage of bays, marshes, tidal flats, and floodplains. In the humid tropics, delta plains are likely to be under a luxuriant vegetation cover of freshwater swamps and mangroves. In the Mahakam Delta of humid Kalimantan, the interdistributary areas between the levees are under mangroves. *Sonneratia*, *Avicennia*, and *Rhizophora* grow near the sea. An extensive zone of nipa palms (*Nypa fruticans*) and *Heritiera* is found behind these mangroves, which is replaced landward by a swamp forest. (Woodroffe 2005 and references therein). In contrast, vegetation is scarce in arid areas, being replaced by a saline crust of gypsum and halite on the surface. These interdistributary areas are flooded during the wet season but emerge as dry land at other times. Overbank flooding and crevasse splays are common during high flows, and so is channel avulsion.

Various other processes also characterise delta morphology. If the sand supply is high, dunes may appear as in the delta of the Sâo Francisco River. The effect of the tidal passage of water typifies the lower part of a tide-dominated delta. There the channels have funnelled shapes resembling estuaries and are of low sinuosity. Bedforms and sand bars are common in the channel. For example, sand bars in low sinuosity channels occur in the Mahakam Delta in eastern Kalimantan. Channels in a river-dominated delta with negligible tidal influence tend to be sinuous instead. In arid areas with discrete high flows and coarse bedload, the distributary channels may braid or anastomose (Elliott 1986).

The subaerial delta plain carries a range of sediment that varies in both texture and structure. Levees, crevasses splays, and beach ridges commonly are in sand, whereas the interdistributary lowlands are in clay. In front of it, interlaminated or thinly interbedded alternate mud and sand layers build the delta front. The variation in beds has been associated with a change in processes such as changing tides. Relatively thicker beds (about 2 cm) as in the delta of the Fly River, Papua New Guinea have been attributed to seasonal changes in wind pattern.

The delta front occurs in the subaqueous area where the sediment-laden river water meets the saline sea water and sediment is actively deposited as a clinoform. Delta slopes steepen towards the sea and form the subaqueous delta front. Part of the sediment is delivered by traction. Bars are frequently deposited in front of a distributary mouth and the channel bifurcates to flow around the bar, extending the delta into the receiving basin. Tide-dominated deltas thus have wide delta fronts (Nittrouer et al. 1986). Waves may redistribute the sediment, remodelling the shape of the bar according to the direction of wave approach, unless the sediment supply is strong and persistent. Exposure to strong waves tends to make the delta front straighter, building beaches and beach ridges behind them. The delta front advances over the next zone, the prodelta, which has a gentle seaward slope, building a coarsening-upward vertical sequence.

The prodelta is located seaward of the delta front, a layer of clay and silt on the floor of the receiving basin. It is the advancing edge of a delta. The fine sediment is intercalated with silt stringers and thin shell beds. It is highly bioturbated. The presence of shell beds and coarse material within the prodelta sediment is usually attributed to storms.

The succession of prodelta, delta front, and delta plain gives the delta sediment a characteristic vertical sequence. A coarsening-upward change occurs between the sediment of the prodelta and the delta front. This is overlain by a fining-upward sequence from the delta front to delta plain. Coarser and better-sorted sediment indicates a transporting agency with high energy. In such cases the fining-upward sequence may be replaced by a coarsening one if dunes or beach ridges occur on the delta plain. The generalised succession is not found everywhere. Sedimentation in deltas is variable and complex (Hori and Saito 2007).

A structural pattern in delta sediment was recognised by G.K. Gilbert in 1855 around lake margins in the western United States. The concept indicated that deltas start with a near-horizontal deposit of fine material in the receiving basin, called the bottomset. As the delta progrades, a clinoform, called the foreset, advances over the bottomset. This is then overlain by another near-horizontal layer of fine sediment, the topset. The angle of the foreset of the slope of the delta depends on the coarseness of the sediment. This concept works for deltas in lakes and fan deltas in coarse material where foreset beds have an inclination of 10–25°. The majority of the deltas of the world are in fine material. According to Milliman and Meade (1983), the major deltas of the world receive 80–90%

fine-grained suspended load and 10–20% coarse-grained bed load. In sum, deltaic systems reflect the complex interaction of sediment availability, accommodation space, and the efficacy of coastal processes – each varying over space and time.

6.5 The Ganga-Brahmaputra Delta: An Example of a Major Deltaic Accumulation

Deltas of large rivers can be complex. Such a delta may extend over a large area and display varied morphology, multiple delta-forming processes, and the contribution of water and sediment discharges of the large river. The latter may even override the effect of a changing sea level. The Ganga-Brahmaputra Delta demonstrates the complex nature of major deltas. These deltas act as sinks at the end of a long conduit that carries water and sediment. The Ganga and Brahmaputra, two large rivers that rise on opposite sides of the Himalaya Mountains, together bring down about a billion tonnes of sediment annually to form one of the largest deltas of the world, with a subaerial extent of about 111 000 km^2 in Bangladesh and India. If the present prograding clinoform is included, the total subaerial and subaqueous area stretches for more than 250 km across the continental shelf and extends about 125 km from land, measuring approximately 140 000 km^2. In this humid monsoon setting, the densely populated delta, in spite of considerable forest destruction, still carries a dense mangrove forest (the Sundarbans) near the delta face (Kuehl et al. 2005). The Ganga and Brahmaputra meet at Aricha (Figure 6.6). The combined waterway, known as the Padma, flows southeast to meet another huge river, the Meghna, draining the Sylhet Basin. Their combined water flows into the Bay of Bengal past Noakhali.

6.5.1 The Background

Their sink is tectonically active, and the delta has been divided into subsiding basins and uplands by tectonics (Figure 6.6). The Bengal Basin has been subsiding since the Holocene, and beyond a hinge zone that marks the boundary between the Indian continent and the oceanic crust provides space for a very thick volume of deltaic sediment (Sengupta 1966). It is a low delta, with a surface elevation of 10–15 m in the northwest from where it slopes southeast to the sea. The relief is low, the only uplands being the Barind and Madhupur Terraces which rise to about 15 m above the level of the delta.

The offshore surface dips gently. The only exception is found in the east near Chittagong, where a series of north–south structural ridges and lows are found offshore. Towards the west, near the India–Bangladesh border, a submarine canyon (the Swatch of No Ground) starts within 30 km of the coast. The canyon is supposed to have been incised during the low stand of the huge river in the Pleistocene. It provides a link to the Bengal Deep Sea Fan which extends under water almost to the Equator. Elsewhere, the offshore surface is covered by the Holocene sediment of the Ganga and Brahmaputra.

The seasonal monsoon system controls the hydrology of the two basins. Rain falls between late June during June–November and May–November, respectively. The low flows are during January–April. The joint flow of the rivers to the northern Bay of Bengal increases from 10^4 to 10^5 m^3 s^{-1}. Most of the sediment is also released during this period. Tropical storms, several reaching the level of tropical cyclones, also inundate the

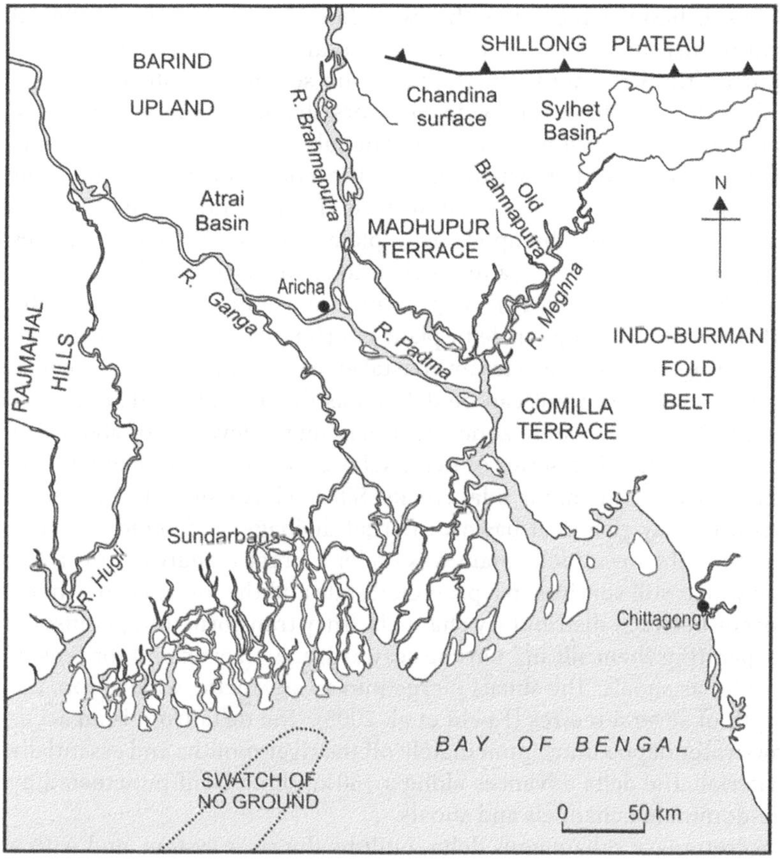

Figure 6.6 Physiographic and tectonic map of the Ganga-Brahmaputra Delta. Adapted from Kuehl et al. 2005, Sengupta 1966 and Bangladesh geological maps.

lower delta, mobilising a huge amount of offshore sediment and raising the coast and islands. This is a tidal delta, and tides may be as high as 5 m in the estuaries (Barua 1990). A synchronised combination of tropical cyclones and high tides significantly erode the coast.

6.5.2 Morphology of the Delta

Kuehl et al. (2005) divided the Bengal Delta into three morphological compartments: upper delta plain and flood basins; lower delta plain and delta front; and subaqueous delta. The upper delta plain is a 200 km wide zone of the extreme landward part of the delta. The landward boundary has been recognised as the initiation of the deltaic distributary of the Ganga River and where frequent avulsion of the channel of the Brahmaputra has been recorded. The dry season inland extension of salt water marks its lower boundary. This is an area of river floodplains bounded by uplands. The uplands consist of broad fluvially dissected surfaces of Barind and Madhupur Terraces, probably Pleistocene in age; and less distinct, terrace-like features of younger Chandina and Comilla surfaces (Figure 6.6). The upper delta is partitioned by the uplands into individual and lower sub-basins which are either wide flood basins or narrow, alluvial corridors. The flood basins such as the Sylhet and Atrai are inundated under several metres of water during the wet monsoon, trapping fine sediment. Both the Ganga and Brahmaputra flow through 40–80 km wide corridors characterised by braid belts, overbank floods, and channel avulsions. The dominant subsurface material is channel sand. The sediment is transferred downstream via the corridors but a significant part gets trapped building a low and very flat surface. These rivers are mobile and change their courses repeatedly on the upper delta plain. The Brahmaputra provides a classic example, the huge river avulsing alternately into channels east and west of the Madhupur Terrace (Pickering et al. 2013). At present, the Brahmaputra is using the western channel (Figure 6.6). The upper delta plain is primarily controlled by fluvial processes, influenced by tectonics, downstream coastal evolution, and sea-level change.

The lower delta plain and delta front can be defined as the part of the delta always affected by salt water. This 100 km wide zone has a maximum elevation of about 3 m at its inland limit. The coastline is a series of peninsulas between major channels, the peninsulas being criss-crossed by minor tidal creeks. Silt to clayey silt with little sand is found on the surface. Clay-rich peat basins, internal drainage, and standing water conditions are common. The lower delta plain was under a larger mangrove forest, the Sundarbans, part of which still survives in a protected state and the forest is now much depleted. Islands occur between distributary channels. They transform into peninsulas as tidal channels separating them silt up. On the seaward side, they extend for tens of kilometres under water as shoals. The shoals merge into an advancing wide apron, the delta front, at a depth of several metres (Kuehl et al. 2005). The delta front is an accumulation of shallow-water depositions immediately off the river mouths and essentially made of coarse material. The delta advances along a 150 km long front punctuated by ebb- and flood tide-dominated channels and shoals.

Downslope, an accretionary subaqueous delta, built by the river system, and with a complete topset, foreset, and bottomset, occurs on the continental shelf. This large delta, like the deltas of several large rivers, has two sets of clinoforms, both of which can be identified as a delta fronts. Beyond the subaqueous bottomset is the outer shelf. Submerged former deltas, dating back to glacial low stands, are evident on the outer shelf. Oolitic ridges occur on the outer shelf indicating a low supply of silica-rich sediment following a quieter river system during the cold times of the Pleistocene (Goodbred 2003).

A submarine canyon, knows as the Swatch of No Ground, dissects the shelf, starting from about 30 km off the present coast. The canyon is 20–30 km wide with a steep eastern wall and is believed to be filled with deltaic sediment. Such a fill is periodically emptied by turbidity currents that transfer the sediment to the enormous 2000 km long Bengal Fan.

6.5.3 Late Glacial and Holocene Evolution of the Delta

The evolution of the delta is related to changes in the river system, climate, and sea level during the Late Pleistocene and Holocene (Goodbred and Kuehl 2000a; Goodbred 2003). During the low stand of the sea around 18 000 years ago, both major rivers flowed through incised valleys. Most of the surface of the basin outside such valleys consisted of broad lateritic uplands which were about 45–55 m below the current sea level (Goodbred and Kuehl 2000b). Lateritic remnants have been found on the outer shelf.

The oolitic ridges on the surface indicate low terrigeneous outputs from rivers of glacial time. Around 15 000 years ago, the summer monsoon became stronger with increased precipitation which led to the transfer of an enormous amount of sediment to the basin. This enhanced summer monsoon continued until the early Holocene with extremely high sediment discharge which was 2.5 times the modern volume. Delta formation thus started early, and the high sediment discharge continued for several thousand years (Kuehl et al. 2005 and references therein).

The rising sea transgressed the low areas, trapping the sediment on top of the delta which aggraded at an enormous rate between 11 000 and 7000 years BP. The rapidly rising sea level was offset by the huge amount of sediment arriving at the delta due to the enhanced summer monsoon. The maximum transgression of the sea, as indicated by pollen abundance, happened near the end of this time frame (Banerjee and Sen 1988 referenced in Kuehl et al. 2005; Umitsu 1993). Subsequently, after 7000 years BP, deltaic deposition became progradational and the delta primarily extended seaward as the beginning of the subaqueous delta. Progradation was most prominent in the middle Holocene in the western delta where the Ganga once used to discharge actively to the sea.

The surface of the delta was selectively filled by the avulsion of large rivers, mainly the Brahmaputra. This huge river avulsed between its western and eastern channels, by passing or filling the Sylhet Basin (Pickering et al. 2013). The main channel of the Ganga migrated or avulsed eastward about 5000 years ago. As a result, the Meghna estuary region to the east has gone through annual shoreline accretion of 5.5–16 km^2 over the past several hundred years (Kuehl et al. 2005 and references therein).

The Ganga-Brahmaputra Delta demonstrates the complexity of surface forms, stratigraphy, and evolution of the delta of a large river. It also indicates that deltas could be influenced not only by sea level changes but also by varying discharges of huge amounts of water and sediment.

6.6 Conclusion

Evolution of a large delta primarily requires deposition of huge discharges of water and sediment in a depositional basin. A delta has a characteristic physiography and sedimentary sequence. Large deltas are often built by a combination of processes and a display a complexity of forms. In general, the present deltas started in the Early Holocene, although variations did happen. Deltas are morphologically fragile and change over time. The recent spate of dam construction on major rivers has led to considerable decreases in water and sediment discharge. Such anthropogenic activities and global warming have led to alterations in regional wave climate and storm patterns (Giosan and Bhattacharya 2005) besides affecting the river system. Deltas of large rivers are commonly indicative of past sea-level changes.

Questions

1 Why do deltas vary in their morphology and constituting sediment?

2 Deltas are of importance because of resource availability and hazard maintenance. Discuss.

3 How is a delta formed?

4 Is it possible to recognise the related delta-forming processes from the morphology of individual delta?

5 Is the Orinoco Delta built by the sediment transported by the Orinoco River?

6 Is a delta formed by a single process or a combination of processes?

7 How old are the present deltas?

8 Stanley and Warne (1994) have dated the beginning of modern deltas at 4500–6500 BP. The base of the current Ganga-Brahmaputra Delta has been dated at 10 000–11 000 BP. Why is there a difference?

9 What vertical sequence indicates the advance of the prodelta, delta front, and delta plain? Why?

10 Why do large rivers have large deltas?

References

Banerjee, M. and Sen, P.K. (1988). Paleobiology and environment of deposition of Holocene sediments of the Bengal basin, India. In: *The Palaeoenvironment of East Asia from the Mid-Tertiary, Proceedings of the 2nd Conference*, 703–731. Hong Kong: Centre of Asian Studies, University of Hong Kong.

Barua, D.K. (1990). Suspended sediment movement in the estuary of the Ganges-Brahmaputra-Meghna system. *Marine Geology* 92: 243–254.

Coleman, J.M. (1982). *Deltas. Processes and Models of Deposition and Exploration*, 2e. Boston: International Human Resources Development Corporation.

Correggiari, A., Cattaneo, A., and Trincardi, F. (2005). Depositional patterns in the late Holocene Po Delta system. In: *River Deltas: Concepts, Models and Examples*, vol. 85 (eds. L. Giosan and J.P. Bhattacharya), 365–392. Tulsa: SEPM (Society for Sedimentary Geology).

Elliott, T. (1986). Deltas. In: *Sedimentary Environments and Facies* (ed. H.G. Reading), 113–154. Oxford: Blackwell Scientific Publications.

Fielding, C.R., Trueman, J., and Alexander, J. (2005). Sedimentology of the modern and Holocene Burdekin River delta of North Queensland, Australia – controlled by river output, not by waves and tides. In: *River Deltas: Concepts, Models and Examples*, vol. 85 (eds. L. Giosan and J.P. Bhattacharya), 467–496. Tulsa: SEPM (Society for Sedimentary Geology).

Giosan, L. and Bhattacharya, J.P. (2005). New directions in deltaic studies. In: *River Deltas: Concepts, Models and Examples*, vol. 85 (eds. L. Giosan and J.P. Bhattacharya), 3–10. Tulsa: SEPM (Society for Sedimentary Geology).

Goodbred, S.L. Jr. (2003). Response of the Ganges dispersal system to climate change: a source-to-sink view since the last interstade. *Sedimentary Geology* 162: 83–104.

Goodbred, S.L. and Kuehl, S.A. (2000a). Enormous Ganges-Brahmaputra sediment discharge during strengthened early Holocene monsoon. *Geology* 28: 1083–1086.

Goodbred, S.L. and Kuehl, S.A. (2000b). The significance of large sediment supply, active tectonism and eustasy on margin sequence development: late Quaternary stratigraphy and evolution of the Ganges-Brahmaputra delta. *Sedimentary Geology* 133: 227–248.

Hori, K. and Saito, Y. (2007). Classification, architecture, and evolution of large-river deltas. In: *Large Rivers: Geomorphology and Management* (ed. A. Gupta), 75–96. Chichester: Wiley.

Kuehl, S.A., Allison, M.A., Goodbred, S.L., and Kudrass, H. (2005). The Ganges-Brahmaputra Delta. In: *River Deltas: Concepts, Models and Examples*, vol. 85 (eds. L. Giosan and J.P. Bhattacharya), 413–434. Tulsa: SEPM (Society for Sedimentary Geology).

Kulp, M., Fitzgerald, D., and Penland, S. (2005). Sand-rich lithosomes of the Holocene Mississippi River delta plain. In: *River Deltas: Concepts, Models and Examples*, vol. 85 (eds. L. Giosan and J.P. Bhattacharya), 279–293. Tulsa: SEPM (Society for Sedimentary Geology).

Meade, R.H. (1996). River-sediment inputs to major deltas. In: *Sea-Level Rise and Coastal Subsidence* (eds. J.D. Milliman and B.U. Haq), 63–85. Dordrecht: Kluwer Academic Publishers.

Meade, R.H. and Moody, J.A. (2010). Causes for the decline of suspended-sediment discharge in the Mississippi River system, 1940–2007. *Hydrological Processes* 24: 35–49.

Milliman, J.D. and Farnsworth, C.L. (2011). *River Discharge to the Coastal Ocean*. Cambridge: Cambridge University Press.

Milliman, J.D. and Meade, R.H. (1983). World-wide delivery of river sediment to the ocean. *Journal of Geology* 91: 1–21.

Nittrouer, C.A., Kuehl, S.A., and DeMaster, D.J. (1986). The deltaic nature of Amazon shelf sedimentation. *Geological Society of America Bulletin* 97: 444–458.

Pickering, J., Goodbred, S., Reitz, M. et al. (2013). Late Quaternary sedimentary record and Holocene channel avulsions of the Jamuna and Old Brahmaputra River valleys in the upper Bengal delta plain. *Geomorphology* 227: 123–136.

Rao, K.N., Sadakata, N., Hema Malini, B., and Takayasu, K. (2005). Sedimentation processes and asymmetric development of the Godavari delta, India. In: *River Deltas: Concepts, Models and Examples*, vol. 85 (eds. L. Giosan and J.P. Bhattacharya), 435–451. Tulsa: SEPM (Society for Sedimentary Geology).

Sengupta, S. (1966). Geological and geophysical studies in the western part of the Bengal Basin, India. *American Association of Petroleum Geologists Bulletin* 50: 1001–1017.

Stanley, D.J. and Warne, A.G. (1994). World initiation of Holocene marine deltas by deceleration and sea-level rise. *Science* 265: 228–231.

Syvitski, J.P.M., Vorösmarty, C.J., and Kettner, A.J. (2005). Impacts of humans on the flux of terrestrial sediment to the global coastal ocean. *Science* 308: 376–380.

Ta, T.K.O., Nguyen, V.L., and Tateishi, M. (2005). Holocene delta evolution and sediment discharge of the Mekong River Delta, southern Vietnam. In: *River Deltas: Concepts, Models and Examples*, vol. 85 (eds. L. Giosan and J.P. Bhattacharya), 453–466. Tulsa: SEPM (Society for Sedimentary Geology).

Umitsu, M. (1993). Late Quaternary sedimentary environments and landforms in the Ganges delta. *Sedimentary Geology* 83: 177–186.

Warne, A.G., Meade, R.H., White, W.A. et al. (2002). Regional control on geomorphology, hydrology and ecosystem integrity in the Orinoco Delta, Venezuela. *Geomorphology* 44: 273–307.

Woodroffe, C. (2005). Southeast Asian deltas. In: *The Physical Geography of Southeast Asia* (ed. A. Gupta), 228–236. Oxford: Oxford University Press.

7

Geological History of Large River Systems

7.1 The Age of Large Rivers

A number of large rivers are also long-lived rivers. The Mississippi, for example, has been a river system since at least the late Jurassic, transporting large amounts of sediment to the Gulf of Mexico since the Cretaceous (Coleman 1988). Global climatic and sea-level changes have repeatedly altered major rivers worldwide during the Quaternary. In recent years, anthropogenic alterations – widespread land use changes, impoundments, and reservoirs, engineered levees, etc. – have impacted many rivers. A large river may reflect all these changes.

All rivers have a history. Tectonic processes commonly influence the origin, location, and modification of major rivers. It is probably safe to assume that the present large rivers in the world were formed at different times after the break-up of Pangea in the Cretaceous. For example, the collision of the Indian Plate with the Eurasian Plate built the Himalaya Mountains, and created the present location and network of the major rivers of South and Southeast Asia (Tapponier et al. 1986; Brookfield 1998). It has been suggested that this collision led to disintegration of an earlier drainage net of the Sông Hóng (Red River) due to progressive river capture after the uplift of the Tibetan Plateau. The former drainage network then disintegrated into separate headwaters which developed into several major rivers: Yangtze, Mekong, Salween, and Tsangpo (Clark et al. 2004). River networks change over time.

As described in Chapter 2, the course of a large river is commonly controlled by one of several types of tectonic settings. A large river can extend from the active side of a continent to its passive coast (the Amazon), follow a foredeep along a continental collision belt (the Ganga), be confined in a rift (Rio Grande), run on a dome flank following the rise of a mantle plume (the Paraná), and drain an intraplate passive margin setting (the Murray-Darling) (Inman and Nordstrom 1971; Potter 1978; Cox 1989; Tandon and Sinha 2007). The history of a large river should include the formation of a structural low, considerable sediment accumulation in this linear depression, periodic morphological and sedimentological changes during the Quaternary, and large-scale anthropogenic alterations over the last hundred years or more. The Quaternary changes and the anthropogenic alterations happened worldwide and were superimposed on different geological characteristics of individual effects of the rivers.

For example, the drainage system of South-Central Africa illustrates major reorganisations following the break-up of Gondwana. The area is now drained by two major rivers: the Zambezi and Limpopo (Figure 7.1). But during the Upper

Introducing Large Rivers, First Edition. Avijit Gupta.

Jurassic and Cretaceous, three major systems used to operate in this area. The Okavango–Cuando–Zambezi-Luanga river system formed the headwaters of an ancestral Limpopo; the lower Zambezi–Shire formed a separate system that was graben-bound and emptied in the Indian Ocean near the mouth of the present Zambezi; and lastly, a large system discharged into the Indian Ocean near the mouth of the present Save River. A Cretaceous uplift changed the old network and (i) created

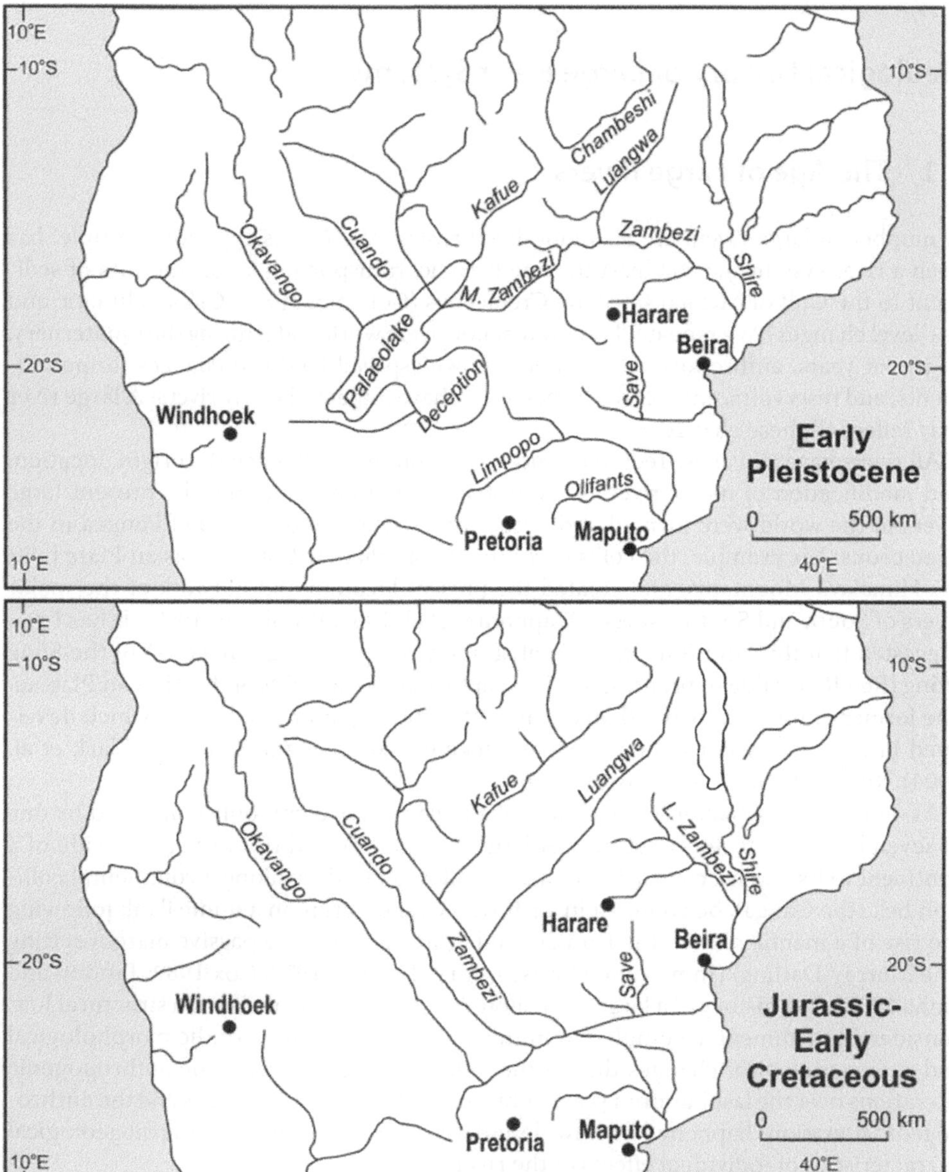

Figure 7.1 Simplified diagram showing drainage reorganisation in South Central Africa, Jurassic–Cretaceous to Upper Pleistocene. Source: Adapted from Moore and Larkin 2001.

a senile endoreic drainage towards the Kalahari Basin, and (ii) rejuvenated the lower Zambezi which captured several streams to build a new major network. Thus, the present pattern of two river systems to the coast (the Zambezi and Limpopo) as shown in Figure 7.1 came into being (Moore and Larkin 2001). A series of large dams and reservoirs now operate on the Zambezi which alternates between gorges and alluvial basins. The current river reflects all such developments.

7.2 Rivers in the Quaternary

We start this account of the geological history of large rivers from the Quaternary, as not much is known about earlier times.

7.2.1 The Time Period

The Quaternary period includes the Pleistocene and Holocene epochs. The Pleistocene extends from 2.59 million to 11 500 radiocarbon years ago, when the Holocene began. The Pleistocene was a period of repeated glaciations separated by warmer times. Land-based glacial sediment and relict landforms have been recognised from the middle of the nineteenth century as related to glaciations in the Pleistocene. Further details, such as sequences and dates, however, were worked out much later in the twentieth century from (i) sea-floor cores and (ii) ice cores from Antarctica and Greenland. Oxygen isotopes examined vertically in the sea floor cores show a series of alternating changes in their ratio over time, which were related to rise and fall in the global temperature. ^{16}O evaporates at a higher rate than other oxygen isotopes, and following evaporation from the ocean, is locked up in ice during colder periods. Thus, a higher value of ^{18}O in the skeletons of marine organisms from sea-floor cores indicates a colder climate, whereas that of ^{16}O suggests a warmer climate, melting of ice, and draining of meltwater back to the oceans. Each climatic directional change, starting from the top of the core, has been given a Marine Isotope Stage (MIS) number (Figure 7.2). The Holocene is MIS 1, the previous interglacial maximum is MIS 5e (130–118 ka ago). Several glacial stages (stadials) and warmer events (interstadials) have occurred between MIS 1 and MIS 5e, but not a full interglacial. The base of the Pleistocene is associated with MIS 103. Such a high number indicates a large number of changes in the Early Pleistocene (Murray-Wallace and Woodroffe 2014 and references therein).

These changes have been related to periodic modifications in the elliptical orbit of the Earth around the Sun which altered the amount of solar intensity reaching the Earth's surface which in turn caused a temperature change. The alterations in orbital configuration of the Earth around the Sun were due to three factors: changes in eccentricity (which occurred at 100 ka cycles), in obliquity (at 41 ka cycles), and in precession (at 23 ka cycles). Our understanding of the association between orbital forcing and Pleistocene climatic changes came from work by Joseph Adhémar, James Croll and Milutin Milankovitch. Milankovitch reviewed in detail the association of the three types of orbital changes with glaciations in the Pleistocene.

Climate in the Pleistocene thus repeatedly went through a number of cold glacial and warm interglacial times. Some of the interglacials were comparable with the present global climate but the average temperature of each glacial and interglacial varied. The

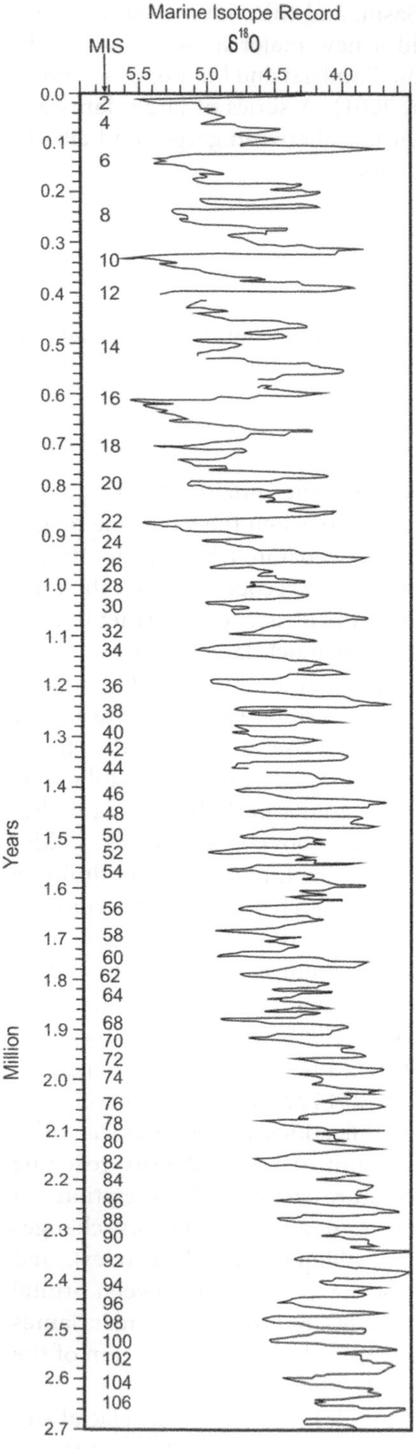

Figure 7.2 Global correlation time chart of the Quaternary showing the Marine Isotope Record (MIS). Source: Murray-Wallace and Woodroffe 2014.

differences between average annual temperatures of the cold (glacial) and warm (interglacial) periods were small (only 5–10 °C) but were enough to alter a glacial to an interglacial time or vice versa. The deep-sea cores indicated that the change from warm interglacial times to cold glacials was slow and gradual but the change from cold to warm was faster. The average temperature curve for the Pleistocene thus has a saw-toothed appearance (Figure 7.2).

Climate changes happened also before the Pleistocene. Certain events that occurred in the Tertiary helped at times to create favourable conditions for a colder climate. For example, the closing of the Panama isthmus prevented warm water from the Pacific from entering the Atlantic to keep the North Atlantic warm. Other developments that cooled the surface of the Earth were the expansion of ice over Antarctica, the building up of ice sheets in the temperate latitudes, the expansion of tropical grasslands separating forests, and development of high plateaux (Ruddiman and Kutzbach 1991; Raymo and Ruddiman 1992; Quade et al. 1995; Derbyshire 1996; Williams et al. 1998). It is the series of climate changes during the Pleistocene that has left significant marks globally, and certainly on large rivers.

7.2.2 The Nature of Geomorphic Changes

Alterations to the global climate during the Pleistocene led to a set of geomorphic changes, mainly (i) the advance and retreat of mountain glaciers and ice sheets and (ii) the related fall and rise of the sea. The nature and intensity of these changes varied across the globe, depending on the latitude and elevation of the place concerned. Both temperature and precipitation of places changed, and so did the nature of storminess and precipitation intensity. Climatic and geomorphic processes changed not only over space but also over time. To illustrate, rivers commonly ran low at the peak of cold timespans as the growth of glaciers locked water into ice, but the same rivers might have carried huge floods later, when the temperature rose and ice melted. A river that persisted through the Pleistocene possibly went through such changes repeatedly, although all the evidence did not survive clearly, and the more recent alterations are usually the ones best recognised in the field.

The Pleistocene has been studied in detail by experts from many academic fields, and we can take advantage of these studies to understand its impact on large rivers. It is worth remembering five major factors while discussing such changes:

1. A large river commonly runs from the mountains to the sea, and therefore would be impacted by both (i) the advance and retreat of glaciers and (ii) the fall and rise of the sea.
2. The hydrology of the river would change over time due to modification in precipitation over its basin and differential melting of glaciers. Floods and possibly large storms would dominate river hydrology at the time of ice melt, whereas the same river may run low during peak glaciation.
3. The supply of sediment from the slopes and tributary channels would vary over time owing to changes in climate and glaciation which would modify land cover and sediment yield (Figure 7.3).
4. The morphology of the channel, floodplain, and delta is likely to be reconstructed repeatedly, following changes in glaciation in the mountains, basin climate, and base level.
5. Theoretically, all such changes should be repeated for glacial and interglacial stages, although not necessarily at the same level; not much is known about this.

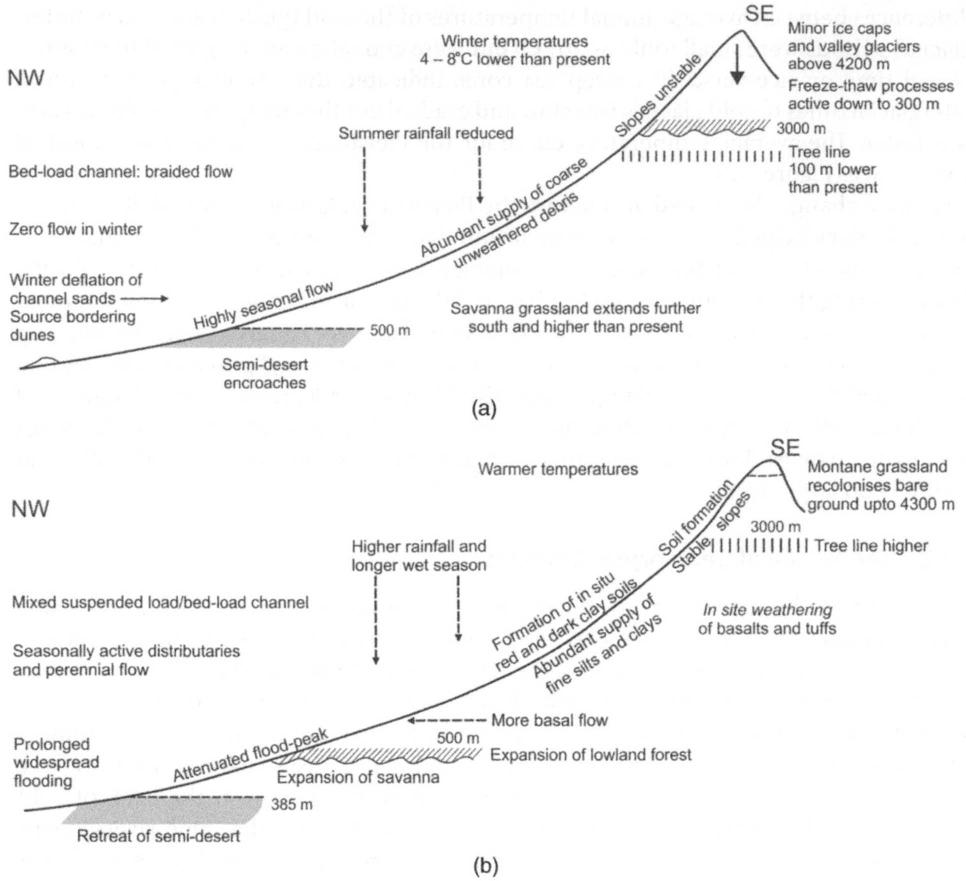

Figure 7.3 Landforms and geomorphological processes in the Blue Nile Basin during (a) the Last Glacial Maximum and (b) the early Holocene. Source: Woodward et al. 2007.

7.2.3 The Pleistocene and Large Rivers

We start with a simple example. Imagine a large river running from high mountains to the sea. The upper mountainous part of this river was affected by glaciation and climate change, the middle part by changing climate and sediment supply, and the lower part mostly by changes in sea level. This set of changes of different types affected various parts of the river which as a system had to adjust to all changes multiple times throughout the Quaternary. We review this hypothetical river in three stages: a full glacial, a full interglacial, and a transition stage between the two. Each stage may leave its impact on the channel, floodplain, and basin of the river.

7.2.3.1 The Glacial Stage

In a glacial stage, ice would accumulate in the mountains and advance downslope, especially along the headwater valleys as glaciers, or in places as part of an ice sheet. Extreme cold would prevent the ice from melting, thereby reducing the discharge in the river. Physical weathering and glacial advance would actively erode the surface material, and coarse sediment would cover the slopes and valley floors, chiefly as moraines. The reduction in discharge may lead to morphological transformation of the river, a narrower form eroded into the previous channel bed could be expected. Small terraces may evolve along the main river, and older tributary mouth fans would become inactive. The valley of the major river downstream is expected to adjust to both a changing climate and the altered nature of the tributaries joining it. The sea level was lowered about 125 m in the Last Glacial Maximum (LGM) 21 ka ago. The lower end of the river would adjust to this regression of the sea level which may have several effects.

The lowering of the base level may cause incision in the channel, usually for a limited distance upstream. For example, headward incision in the Mississippi has been recognised only for a few hundred kilometres (Blum 2007 and references therein). In the Amazon, which has a very low channel gradient, it has been estimated to be 1700 km (Mertes and Dunne 2007). As the sea level fell, the pre-existing old delta became inactive, and our hypothetical river would extend an incised channel across the coastal shelf, probably ending in a fan-shaped deposit. If several rivers used to converge into a shallow sea, the floor of which became exposed due to the fall in sea level, the extended channels may come together to form a bigger drainage network. Drowned drainage networks have been traced on the floor of the South China Sea (Sunda Shelf) between (i) eastern Malay Peninsula, Java, and Borneo, (ii) between Java and Borneo, and (iii) on the sea floor of the Malacca Strait between Sumatra and the Malay Peninsula (Emmel and Curray 1982). Similar extension of rivers from land to exposed sea floor and building of a network has been recognised in the English Channel. The Rhine extended for about 800 km along the floor of the present English Channel into the Atlantic Ocean, collecting the Thames as a tributary (Blum and Törnqvist 2000). Evidence of catastrophic flooding also has been recognised on the floor of the English Channel (Gupta et al. 2007).

In brief, the headwaters in the mountains are likely to be altered with glaciation and valley aggradation; the tributary fans would become inactive; the main channel slightly incised in the valley alluvium while carrying a reduced volume of discharge and sediment; a quiet period in the old delta; and an extension of a slightly incised channel over the exposed coastal shelf that probably joined another channel or ended in a fan-shaped deposit.

7.2.3.2 The Transition

The ice has a history of rapid melting between the glacial and interglacial periods, producing huge floods of short duration which ultimately travelled down an existing river channel. Immense fluxes of meltwater were released from continental ice sheets at the end of each glacial period; the deglaciation from the last event (MIS 2) which happened about 20 000–11 000 years ago is probably the best known. The meltwater volume was so huge that it tends to be estimated in units of 1 million $m^3\ s^{-1}$ (sverdrups, units used for measuring ocean currents).

Large lakes were formed near the boundary of continental ice sheets, the water blocked between the ice and local high relief. A well-known example was Lake Missoula which existed south of the Cordilleran Ice Sheet, in the northwestern United States. The significance of the floods from Missoula was first recognised by Bretz (1923). Baker (1981, 2007) determined the maximum extent of the lake to be 7500 km^2 with a volume of about 2500 km^3. Large-scale failure of its ice dam led to a cataclysmic volume of water escaping west, creating a devastated landscape aptly described as the scablands in the state of Washington, USA, and ending as an immense discharge into the Columbia River system. The biggest flood was of about 20 million $m^3 s^{-1}$. There were also other floods, of smaller but still of huge magnitude, from Lake Missoula. Spilling of Lake Bonneville in the late Pleistocene, ultimately forming a canyon along the Snake River, is another example from the region. These floodwaters, on reaching the Pacific Ocean, submerged as hyperpycnal flows and deposited an estimated 5000 km^3 of sediment on the sea floor (Baker 2007 and references therein). Retreat of the Laurentide Ice Sheet in central and eastern Canada gave rise to huge meltwater lakes to its south and west which spilled as similar megafloods, eroding a number of large channels that drained into the Mississippi, St Lawrence and Mackenzie rivers (Keshew and Teller 1994; Teller et al. 2002).

In northern Eurasia, similar huge proglacial lakes and spillways have been reported (Figure 7.4), diverting the drainage of the time. Cataclysmic floods impacted the south-flowing rivers (Dneiper and Volga) while the course of the north-flowing ones (Irtysh, Ob, and Yenisei) was impounded by ice sheets. Huge lakes spilled south to the basin of inland seas such as the Aral or the Caspian. Landscapes delineated by cataclysmic floods can also be identified in the Central Asian Mountains such as the Altai, from modified river channels. Evidence of Late Quaternary palaeofloods was seen

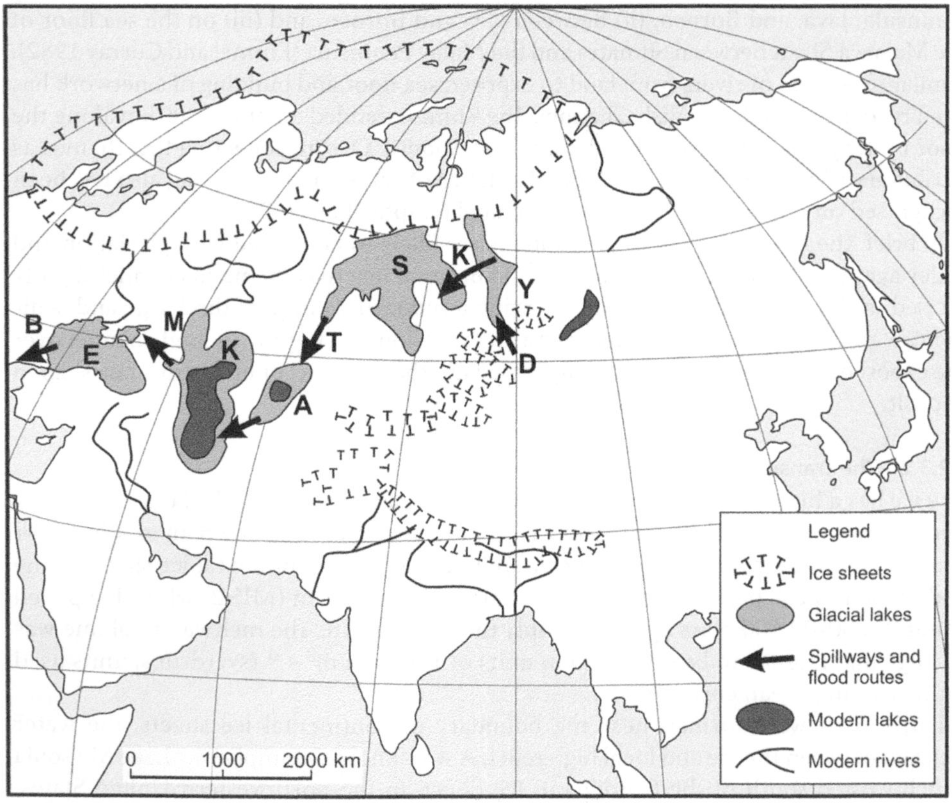

Figure 7.4 Sketch map of the late Pleistocene glacially diverted drainage system of Central Asia. Lakes (expanded and ice-dammed): the Yenisei Lake (Y), Lake Mansi (S), the Aral Sea (A), the Kvalynian Lake (K), and the new Euxine Lake (E). These were connected by a series of spillways: Kas-Ket (K), Turgai (T), Manych (M), Bosporous (B), and Dardanelles (D). Source: Baker 2007.

also in north-central Mongolia and Kirgizstan. Amongst other rivers, the upper Yenisei was modified by such palaeofloods. Many rivers of the middle and higher latitudes in the temperate zone were impacted (Baker 2007). In the tropics, large floods modified river channels in the mountains but as the world became warmer, their middle and lower courses were probably more affected by a changing climate than by meltwater.

7.2.3.3 The Interglacial Stage

After the ice-melt, a full interglacial stage carried rivers more comparable with present rivers. The effect of mountain glaciers became limited and the sea level rose higher, transgressing the coastal plains and disintegrating the lower part of existing river networks. This rise in sea level also provided accommodation space at the end of the river, and enhanced sedimentation leading to formation of deltas. Temperature, precipitation, and storminess may be expected to increase over the basin. Glaciers retreated up the headwaters and the meltwater transferred channel and valley sediment downstream. In certain rivers, the channel morphology possibly changed from the braided condition of an earlier colder time to a meandering phase. Unless the climate was going through a brief very warm (hypothermal) period, sedimentation on tributary-mouth fans and deltas would have been slow. In contrast, hypothermal periods were associated with rapid meltwater production and increased storminess, leading to very high water and sediment discharges down the river which would erode and transfer the valley fill, display flood morphology in the channel, widely and rapidly erode tributary-mouth fans and alluvial plains, build deltas speedily, and accumulate sediment on the coastal shelf. The exact morphology and behaviour of the river would depend on the degree of warmth and wetness of the climate. The nature of the river would thus depend on the pattern of warming and planetary location of the river basin.

7.3 Changes During the Holocene

Over the next 11 500 years after the Pleistocene, the rivers of the world adjusted to a series of different types of natural and anthropogenic changes. Besides the general warming at the end of the cold MIS 2, the regional climate of different parts of the world went through various smaller changes, mainly with variations of wet and dry times. Differential changes in climate may have happened across the huge basins of large rivers. Enormous amounts of glacial meltwater were discharged to the oceans as in previous interglacial times which modified the pattern of ocean currents and ultimately coastal climates in the early Holocene. The sea level rose, until after some fluctuation it reached its present position and the present deltas were built at river mouths over several thousand years. Changes in land use and growth of settlements started to modify hydrology and sediment yield of smaller tributary drainage basins. Over the last three hundred years, as will be described in Chapter 8, engineering work such as construction of levees, dredging of channels, and building of impoundments with reservoirs modified larger rivers. Currently, anthropogenic climate change is also impacting rivers (Chapter 12), although our understanding of it is incomplete.

The early Holocene was often wetter than the present climate in places. This has been demonstrated in a number of rivers by their alluvial stratigraphy and channel morphology. The Amazon, Nile, Narmada, and Murray-Darling all demonstrate this pattern of climate change. Goodbred and Kuehl (2000a) estimated from deltaic cores that the sediment released by the Ganga-Brahmaputra system in the 4000 years between 11 000 and 7000 years BP was double that of the current enormous amount. The monsoon system was stronger than the present, causing accelerated erosion. Climate changes in northwest India during the Holocene have been demonstrated by Enzel et al. (1999) from the analysis of sediment from the dry lake at Lunkaranasar, Rajasthan. This shallow lake reached its highest level around 6300 radiocarbon years ago but dried out completely around 4800 years BP and subsequently a series of dunes were formed. Wasson (1995) has concluded that the available moisture peaked in many regions approximately between 8000 years and 6000 years ago. His list of such areas includes monsoon Australia, India, Arabia, China, Japan, Taiwan, and Thailand but not southern India and eastern China. Morphology and behaviour of rivers had to adjust to the shifting of their hydrology and sediment production more than once during the Holocene. The climatic system shows many changes and clustering of floods over a short period. Kale (1998) has referred to several palaeoflood periods for the Narmada River of central India. For example, two sequences of frequent but moderate floods, 0–400 and 1000–1400 CE, were separated by a period (400–1000 CE) of fewer but larger floods.

The present deltas were formed in the early Holocene, generally about 6000–7000 years ago, as detailed in Chapter 6. Three factors were involved: a rising sea level; the availability of accommodation space for storing riverine and coastal sediment; and considerable sediment production by rivers. The geomorphic history of rivers of the world, even of the major rivers, did not end there. Their flows, sediment loads, channel and valley floor morphologies, water quality, etc. were modified by anthropogenic practices especially over the last hundred years (Chapter 8). The on-going anthropogenic global warming of current times is expected to raise the sea level and change the climate, producing a different pattern of precipitation and snow melt. All these will impact on rivers (Chapter 12).

This historical summary of global rivers implies the dynamic nature of their adjustment to a series of changing environments. A generalised world-wide account is difficult to produce because of the variable nature of river channels and basins. Instead, we summarise the geological history of three major rivers, the Mississippi, Ganga-Brahmaputra, and Amazon, in order to illustrate the range over which the major rivers have evolved. The Mississippi demonstrates the effects of three historical factors: the control of its location by a structural embayment; the events in the Pleistocene; and the modification of its form and function by extensive engineering over the last two hundred years. A changing hydrology has considerably modified the Ganga-Brahmaputra Basin during the Pleistocene and early Holocene. Unlike the Mississippi this drainage basin has been extensively cultivated for a long time. People have settled in the valley for more than several thousand years, although the demand for water and transport led to construction of engineering works much later. The Amazon has stayed more of a natural river, displaying the effect of underlying geological structure and Quaternary changes in climate and sea-level alterations without much anthropogenic modification.

7.4 Evolution and Development of the Mississippi River

The 6000 km long Mississippi drains a huge basin of 3.22×10^6 km^2 (Figure 7.5). The Mississippi River has been active since at least the late Jurassic. Riverine sediment of

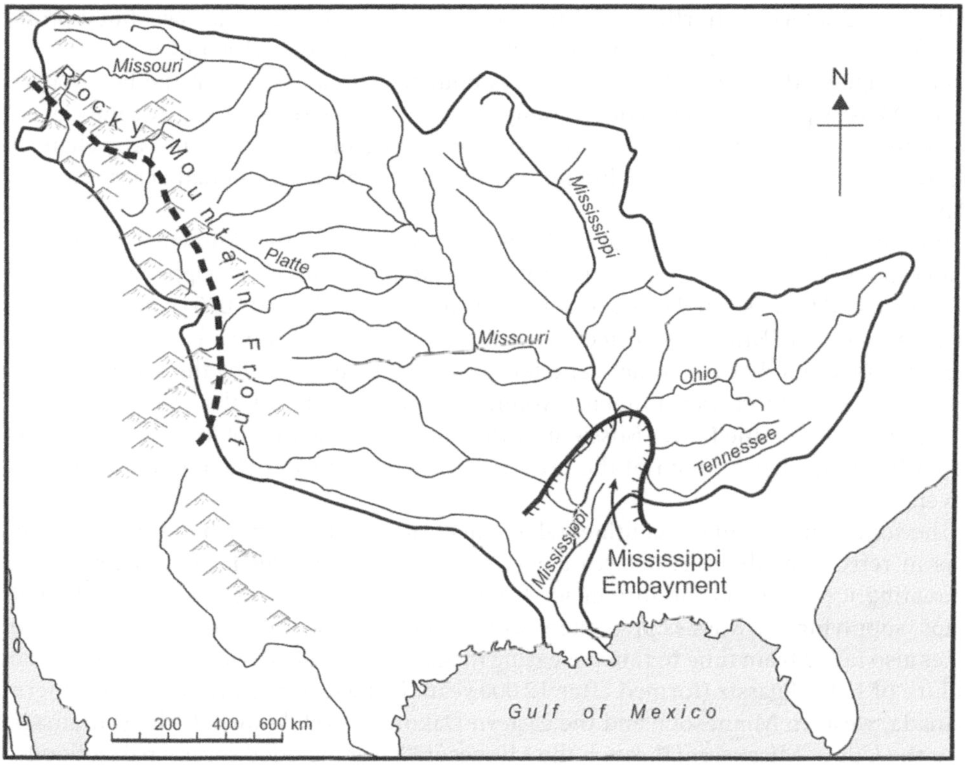

Figure 7.5 The Mississippi Basin.

Cretaceous age has been found in the Gulf of Mexico (Coleman 1988). The river has been strikingly modified during the Pleistocene. The modern surface configuration of the basin north of its two major tributaries, the Ohio from the east and Missouri from the west, is due to multiple Pleistocene glaciations (Knox 2007 and references therein) whereas its lower course has been significantly impacted by the changing sea levels of the Pleistocene. Blum (2007) has summarised this large river as a complex polyzonal system that discharges to a passive continental margin.

South of the confluence with the Ohio, the Mississippi flows over a broad crustal downwarp of late Jurassic age. This embayment is considered to originate from the late Precambrian (Ervin and McGinnis 1975). However, many incised bedrock tributary valleys in the basin were formed probably after the late Pliocene (Knox 2007). The basin has been uplifted over the last few million years, but variably, the maximum uplift happening near the Rocky Mountains on the western edge of the basin (Knox 2007). The resulting pattern of sedimentation from the west towards the central valley of the Mississippi indicates that the present configuration of the Missouri was established in the Pleistocene during MIS 6–8. A deeply entrenched large river of the past, the Teays-Mahomet proglacial valley system, used to flow towards the former Mississippi from the east. These two lines of drainage joined the ancestral Mississippi near the head of the Mississippi Embayment, indicating that the middle and lower parts of the main Mississippi used to be approximately in the same location as now.

The pre-Quaternary northern boundary of the basin is not clearly known, but it probably followed the existing east–west relief heights such as the Niagara Cuesta and the Sioux Ridge to the Rocky Mountains. This boundary area along with the headwaters of the Mississippi have been considerably modified by continental Pleistocene glaciations, in places creating anomalous relations between geological structures and channel locations of the upper river, including incised narrow gorges extended across cuestas (Anderson 1988; Knox 2007).

The history of the Quaternary Mississippi is much better known. Continental glaciation in the Pleistocene impacted the channels and basins of the Upper Mississippi, Missouri, and Ohio several times, of which events over the last 300 000 years are known in greater detail (Knox 2007). Ice sheets blocked rivers, channels became diverted, meltwaters drained new channels, and terraces were formed. During the last glaciation, rivers of the basin carried immense volumes of meltwater and glacial outwash with coarse sediment to the Lower Mississippi and finally into the Gulf of Mexico which then had a lower sea level. Much of the pre-MIS 4 glaciated basin is hidden under a loess cover.

The southern edge of the continental ice sheet (MIS 2) to the north of this land mass was in retreat by about 17 000 years ago. Huge proglacial lakes formed between the retreating ice edge and the former ice front positions. These lakes acted as sediment traps, supplying the Mississippi system with water of low sediment concentration. The lakes also failed from time to time, releasing megafloods to the Mississippi network. The failure of Lake Agassiz (formed after 12 000 years BP) which used to occupy southern Canada, western Minnesota, and the eastern Dakotas is well known. Its huge drainage into the Upper Mississippi through the Minnesota River included several palaeofloods, the biggest of which has been estimated to carry a discharge between 50 000 $m^3 s^{-1}$ and 1 000 000 $m^3 s^{-1}$. A number of palaeoflods of comparable volume drained several other palaeolakes located in the north (Knox 2007 and references therein). Both episodic megafloods and sustained high meltwater discharge, low in sediment, eroded the river channels.

The lowland of the Lower Mississippi Valley through which the main river flowed, extends from the Mississippi–Ohio confluence in the north to the Gulf of Mexico in the south, a distance of nearly 1000 km. During the last glaciation (MIS 2), the Lower Mississippi was a braided river along the western side of the lowland, was fed by extensive proglacial outwash, and transported sand and gravel to the Gulf. An increase in meltwater discharge due to the collapse of upstream glacial lakes apparently resulted in avulsions that incised through low ridges to shift the river from the west to the eastern side of the lowland and, along with a rise in sea level, transformed the river to the present meandering system. The channels were incised between 65 and 25 ka ago while the river was in the western lowland, south of the embayment, but the river abandoned the western side of the valley for the eastern side at about 25–20 ka ago; shifting completely to the east by about 11.5 ka ago (Blum 2007 and references therein).

The effect of the sea-level changes at the end of Pleistocene was limited to the last few hundred kilometres; Blum and Törnqvist (2000) estimated a length of 300–400 km. The gradient of the river was also reduced after the last glacial time. The shift of the lower river from the braided pattern to a meandering one probably started about 14 ka ago near the river mouth and reached near the junction with the Ohio River in another

2000 years (Knox 2007 and references herein). Subsequently, the lower river developed a series of meander belts in the Holocene. Large volumes of coarse sand and gravel were eroded and transferred to the Gulf of Mexico, being replaced over time by clay, silt, and fine sand forming floodplains, levees, inset point bars, abandoned channels and backswamps in the river valley. The Mississippi is an excellent example of a large river in the middle latitudes, significantly shaped by changes during the Quaternary.

The Mississippi River and its sediments have been strongly influenced by adjustments to Pleistocene activities and the subsequent climate and vegetation changes in the Holocene. Knox (2007) suggested that alluvial deposits have been redistributed along the Upper Mississippi and the Ohio during the Holocene. The vegetation that the early settlers saw in the Basin probably arrived in those locations about 10 ka ago or later, moving north from the warmer south and replacing the colder environment. A forest cover grew up over the eastern and southern basin and over parts of Wisconsin and Minnesota in the north. The other parts of the north and the entire western basin became grassland. The Holocene climate went through periods of dry and wet spells which at times, depending on the prevalent climate, accelerated sediment production and large floods in the Mississippi. Rates of erosion and sedimentation went through significant changes in the grasslands, leading to remobilisation of sediment from small valleys, accumulation of sediment in the main valleys, and development of alluvial fans. The Mississippi adjusted by moving laterally opposite tributary-mouth fans. Channels ranging from small valleys to the Upper Mississippi were impacted by Holocene floods of varying magnitude. Knox has commented that Upper Mississippi floods of the Holocene suggest a very non-stationary behaviour for the means and variances in the flood series (Knox 2007). The larger floods deposited considerable sand over the floodplain but the others were less efficient.

The Lower Mississippi River, downstream of the embayment, receives drainage and sediment from more than 75% of the upper basin. In the late nineteenth century, before construction of dams and reservoirs, about 84% of the sediment used to be delivered by only the Missouri from its semiarid basin to the west. This resulted in the Holocene meander belts of the Lower Mississippi and five delta lobes. According to Coleman (1988), each lobe on average covers 3000 km^2 with a thickness of about 35 m, switching to a new one approximately every 1500 years. Both the Upper and Lower Mississippi have been divided into multiple reaches with variations in their long profile and morphology following adjustments to local discharge and sediment characteristics (Knox 2007).

Over the last two hundred years or so, the river has been remodelled for better navigation and flood control. This required snag removal, channelization, meander cutoff, levee building, etc. A number of dams and reservoirs have been built in many of its tributary basins, the Missouri and Tennessee being striking examples. The Upper Mississippi was modified early, mainly for flood control and improved navigation for loaded river barges. The US Corps of Engineers was instructed by the passing of the 1866 Rivers and Harbors Act of the US Congress to inspect part of the upper river. In 1881, a comprehensive plan was approved by the Congress to continuously improve the middle part of the Mississippi River by reducing its width to about 760 m and keeping the channel deep. This was the beginning of significant modification of the upper river by sand dredging, construction of wing dams (rock and brushwood dams) along channel margins, and prevention of water escaping into side channels. Finally, a system of 29 locks and dams was constructed between St Paul, Minnesota and St Louis, Missouri to maintain a targeted depth of 2.75 m in the channel (Knox 2007).

Large-scale modification of the Lower Mississippi started in the early eighteenth century with construction of levees at New Orleans for protection from floods. Starting from 1 m in height, these levees were increased to more than 10 m for flood protection and maintained a self-scoured navigation channel (Winkley 1994). The repeated flooding on the Lower Mississippi prompted the US Congress to create the Mississippi River Commission in 1879, which resulted in the building of 2400 km of levees along the river. This separated the channel from its floodplain, raised flood stages, and led to disasters when the levee broke as in the Great Mississippi Flood of 1927. Dams and reservoirs on the tributaries and spillway structures for flood diversion and storage were subsequently added. Kessel (2003) has pointed out the limited nature of sediment supply from the adjacent floodplain because of artificial revetment structures that prevented bank caving and also accelerated storage of sediment on channel bars. The Corps of Engineers attempted to protect river banks to prevent meander cutoffs, dredged crossings between bends, and in general shortened channel reaches for improved navigation. This led to instability in certain reaches of the river. In spite of enormous investments many reaches remained unstable, and the Mississippi did not transfer its full sediment load down the channel (Winkley 1994).

The Mississippi has also been considerably modified by the construction of numerous dams and reservoirs on its tributaries. Such engineering projects are perhaps best known for the set of dams on the Missouri and Tennessee rivers, both of which dramatically reduced the sediment load of the Mississippi. Earlier, erosion and transfer of a large amount of sediment was common along the western tributaries of the Mississippi River such as the Missouri and its tributary the Platte. Milliman and Farnsworth (2011) estimated the natural sediment yield from these tributary basins to be about 200 $t km^{-2} yr^{-1}$. Episodic storm-derived peaks were also common. The Mississippi River, about three hundred years ago, when the early European settlers arrived, was still adjusting to such sediment-laden runoff events which started to arrive after the last glaciation (Milliman and Farnsworth 2011). Construction of numerous dams, totalling tens of thousands and including some very large ones, blocked sediment and regulated the water and sediment discharges of the western tributaries (Meade and Moody 2010). From the 1940s, accelerated soil conservation measures also controlled supply of sediment to the river. The river therefore had to adjust to a new pattern of regulated discharge, much lower sediment volume, and a levee-guided channel – all due to anthropogenic alterations. The effect of these dams extended also to the channel and floodplain of the lower river; even the Mississippi wetlands became vulnerable to water erosion. Disasters still happen when levees break as in the Category 5 Hurricane Katrina which destroyed New Orleans and surrounding area in 2005.

The long history of the present Mississippi therefore records a series of controls. Its location was broadly determined by the underlying geological structure; the river and its basin went through large and repetitive changes during the Pleistocene, at times resulting in diversions and incisions across uplands of bedrock roughening the profile of the channel; it needed to cope with changes in the climate and vegetation patterns in the Holocene; and the current anthropogenic modifications considerably modified the river, its tributaries, and most of its basin. Not all large rivers demonstrate such changes so clearly, but they all have a history which explains their present configuration and behaviour.

7.5 The Ganga-Brahmaputra System

The Ganga and Brahmaputra rise on the opposite sides of the Himalaya Mountains, traverse a couple of 1000 km before joining, and then meet with the third large river, the Meghna, to flow into the Bay of Bengal (Figures 4.6 and 4.7). Their combined drainage is over 1 million km^2, the joint discharge being the fourth biggest in the world, and the total average annual sediment load the second highest. One of the main headwaters of the Ganga, the Bhagirathi, begins at 3800 m altitude from the Gangotri Glacier in the Himalaya. The river is called the Ganga after joining with the other main headwater, the Alaknanda, at Devprayag. Descending to 290 m in 300 km, the Ganga leaves the mountains at Haridwar and enters the plains. Its headwaters descend through tectonically active mountains but after emerging from the Himalaya, the Ganga and its tributaries flow through an extensive low gradient alluvial plain, a subsiding continental interior foreland to the south of the Himalaya.

All along its length, as the Ganga flows southeast and then east, it is joined by large Himalayan tributaries from the north bringing in huge volumes of water and sediment. All these rivers, including the Ganga, have built large megafans at the highland–lowland contact. Although its main Himalayan tributary, the Yamuna, joins the Ganga on its southern bank, the other major tributaries that come in from the south drain the old cratonic rocks of the northern wedge of the Indian Peninsula. At the eastern margin of its alluvial plain, the river passes through a gap in the basaltic low hills of Rajmahal, enters its delta, and divides into two major distributaries. One of them, the Hugli (part of which is also known as the Bhagirathi), turns south to flow into the Bay of Bengal though India, collecting further drainage and sediment from the northeastern corner of the Indian Peninsula. The Padma, the main channel, carries most of the river eastwards into Bangladesh where it meets with the Brahmaputra, and then the Meghna. The three major rivers have built the Ganga-Brahmaputra Delta.

The Brahmaputra, rising on the southern slopes of the Kailash Mountains, north of the Main Himalayan Range, flows eastward along the Indus-Tsangpo suture through the Tibetan Plateau on a relatively gentle gradient, about 0.001. After about 1200 km on the Tibetan Plateau, the Tsangpo, as it is regionally known, takes a large U-turn in a 5075 m deep gorge around the Namche Barwa Peak in the eastern Himalaya, a part of the mountain known as the Eastern Syntaxis. The sudden bend and the deep gorge with a gradient of 0.03 are yet unexplained. It has been suggested that the Himalayan orogeny and successive river captures modified a former Yarlung Tsangpo-Irrawaddy-Chindwin system to the present separate river systems of the Yarlung Tsangpo-Brahmaputra and Irrawaddy-Chindwin (Robinson et al. 2014).

At present, the Tsangpo turns south to flow into India, receives a set of tributaries, and enters the Assam Plains as the Brahmaputra. It has now become a wide, deep, braided river carrying enormous amount of sediment, flowing WWS in the reverse direction of the Tsangpo. It then turns south to enter Bangladesh as the Jamuna, and joins the Ganga at Aricha. From the source to this confluence its length is 2900 km. An old channel of the Brahmaputra exists to the east in Bangladesh, each of the two channels periodically been occupied by the main river (Pickering et al. 2013). The huge combined delta is the final product of the sediment eroded and transported mainly from the Himalaya.

The Ganga River system is complex. Its erosional headwaters rise in the actively folded Himalaya Mountains while its aggradational valley runs along a structural low parallel to the ranges. The foreland basin was properly established in the Middle Miocene following lithospheric flexures and basin subsidence. An efficient transport system developed which transferred most of the sediment out to the delta and beyond (Tandon and Sinha 2007 and references therein). Fluvial deposits of about 30 million years ago occurred in the Himalayan foredeep, and a major river system could have flowed to the southeast in the early and middle Miocene (Tandon and Sinha 2007). A considerable amount of sediment derived from the Himalaya by transverse tributary river systems, asymmetric subsidence, and uplift of parts of the foreland displaced the Ganga southwards to the edge of the peninsular craton in the Plio-Pleistocene. The drainage system was rearranged. About 5 million years ago, major tributaries of the northwest (often collectively referred to as the Punjab rivers) changed their courses to leave the Ganga and join the Indus either by river capture or tectonic events (Clift and Blusztajn 2005). The present river system developed south of the Himalaya by the end of the Neogene.

The river and its basin underwent a series of well-studied changes during the Pleistocene which are better known (Goodbred 2003). In association with the changes in the 23 ka precession cycle as Milankovitch indicated, glaciations in the Himalaya varied in extension and intensity which impacted periodically on the river. The river was also significantly affected by the changes in the intensity of the annual monsoon system (Goodbred 2003). Goodbred produced an integrated account of the Ganga for the Late Pleistocene from the glaciated headwaters to the delta (Table 7.1). The length of the river varied with the changing sea level, and the lengthened river of the glacial times became associated with a submarine canyon, known as the Swatch of No Ground, and a large submarine fan extending into the Bay of Bengal beyond about 10° of latitude (Figure 7.6). The climatic indices of temperature and precipitation together affected the intensity of glaciation in the mountains and aggradation and degradation in the river valleys, and the centre of deposition periodically shifted between deltas and deep-sea fans. The configuration of the river and its basin varied in different parts of the late Pleistocene. Presumably such modifications also happened in the earlier parts of the Pleistocene.

As Table 7.1 indicates, the configuration and behaviour of the river system were different throughout the late Pleistocene among the glacial, transition, and interglacial times. Unlike the direct glacial activity in the Mississippi Basin, it was the change in the strength of the monsoon that left its mark on the stratigraphy of the river. Peak glaciation in the southern slopes of the Himalaya did not happen during MIS 2, which is globally considered to be the LGM. Instead, peak glaciation and associated conditions happened during MIS 3 and 4 in the southern slopes of the Himalaya when the monsoon was more intense, and precipitation was higher. It is likely that the Ganga had to adjust to several large floods from meltwater, lake collapse, and storms, some of these probably occurred during a hypothermal period in the early Holocene. It was during the same period that the river carried a high discharge and huge amount of sediment to build the present delta (Goodbred and Kuehl 2000a). Climate signals were not only significant for the river, but the response was also rapid throughout its length (Goodbred 2003).

To recall, the Ganga and its tributaries flow in gorge-like valleys flanked by small discontinuous floodplains and terraces in the Himalaya. Then the river and its major tributaries flow through the Himalayan foreland, a series of consistently low-gradient alluvial plains filling up a subsiding structural low. The northern part of the alluvial plain is a series of megafans built by the major tributaries from the Himalaya. Such fans constitute a large part of the Gangetic alluvial plain. It has been suggested that a large amount of sediment accumulated in the valley during MIS 3 which was cold and

Table 7.1 Source to sink response of the Ganga River since MIS 3.

Time (ka)	Climate	Water discharge	Sediment discharge	Himalayan valleys	Megafans and alluvial plains	Delta and coastal shelf	Deep sea fan
Mid-Holocene (7 to now)	A quieter monsoon than before, less rain	High	High	Retreating glaciers, both sediment removal and aggradation	Low aggradation	Aggradation but slower than before	Less aggradation
Hypothermal (18–7)	Warm and very wet, stronger monsoon	Very high	Very high	Brief glacial advance, erosion of valley fill	Extensive rapid erosion, nearly exposing bedrocks	Very rapid aggradation	Aggradation shifting to delta
MIS 2 LGM (24–18)	Cold and dry	Low	Low	Limited glaciation and valley fill	Largely inactive fans, minor incision, alluvium in southern tributaries	Inactive, possible incision	Probably inactive; river linked with the Bengal Fan via canyon
MIS 3 (58–24)	Slightly colder and drier than present	Moderate	High sediment transfer	Marked glacial advance and valley aggradation	Aggrading megafans, upper Ganga plains, and southern tributaries	Incision and possible local aggradation	Possible aggradation

Source: Based on Goodbred 2003.

wet, sedimentation was inactive during the dry MIS 2 (LGM), but the fans were deeply incised during the enhanced wet monsoon of the late Pleistocene to early Holocene.

The river channels in the wide alluvial plain between the Himalaya and the peninsular craton are entrenched below the surface of the plain. The channel of the Ganga runs within a lowland belt, 10–25 km wide and bounded by alluvial cliffs, several metres high. The cliffs enclose the channel with braid bars, point bars and linear bars, floodplain, terrace-like features, and wetlands. Anecdotally, the highest surface, lying above the alluvial cliffs, is never flooded but high flows modify the rest of the valley features. Gullies and small ravines are common, and abandoned channels and wetlands appear on top of the cliffs on the alluvial plains. The type of bars in the channel depends on the physical configuration of the valley and the size of the seasonal and storm-driven flow. Both the Ganga and Yamuna were mobile in the past as indicated by their stratigraphic records. Gibling et al. (2005) give a slightly different but similar account of events. They suggested that earlier MIS 5 was a period of widespread aggradation on alluvial plains followed by erosion of incised valleys. This was followed by a decrease in flood sizes between ca. 27 ka and 15 ka ago resulting in valleys that were underfit, which were

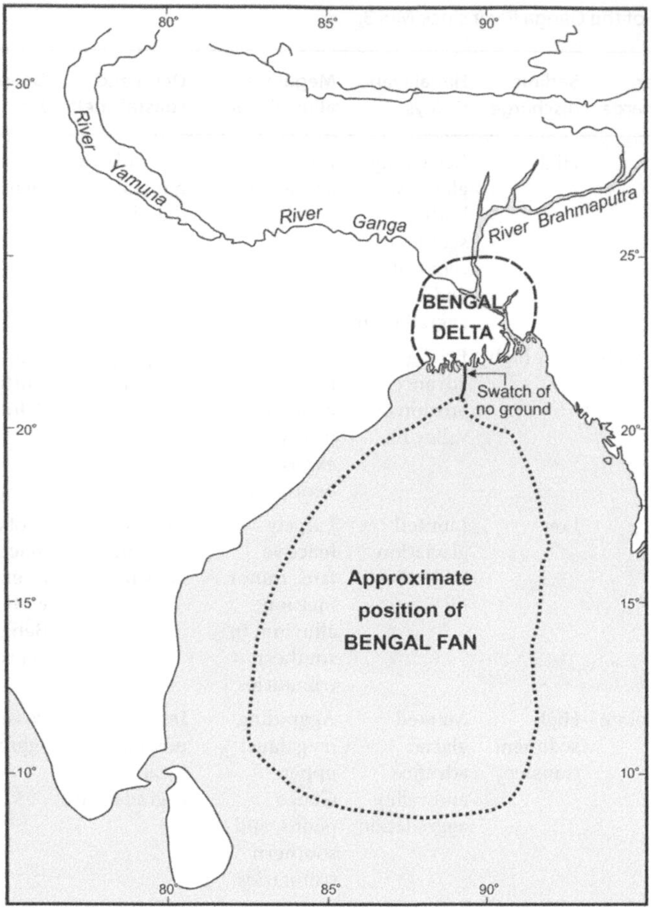

Figure 7.6 Diagrammatic sketch showing the relative position of the major rivers on land, the Bengal Delta, and sediment down the Swatch of No Ground to build a large submerged Bengal Fan. Source: Generalised and adapted from Goodbred 2003.

followed by incision and lateral migration of channels and reworking of the floodplain about 13–9 ka ago, during a hot wet monsoon.

It is likely that there have been older deltas, synchronising with different sea levels, but the basal age of the present one has been dated to 10 000–11 000 years BP (Goodbred and Kuehl 2000a). The Ganga-Brahmaputra-Meghna Delta has been described in Chapter 6 in detail. Briefly, a low stand of the sea occurred before 18 000 years BP, when both the Ganga and Brahmaputra flowed through incised valleys and the rest of the land surface of the Bengal Basin consisted of broad laterite uplands, about 45–55 m below the present sea level (Goodbred and Kuehl 2000b). About 15 ka ago, the summer monsoon started to strengthen and a vastly enhanced precipitation resulted in the transfer of an enormous amount of sediment to the delta. The first batch of sediment reached the Bengal Fan, but subsequently as the sea level rose and transgressed the low areas, the sediment was trapped on top of the delta which between 11 000 years BP and 7000 years BP, aggraded at an enormous rate. Afterwards deltaic deposition became progradational extending seawards. The surface of the delta was selectively filled by avulsing large rivers, principally the Brahmaputra. The main course of the Ganga migrated or avulsed eastwards about 5 ka ago. The Bengal delta therefore consists of wide corridors of low ground used by major channels and separated from each other by blocks of higher ground. A set of peninsulas separated by estuaries and ending in scattered islands extend the delta southwards to the sea.

People have lived in the basin of the Ganga for millennia. Parts of the plain have been cleared for cultivation and settlements but different types of forest still cover the Himalayan foothills, parts of the megafans, and the edge of the peninsula to the south. Considerable erosion has taken place in parts of the cleared land as in the southwest basin, resulting in large ravines. Forests have survived, often close to the river, across the hilly areas to the north, and also over the southern divide of the basin, as described in the accounts of big game hunting by the Mughal Emperors in the sixteenth and seventeenth centuries and the British residents thereafter. We know very little about the effect of land clearing on the Ganga but the condition of the large river deteriorated over time with industrial and residential pollution from numerous urban centres that grew up on its banks. The effect of pollution and sedimentation was heightened by several irrigation canals that were constructed to transfer the water of the river. The effect was heightened specially in the dry season. None of the large rivers yet have dams constructed across them, except the Farakka Barrage on the lower Ganga and the Kosi Barrage below the hills, but numerous dams have been built on the smaller tributaries, some for irrigation, some, especially in the Himalaya, for generation of hydroelectricity. The Farakka Barrage was constructed to ensure entry of the required amount of water for navigation in the Lower Hugli. Pollution is very much a problem for most of the river, especially near the towns and cities of the Gangetic plain, and during low flow.

7.6 Evolution of the Current Amazon

The largest river with the huge drainage basin of about 7 million km^2 does not display marked anthropogenic changes in its configuration or behaviour unlike the Ganga or Mississippi. A few of its tributaries have been dammed and the earlier forest cover has been replaced over a large section of the southern and eastern parts of the basin, but it is primarily a natural system. The river is more a product of its geological history, Quaternary climatic changes, and adjustments to post-MIS 2 changes in sea level (Mertes and Dunne 2007).

The pre-Quaternary stratigraphy of the Amazon and its underlying geological structures are well known from oil explorations. The present configuration of the river came into being after the uplift of the Andes in the Miocene. The river rises in the high fold mountains of the Andes in the west; flows eastward across a vast lowland overlying a foreland basin; passes further to the east along a trough between two low cratonic uplands, the Guyana and Brazil shields; and discharges into the South Atlantic (Figure 4.4). The headwaters of the river in the Andes Mountains flow in steep bedrock and gravel-bedded channels. The river meanders eastwards in the foreland basin and the central trough. At the surface, the meandering Holocene Amazon is a complex pattern of channels of varying dimensions. Scroll bars and levees are common. Lakes occur in various places: tributary mouths, channel cutoffs, and backswamps.

Discontinuous terraces lie above the Holocene floodplain of the current mainstream Amazon. The mouth of the Amazon is an estuary that flows through a delta plain marked by several large islands. Below the Andean headwaters, the very low gradient of this large river is measured in centimetres per kilometre. This configuration of the Amazon is created by its geological history.

The present Amazon Basin lies between the folded Andean arc at the leading edge of the South American Plate and a graben at the trailing edge of the continent that localises the river mouth (Potter 1978). The base of the valley is a 6000 m rift of probable Palaeozoic age, which is likely to have been reactivated in the Triassic when South America drifted apart from the African Plate. However, a significant eastward surface flow started later in the Miocene when the Andes were uplifted. This was a change from westward sedimentation in an earlier marine embayment which was connected to the ocean via the Orinoco foreland and later through the Lake Maracaibo lowland. Marine sediments were subsequently overlain by swamp, lake and fluvial sediments dominated by crevasse splays, and led to a directional change in eastward transfer of water and sediment.

Although the Andes and the adjacent mountains occupy only a part of the basin (Figure 4.4), it is the main source of sediment for the river, providing 500–600 million tonnes of sediment annually throughout the Late Cenozoic. The low slopes of the foreland hinder transport of material coarser than sand, and the central part of the trough is occupied by Neogene and Quaternary lacustrine and fluvial sand and silt, often weathered to clay in a landscape of short hillslopes under thick forest.

Transverse subsurface structural arches in bedrock affect the configuration of the river as it flows eastward along and above the structural trough mentioned earlier. The eastward-flowing Amazon crosses four major structural arches – Iquitos, Jutaí, Purús, and Gurapá - in sequence, and also the Monte Alegre Intrusion (Figure 2.2). These rising features affect the channel of the Amazon. While crossing them the valley becomes narrower and the river flows over a steeper gradient (Mertes and Dunne 2007 and references therein).

Large-scale fractures in rock also affect the drainage patterns of the Amazon tributaries. Deep-seated basement fracturing may be responsible for surface drainage patterns which are oriented northeast and northwest, and location of several lakes. The alignment of the lower Negro River, a major tributary of the Amazon, has been interpreted by Franzinelli and Igreja (2002) as related to a large NW-SE tectonic lineament. Changes in alignment of straight reaches of the middle river have been interpreted by Latrubesse and Franzinelli (2002) as related to activities on a set of fractures. Such mapping had been earlier carried out by Sternberg and Russell (Mertes and Dunne 2007). Latrubesse and Franzinelli also described the realignment of the channels at the confluence of the Purús River with the Solimóes-Amazon. They attributed the present form of the channels to an avulsion of the Amazon main channel associated with a sunken tectonic block, about 1000 years ago. The underlying structure influences the channel and the modern floodplain of even a river of the dimensions of the Amazon.

The Andes Mountains source more than 90% of the sediment load of the Amazon River (Meade 2007 and the references therein). In contrast, a large volume of water but very little sediment is derived from the cratonic old stable rocks of the Guyana and Brazil shields (Figure 4.4). This sediment load of the Amazon annually averages above 1200 million tonnes at Óbidos (Mertes and Dunne 2007). The sediment is transferred from the Andes chiefly by the Amazon mainstream and one of its principal tributaries, the Madeira. A very large part of the sediment is deposited in the Andean foreland, the remainder flowing east to the Atlantic. The eastward transfer, however, often happens in stages, the sediment could remain stored for a period in the large floodplain. This storage and transfer may happen in several stages; the residence time of a sediment grain in the Amazon floodplain is estimated in thousands of years. This sediment either forms features in the channel or is stored in the Holocene floodplain of 100 000 km^2, incised more than 10 m below the regional landscape of Neogene and Quaternary sediment. The floodplain is interlaced by a network of channels with a wide range of size. The lateral movement of sediment, in and out of the floodplain, is at least comparable with the amount that is transported downstream (Dunne et al. 1998). The surface of the floodplain displays a network of channels, lakes, and scroll bars. The floodplain narrows and the river gradient steepens where the Amazon crosses a structural high, but elsewhere displays of anabranches, distributary channels with levees, deltas in floodplain lakes, and a range of other complex features mark a wide floodplain with low gradient. The configuration of the Amazon is a result of tectonic impacts at various scales from continental to local; movement of channels eroding the base of terraces; and transfer and storage of a vast amount of fine sediment by a large discharge.

As glaciers expanded during the colder times in the Pleistocene, the snowline shifted downwards, colder climate operated over the basin, and the sea level dropped. Probably the changes associated with the LGM are best known but different dates have been ascribed to the LGM. The dates primarily come from the Andes and range between 34 000 and 21 000 years BP (Mertes and Dunne 2007 and references therein). Various estimates have been made about a lowered snowline, perhaps 1000 m being the maximum, and the temperature over the basin fell by several degrees. A complex history of changes in precipitation have been suggested including various wet and dry periods. Forest cover probably survived in the Andean foothills and the western lowlands, although the nature of the forest, in the driest period, was probably more open in terms of structure and composition. Uncertainties prevail regarding the precipitation and vegetation in the northern, eastern, and southern parts of the Amazon Basin in the colder times. Savanna conditions have been identified from pollen studies in the eastern and northeastern basin margins, and its subsequent replacement by forest has been suggested with the coming of a wetter and warmer period at the beginning of the Holocene (Van der Hammen and Absy 1994). In general, the glacial times probably saw a reduced rainfall and a reduced rainforest area. From a study of oxygen isotopes in planktonic foraminifera in an offshore core, Maslin and Burns (2000) estimated that the Amazon discharge was about 80% of the present value at 14 000 BP, and at most 60% of the present value 12 ka ago. It increased steadily in the Holocene. Mertes and Dunne (2007) concluded that even if precipitation declined over the Amazon lowland, the discharge of the river probably was not much reduced. They also suggested that more runoff and sediment arrived from the Andes into the foreland basins during the cold period and thinning of vegetation allowed slightly higher erosion in the Amazon lowlands and the surrounding shields. The fall of the sea level, however, had a greater impact on the Lower Amazon.

It has been suggested (Mertes and Dunne 2007) that during the LGM, the sea level was about 120 m lower, and in response the Lower Amazon and regional tributaries incised their beds. These incised valleys were later filled with sediment when the sea level rose in the Holocene, but not entirely, resulting in the formation of river mouth lakes. The origin of these lakes is different from the upstream ones mentioned, which are structure-guided. The lowering of sea level led to a steepening of the river gradient, allowing both incision and passage of a larger sediment load. The incision impacted up to an upstream distance of 1700 km. It was 3000 years BP when the bed of the Amazon attained its present elevation.

The present climate of the Amazon basin is dominated by the shifting location of the Intertropical Convergence Zone (ITCZ) which determines the seasonality of rainfall between the northern and southern basins. The Andean region is also influenced by the South Atlantic Convergence Zone (SACZ) which also shifts position. Precipitation of 2000–2500 mm falls fairly uniformly across the basin, with higher values on the lower slopes of the Andes, where it rises to a maximum value of around 7000–8000 mm. The resulting hydrograph of the Amazon has been summarised as unimodal, damped, and controlled by the size and length of the basin network, the huge floodplain, and the difference in timing between the arrival of water from the north and the south. Precipitation and runoff of the rivers in the basin are affected in well-developed El Niño Southern Oscillation (ENSO) years. This has been demonstrated for the period 1906–1985 for the Amazon at Manaus (Richey et al. 1989). In contrast, higher rainfall and discharge may be expected during the La Niña years. Landslides and floods in the headwaters and tributary basins are not uncommon although the related morphological changes in the channel are difficult to identify (Aalto et al. 2003). Off the mouth of the Amazon, inner-shelf mud deposits have built up the subaqueous part of the delta. Beyond lies the Amazon Canyon and the Amazon Cone, a very large deep-sea fan formed in the Late Pleistocene when a lowered sea level allowed direct sedimentation on the cone (Mertes and Dunne 2007 and references therein).

No direct impoundment or significant anthropogenic structural change has been imposed on the mainstream Amazon. There has been deforestation in the basin, but its effect on the hydrology of the river is doubtful. However, it has been demonstrated that an increase in annual runoff occurred following forest clearing and transforming of forests into pastures in the basin of the Tocantins River, in the drier southeastern area (Costa et al. 2003). The largest river in the world does not seem to have been impacted yet by anthropogenic modification. Its growth and evolution display only the impacts caused by natural events.

Almost the entire sediment still comes naturally from the Andean slopes but only a small part reaches the Atlantic. Several estimates of sediment load and transfer are available. According to Aalto et al. (2006), about 2300–3100 million tonnes are derived annually from the Andes. A very large part of it is stored in the foreland, only 1400 million tonnes reach the lowland Amazon, and only 1240 million tonnes pass Óbidos, the tidal limit. Mertes and Dunne (2007) have referred to a previously estimated deposition of 400–500 million tonnes below Óbidos. According to Kuehl et al. (1986) annually about 610–650 million tonnes reach the continental shelf. The sediment load carried by the Amazon appears to have built an incomplete floodplain as shown by topographical features such as lakes, complex levees and nearly straight scroll bars. Floodplain sedimentation occurs through vertical aggradation and from flood channels in avulsion cutting through the floodplain, building levees and depositing sediment which builds local small deltas in lakes.

In sum, three factors have influenced the configuration and behaviour of the Amazon: tectonic setting; climatic pattern; and sea-level fluctuations. This happens over a range of scales and can be traced not only for the present river but also for the one in the Late Pleistocene (Mertes and Dunne 2007).

7.7 Evolutionary Adjustment of Large Rivers

The three case studies indicate that rivers reflect evolutionary changes. Their geological background usually determines their location, morphology, and behaviour. Such controls hold even for the Amazon. The effect of several episodically reactivated geological structures on its physiography for the Nile is another example, an ancestor of it probably started from the Pan-African orogenic event 550 million years ago. Its present physiography is partially the result of several structural and hydrological factors and the changing level of the Mediterranean Sea (Woodward et al. 2007 and references therein).

Events during the Quaternary impacted the alluvial stratigraphy and channel configuration of large rivers. Approximately over the last two million years such events modified the headwaters, main stems, and the lower channels of large rivers, individually and also as integrated systems, multiple times. A number of rivers, particularly from the middle and upper latitudes, probably had to adjust to a changed sediment load at the end of the Pleistocene which followed a readjustment of natural vegetation across large basins. Anthropogenic activities have subsequently strikingly modified the physiography and behaviour of global rivers.

There have been several very large dams. One single dam, the Aswan High Dam, much bigger than the previous impoundments on the Nile, changed the character of the lower river (Chapter 8). The huge Grand Ethiopian Renaissance Dam (GERD) currently being constructed on the Blue Nile in Ethiopia may also have a considerable impact when closed. The impact of the Three Gorges Dam on the Changjiang may alter the pattern of its sediment storage and transfer. Previously, downstream of the gorges, the gradient of this river lessened markedly, the velocity dropped, and the river meandered in a vast plain with progressive deposition of alluvium on its bed. This may change. The Changjiang Basin already has been impacted by a large number of impoundments. A series of large dams in the Missouri Basin has restricted the Mississippi to a much smaller sediment load than natural.

The current physiography of a large river may be considered to represent a point in time in its evolution. Rivers are rarely at either glacial or interglacial peaks, both of which are rather short time periods. They are more likely to be somewhere in transition (Blum 2007). A period of at least 40 000 years, usually much longer, have been estimated for large systems to adjust to a change (Casteltort and Van Den Driessche 2003; Blum 2007). A geologically averaged large river probably varies from the present ones, being adjusted to a different physical environment, sediment production, and a different base level, midway on the continental shelf. Blum (2007) opined that a long-lived large river system should be expected to have adjusted certain of its categories, such as long profiles, slopes, and sediment production, to an average glacial world and graded to a sea level about halfway up the continental shelf. Rivers are dynamic phenomena but with a slow adjustment to changes. Large rivers are polyzonal, integrating huge drainage basins. Large rivers of the world may in future have to adjust to anthropogenic climate change, resulting in different physiography and varied fluvial processes (Chapter 12).

Questions

1 How old are large rivers? List the factors that control the evolution of their present physiography and behaviour.

2 Imagine a river in temperate latitudes, operating as a long conduit transferring water and sediment from its source in tectonic mountains to its mouth in the passive edge of the continent. What changes would the river undergo from the Last Glacial Maximum to the present time?

3 Imagine that the river in Question 2 is located in the tropics. Would your answer be the same?

4 How would a river that underwent a series of meltwater floods differ from one without any?

5 During the last glacial cycle, the average position of sea level was between 50 m and 60 m below the present sea level. To what extent are rivers still responding to the most recent post-glacial sea-level rise?

6 Discuss the nature of changes that have affected rivers since the beginning of the Holocene.

7 What type of anthropogenic changes may a large river have undergone over the last one hundred years?

8 Select a major river which is not discussed in detail in this chapter. Trace its evolution over time, from its geological beginning to the end of the twentieth century.

9 Is there any large river in the world that exists in a near natural state?

10 The lower part of large rivers is commonly affected by a falling sea level in the last glacial time. How far up the river do such changes impact?

11 To what extent did the relative sea-level changes of the last glacial cycle overprint earlier changes in fluvial landscapes?

12 Are large rivers of the world already adjusted to the Pleistocene impacts on their environment?

References

Aalto, R.E., Maurice-Bourgoin, I., Dunne, T. et al. (2003). Episodic sediment accumulation on Amazonian floodplains influenced by El Niño/Southern Oscillation. *Nature* 425: 493–497.

Aalto, R.F., Dunne, T., and Guyot, J.L. (2006). Geomorphic controls on Andean denudation. *Journal of Geology* 114: 85–99.

Anderson, R.C. (1988). Reconstruction of preglacial drainage and its diversion by earliest glacial forebulge in the upper Mississippi Valley region. *Geology* 16: 254–257.

Baker, V.R. (1981). *Catastrophic Flooding: The Origin of the Channeled Scabland*. Stroudsburg, PA: Hutchinson Ross.

Baker, V.R. (2007). Greatest floods and largest rivers. In: *Large Rivers: Geomorphology and Management* (ed. A. Gupta), 65–74. Chichester: Wiley.

Blum, M.D. (2007). Large river systems and climate change. In: *Large Rivers: Geomorphology and Management* (ed. A. Gupta), 627–659. Chichester: Wiley.

Blum, M.D. and Törnqvist, T.E. (2000). Fluvial responses to climate and sea-level change: a review and look forward. *Sedimentology* 47: 2–48.

Bretz, J.H. (1923). The channeled scabland of the Columbia Plateau. *Journal of Geology* 31: 617–649.

Brookfield, M.E. (1998). The evolution of the great river systems of southern Asia, during the Cenozoic India-Asia collision: rivers draining southwards. *Geomorphology* 22: 285–312.

Casteltort, S. and Van Den Driessche (2003). How plausible are high-frequency sediment supply-driven cycles in the stratigraphic record? *Sedimentary Geology* 157: 3–13.

Clark, M.K., Schoenbohm, I., Royden, L.H. et al. (2004). Surface uplift, tectonics, and erosion of eastern Tibet from large-scale drainage patterns. *Tectonics* 23: TC1006. https://doi.org/10.1029/2002TC001402.

Clift, P.D. and Blusztajn, J.S. (2005). Reorganisation of the western Himalayan river system after five million years ago. *Nature* 438: 1001–1003.

Coleman, J.M. (1988). Dynamic changes and processes in the Mississippi River delta. *Geological Society of America Bulletin* 10: 999–1015.

Costa, M.H., Botta, D., and Cardillo, J.A. (2003). Effects of large-scale changes in land cover on the discharge of the Tocantins River, Southeastern Amazonia. *Journal of Hydrology* 283: 206–217.

Cox, K.C. (1989). The role of mantle plumes in the development of continental drainage patterns. *Nature* 142: 873–877.

Derbyshire, E. (1996). Quaternary glacial sediments, glaciation style, climate and uplift in the Karakoram and northwest Himalaya: review and speculations. *Palaeogeography, Palaeoclimatology, Palaeoecology* 120: 147–157.

Dunne, T., Mertes, L.A.K., Meade, R.H. et al. (1998). Exchanges of sediment between the flood plain and channel of the Amazon River in Brazil. *Geological Society of America Bulletin* 110: 450–467.

Emmel, F.J. and Curray, J.R. (1982). A submerged late Pleistocene delta and other features related to sea level changes in the Malacca Strait. *Marine Geology* 47: 197–216.

Enzel, Y., Ely, L.L., Mishra, S. et al. (1999). High-resolution Holocene environmental changes in the Thar Desert, northwestern India. *Science* 284: 125–128.

Ervin, C.P. and McGinnis, L.D. (1975). Reelfoot rift: reactivated precursor of the Mississippi Embayment. *Geological Society of America Bulletin* 86: 1287–1295.

Franzinelli, E. and Igreja, H. (2002). Modern sedimentation in the Lower Negro River, Amazonas State, Brazil. *Geomorphology* 44: 259–271.

Gibling, M.R., Tandon, S.K., Sinha, R., and Jain, M. (2005). Continuity-bounded alluvial sequences of the southern Gangetic Plains, India: aggradation and degradation in response to monsoonal strength. *Journal of Sedimentary Research* 75: 369–385.

Goodbred, S.L. (2003). Response of the Ganges dispersal system to climate change: a source-to-sink view since the last interstade. *Sedimentary Geology* 162: 83–104.

Goodbred, S.L. and Kuehl, S.A. (2000a). Enormous Ganges-Brahmaputra sediment discharge during strengthened early Holocene monsoon. *Geology* 28: 1083–1086.
Goodbred, S.L. and Kuehl, S.A. (2000b). The significance of large sediment supply, active tectonism and eustasy on margin sequence development. Late Quaternary stratigraphy and evolution of the Ganges-Brahmaputra delta. *Sedimentary Geology* 133: 227–248.
Gupta, S., Collier, J.S., Palmer-Felgate, A., and Potter, G. (2007). Catastrophic flooding origin of shelf valley system in the English Channel. *Nature* 448: 342–345.
Inman, D.L. and Nordstrom, C.E. (1971). On the tectonic and morphological classification of coasts. *Journal of Geology* 79: 1–21.
Kale, V.S. (1998). Monsoon floods in India: a hydro-geomorphic perspective. In: *Flood Studies in India* (ed. V.S. Kale) Memoir 41, 229–256. Bangalore: Geological Society of India.
Keshew, A.E. and Teller, J.Y. (1994). History of the late glacial runoff along the southwestern margin of the Laurentide Ice Sheet. *Quaternary Science Reviews* 13: 859–877.
Kessel, R.H. (2003). Human modifications to the sediment regime of the lower Mississippi River flood plain. *Geomorphology* 56: 325–334.
Knox, J.C. (2007). The Mississippi River system. In: *Large Rivers: Geomorphology and Management* (ed. A. Gupta), 145–182. Chichester: Wiley.
Kuehl, S.A., DeMaster, D.J., and Nittrouer, C.A. (1986). Nature of sediment accumulation on the Amazon continental shelf. *Continental Shelf Research* 6: 209–225.
Latrubesse, E. and Franzinelli, F. (2002). The Holocene alluvial plain of the middle Amazon River, Brazil. *Geomorphology* 44: 241–257.
Maslin, M.A. and Burns, S.J. (2000). Reconstruction of the Amazon Basin effective moisture availability over the past 14 000 years. *Science* 290: 2285–2287.
Meade, R.H. (2007). Transcontinental moving and storage: the Orinoco and Amazon Rivers transfer the Andes to the Atlantic. In: *Large Rivers: Geomorphology and Management* (ed. A. Gupta), 45–63. Chichester: Wiley.
Meade, R.H. and Moody, J.A. (2010). Causes for the decline of suspended-sediment discharge in the Mississippi River system, 1940–2007. *Hydrological Processes* 24: 35–49.
Mertes, L.A.K. and Dunne, T. (2007). Effects of tectonism, climate change, and sea-level change on the form and behaviour of the modern Amazon River and its floodplain. In: *Large Rivers: Geomorphology and Management* (ed. A. Gupta), 115–144. Chichester: Wiley.
Milliman, J.D. and Farnsworth, K.L. (2011). *River Discharge to the Coastal Oceans: A Global Synthesis*. Cambridge: Cambridge University Press.
Moore, A.E. and Larkin, P.A. (2001). Drainage evolution in south-central Africa since the breakup of Gondwana. *South African Journal of Geology* 104: 47–68.
Murray-Wallace, C.V. and Woodroffe, C.D. (2014). *Quaternary Sea-Level Changes: A Global Perspective*. Cambridge: Cambridge University Press.
Pickering, J., Goodbred, S., Reitz, M. et al. (2013). Late Quaternary sedimentary record and Holocene channel avulsions of the Jamuna and Old Brahmaputra River valleys in the upper Bengal delta plain. *Geomorphology* 227: 123–136.
Potter, P.E. (1978). Significance and origin of big rivers. *Journal of Geology* 86: 13–33.
Quade, J., Cater, J.M.L., Ojha, T.P. et al. (1995). Late Miocene environmental change in Nepal and the northern Indian subcontinent: stable isotopic evidence from paleosols. *Geological Society of America Bulletin* 107: 1381–1397.

Raymo, M.E. and Ruddiman, W.F. (1992). Tectonic forcing of late Cenozoic climate. *Nature* 359: 117–122.

Richey, J.E., Nobre, C., and Deser, C. (1989). Amazon River discharge and climate variability: 1903–1985. *Science* 246: 101–103.

Robinson, R.A.J., Brezina, C.A., Parrish, R.R. et al. (2014). Large rivers and orogens: the evolution of the Yarlung Tsangpo-Irrawaddy system and the eastern Himalayan syntaxis. *Gondwana Research* 28: 112–121.

Ruddiman, W.F. and Kutzbach, J.E. (1991). Plateau uplift and climate change. *Scientific American* 264: 42–50.

Tandon, S.K. and Sinha, R. (2007). Geology of large river systems. In: *Large Rivers: Geomorphology and Management* (ed. A. Gupta), 7–28. Chichester: Wiley.

Tapponier, P., Peltzer, G., and Armijo, R. (1986). On the mechanics of collision between India and Asia. In: *Collision Tectonics*, vol. 19 (eds. M.P. Coward and A.C. Ries), 115–157. London: Geological Society of London.

Teller, J.T., Leverington, D.W., and Mann, J.D. (2002). Freshwater outbursts to the ocean from glacial lake Agassiz and their role in climate change during the last deglaciation. *Quaternary Science Reviews* 21: 879–887.

Van der Hammen, T. and Absy, M.L. (1994). Amazonia during the last glacial. *Palaeogeography, Palaeoclimatology, Palaeoecology* 109: 247–261.

Wasson, R.J. (1995). The Asian monsoon during the Late Quaternary: a test of orbital forcing and palaeoanalogue forecasting. In: *Quaternary Environments and Geoarchaeology of India. Essays in Honour of Professor S.N. Rajaguru* (eds. S. Wadia, R. Korisettar and V.S. Kale), 22–35. Bangalore: Geological Society of India, Memoir 32.

Williams, M., Dunkerley, D., De Dekker, P., and Kershaw, A.P. (1998). *Quaternary Environments*. London: Arnold.

Winkley, B.R. (1994). Response of the Lower Mississippi River to flood control and navigation improvements. In: *The Variability of Large Alluvial Rivers* (eds. S.A. Schumm and B.R. Winkley), 45–74. New York: American Society of Civil Engineers.

Woodward, J.C., Macklin, M.C., Krom, M.D., and Williams, M.A.J. (2007). The Nile: evolution, Quaternary river environments and material fluxes. In: *Large Rivers: Geomorphology and Management* (ed. A. Gupta), 261–292. Chichester: Wiley.

8

Anthropogenic Alterations of Large Rivers and Drainage Basins

8.1 Introduction

People have used rivers as a valuable resource for a very long time. Early centres of civilization developed in basins of large rivers such as the Nile, Tigris-Euphrates, Indus, Huanghe, and the Changjiang. The earliest of these centres started probably as far back as 8000 years ago, and they all lasted for thousands of years. Farming began about 12 000–11 500 years BP in southeastern Turkey and northern Syria from where it spread out to other major valleys, increasing the demand for water. Rivers have been used for a long time as a source of water for consumption, navigation, or crop irrigation. As settlements grew up on riverbanks, it became necessary to control floods and to prevent channel migration. We do not have a complete and organised account of anthropogenic alteration of rivers in historical times. The basic story is frequently derived from field evidence as carried out for the Indus Valley. Recorded accounts and flood measures provide a fuller narration for only a few large rivers, such as the Nile or Changjiang.

Two alterations have been common: (i) direct modification of the river itself, and (ii) changes in ground cover and slopes of the drainage basin. Later, accelerated dam building was added along with impoundments and related diversion of river water. About three decades ago, a World Bank report indicated that about 50 km^3 of sediment was being trapped globally behind dams each year by the 1980s (Mahmood 1987). Such activities not only modified the morphology and behaviour of numerous small rivers, but also of large rivers such as the Nile or the Huanghe. The majority of large rivers in the world have been affected by impoundments or diversions (Ramankutty and Foley 1999; Syvitski et al. 2003; Vörösmarty et al. 2003; Nilsson et al. 2005).

A growing demand for water and advances in engineering led to increasingly efficient modification of rivers by water impoundment, diversion, and extensive changes in the land use of river basins. The loss of forests and spread of agriculture in drainage basins led to increased erosion and sediment production. This sediment was added to the natural sediment in the river and conveyed downstream. The dams, as they became larger in the twentieth century, changed river morphology by holding water and sediment in upstream reservoirs. The hydrology, sediment transfer and storage, and channel morphology of the lower river changed. The impoundments also modified the ecology of the river. Such anthropogenic alterations did not significantly affect all large rivers but demonstratively impacted many of them. A brief account of such changes is given in this chapter, illustrating that many large rivers are no longer natural.

Introducing Large Rivers, First Edition. Avijit Gupta.

8.2 Early History of Anthropogenic Alterations

The earliest accounts of the organised utilisation of large rivers came from the valleys of the Tigris and Euphrates in ancient Mesopotamia and the Nile in Egypt. Both regions required flood management and water for irrigation. Both regions demonstrated skills in (i) controlling floodwater and (ii) use of irrigation water by lifting it from channels and distributing it to the fields at higher level. Temporary earth and wood structures were used for diverting floodwater. About 5000 years ago, during the realm of King Menes in Egypt, clay bunds were used to hold back floodwaters of the Nile for basin irrigation which inundated croplands. This was a very early example of state-supported river impoundment. The practice of using seasonal floodwaters of the Nile for irrigation continued for thousands of years. Instruments were innovated for lifting water out of rivers or canals. About 4000 years ago, the *shadoof*, which is essentially a bucket and a counterweight at the opposite ends of a balanced pole, was used for lifting water from rivers and irrigation channels to higher levels. Other instruments for lifting water and pouring it down a channel to reach agricultural fields by gravity were the Archimedean screw and *noria* or Persian wheel. As the agriculture of ancient Egypt depended on the arrival of Nile floods bearing fertile silt, records were kept of the level of the river using nilometers which are basically structures in the river that measure the height and clarity of floods. A long record of the Nile floods therefore exists (Hurst 1952).

The Sumerian civilisation of Mesopotamia irrigated the land between the Tigris and Euphrates rivers by using skilfully constructed impoundments whereby water was diverted to the canals from the rivers and fields. McCully (1996) mentioned the remains of an early damming system, dated 3000 BCE, from the current area of Jordan. The system involved a long weir diverting water into a canal that fed a number of reservoirs blocked by dams of rock and earth. Weirs are low walls across rivers that allows the river to flow but at the same time enhance the storage of water upstream. They are often described as run-of-the-river dams. By the late 1st millennium BCE, dams of stone and earth were in use in many parts of the world.

In the dry environment of Mesopotamia, this, however, resulted in salinisation of the fields. High rates of evaporation of the canal and field water brought up the saline groundwater to the surface by capillary action. The water then evaporated leaving behind a layer of salt. As a result, the major crop wheat had to be replaced by the salt-tolerant barley, and eventually the once cultivated fields were abandoned. This is an early example of anthropogenic alteration of rivers resulting in a degraded environment that still happens. The Indus River and its tributaries supported a civilisation of cities, known as the Indus Valley or Harappan Civilisation that started about 5000 years ago and lasted for about 2000 years. It was based on careful water management to serve canals, urban drainage systems, and navigation.

Water engineering in China started later, about 4000 years ago, but it was extremely efficient, and hydrological records were carefully kept. It included utilisation of the flood prone rivers of the north, especially the Huanghe; the use of the Changjiang River for irrigation and navigation; and the construction of canals including the Grand Canal that connected the north with the Changjiang Valley. The Grand Canal was constructed in stages, starting in the fifth century BCE. The nearly 1800 km long canal from Beijing to Hangzhou was finally completed early in the seventh century, and provided a passage from the Huanghe to the Changjiang. The Dujiangyan Irrigation System in Sichuan is the other large-scale and impressive example of early river engineering. It was supervised by Li Bing on the flood prone Min River, a major tributary of the Changjiang. The project was completed in 256 CE but it is still functional 2000 years later. This engineering marvel controlled the devastating floods of the Min and transferred the water via canals to the drier Chengdu Plains of Sichuan for irrigating crops. The amount of water extracted from the river was carefully controlled by its level, and excess water was returned to the Min downstream of the diversion point.

Aquaducts, navigation canals, small impoundments, and more efficient versions of water wheels continued to modify smaller rivers. Large-scale engineering works started to impact bigger rivers probably from the nineteenth century. A number of canals were engineered in northern England to generate water power and transfer goods of the early industrial revolution. These modifications continued on many rivers in different parts of the world, mainly in Europe, and later in North America. The Erie Canal was in place by 1825, transporting freight between the state of New York and Lake Erie using the Hudson and Mohawk rivers. The factories and mines of Europe used water power from small dams on swift streams towards the second half of the nineteenth century and early part of the twentieth. Dams were essential in opening up the dry western part of the United States. The next major advance in technology involved utilising the large rivers. An excellent historical account concerns the Mississippi.

8.3 The Mississippi River: Modifications before Big Dams

The history of alterations to the Mississippi River (Chapter 7) is easy to follow as (i) such changes are extremely well documented; and (ii) as the river has been affected by human activities for only about three hundred years, the natural pre-change state of the river and its basin was known. The part of the basin on the eastern side of the river used to be predominantly under forest and the western side under grassland, although variations were present on both sides. Between 1750 and 1850, the natural cover was dramatically changed by the westward advance of the Euro-American settlers. Towards the end of the twentieth century, only about a third of the entire basin was left in forest and woodland, although the pattern of land use varied significantly between the tributary basins (Knox 2007). Sediment in the rivers increased strikingly due to this change in land use and peaked in the late nineteenth and early twentieth centuries before extensive application of soil conservation techniques and improved cropping practices brought in some control (Knox 2002).

The Upper Mississippi River was also directly modified by river engineering. This involved clearing of logjams (regionally known as snags), removal of sandbars, and straightening of the channel by cutting across bends – all to improve navigation for loaded river barges. In addition, levees were constructed on the banks of the Mississippi for flood control. In 1866, the US Congress passed the River and Harbors Act which directed the Corps of Engineers to modify the river for navigation. Dredging and snag removal for improving the navigability of the river started in 1868. Removal of the sand bars was supplemented with rock and brush dams along the sides of the channel, and barriers were constructed to prevent water escaping to side channels and backwaters. Banks and nearshore areas were covered locally with mats of willow and rocks to prevent erosion. The channel was repeatedly dredged to maintain a navigable waterway in spite of disputes and controversies about the engineering techniques used. It was 1940 when the final structure became operational.

Work on the Lower Mississippi River, downstream from its confluence with the Ohio, started in the early eighteenth century. Levees were constructed to protect New Orleans but large floods continue to overflow or break through levees. Consequently, levees were raised from an initial 1 m height to locally more than 10 m (Winkley 1994). By the mid-nineteenth century, levees about 2 m high on average expanded along the Lower Mississippi for farmland protection. The disastrous flooding of 1927 was probably due to levee breaks in a large flood in many locations on the river. Post-flood engineering works along the river also included spillway structures to divert water from the main river and construction of dams and reservoirs in tributaries for flood control. The Lower Mississippi was stabilised for navigation from the 1930s by levee construction, bank revetments, and meander cutoffs. These changes improved navigation in certain reaches but not everywhere, and different problems, discussed in the next section, arose and needed to be solved. The Mississippi describes various attempts at modifications of a large river, with or without success.

Part of the river has been allowed to deepen for navigation by decreasing its width by rock and pile dykes. The Mississippi thus became narrower in places and carried a deeper flow between dykes. For example, the channel at St Louis was 1130 m wide and 9 m deep at bankfull in 1837. The width was later constrained by dykes to 640 m, which proportionally increased the depth of the flow. The 1844 peak at St Louis was 36 800 $m^3\ s^{-1}$ with a stage of 12.6 m. Over a hundred years later the smaller 1973 discharge of 24 122 $m^3\ s^{-1}$ had a higher stage of 13.2 m (Schumm 2007). Levees isolated the river from its floodplain, sending the entire flood through the channel.

Erosion and transfer of a large amount of sediment is natural for the western tributaries of the Mississippi River (including the Missouri) that drain east from the mountains. The natural sediment yield for these tributary basins could be about 200 t $km^{-2}year^{-1.}$ (Milliman and Farnsworth 2011). Episodic storm-derived peaks were also common. When the early European settlers arrived about three hundred years ago, the Mississippi River was still adjusting to sediment-laden runoff events that started after the last glaciation (Milliman and Farnsworth 2011). It used to be a broadly meandering river of variable and episodic discharge, adjusting to post-glacial sediment (Knox 2007). Construction of numerous dams, in the tens of thousands, some being very large, impounded sediment and regulated the discharge of the western tributaries (Meade and Moody 2010). Accelerated soil conservation measures from the 1940s also controlled supply of sediment to the river. The Mississippi River therefore had to adjust to a new pattern of regulated discharge, lower sediment volume, and levee-guided channel, all due to anthropogenic alterations. Not only the river changed, a corresponding loss of wetlands by erosion occurred in the Mississippi Delta.

Human alterations are capable of changing even a huge river such as the Mississippi. The river would have been different otherwise and more natural. Wohl (2002) has demonstrated that even the pristine-looking river systems of the Front Ranges, Colorado, USA in reality are channels altered in historical time. She called such streams 'virtual rivers'.

8.4 The Arrival of Large Dams

In the twentieth century, starting in the developed countries, large river projects rapidly increased in number. These generally involved construction of a large dam across a river with a set of related multipurpose projects that met a variety of needs such as flood control, irrigation, and production of hydroelectricity (Figure 8.1). The 221 m Hoover Dam over the Colorado River was constructed in the 1930s with Lake Mead, a storage reservoir with a capacity of 38 million m^3, behind it (Figure 8.2). Hoover remained the highest dam in the world until 1957, when several even higher dams were constructed subsequently in Switzerland, Italy, and California (USA). In the middle of the twentieth century other huge dams also were built on major rivers in the United States, including the Grand Coulee and Bonneville on the Columbia, the Shasta on the Sacramento, and

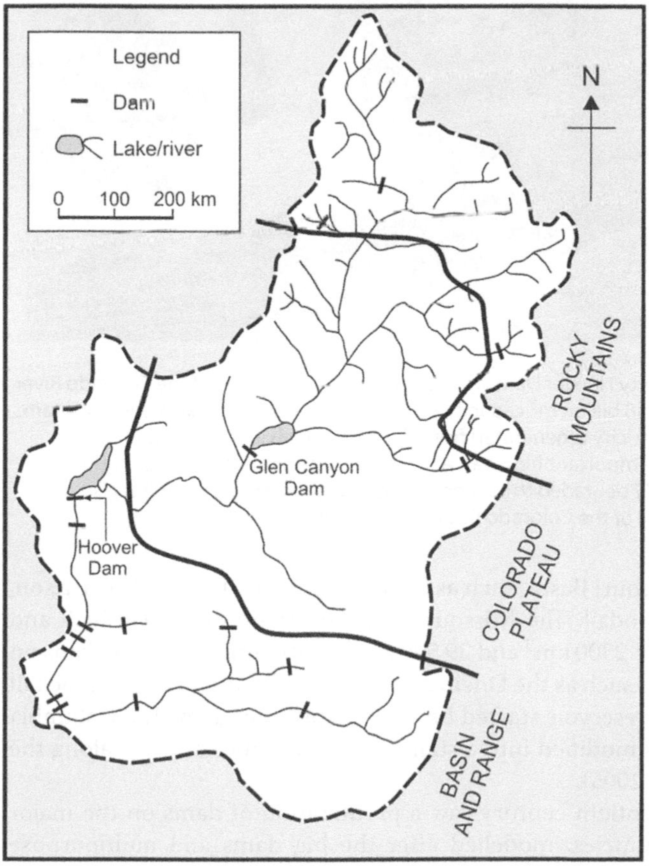

Figure 8.1 Schematic map of the Colorado River Basin showing the major dams on the main river and its tributaries. The two large reservoirs shown are Lake Powell behind the Glen Canyon Dam impacting the pre-dam hydrology, sediment transport and bar deposition in the Grand Canyon downstream (see text), and Lake Mead behind the Hoover Dam.

Figure 8.2 The concrete arch-gravity Hoover Dam was built in the Black Canyon of the Colorado River between 1931 and 1936, with the 38 billion m^3 capacity of impounded water of Lake Mead upstream. An immense amount of hydroelectricity is generated from the dam, which located at the Arizona–Nevada border carries an important highway, and has become a tourist attraction. The dam and reservoir have environmentally degraded the Colorado Delta and its ecosystem. It has also impacted the unique native fishery of the Colorado Basin. Source: Wikipedia.

a set of large dams in the Missouri Basin, such as the Canyon Ferry, Fort Peck, Garrison, Oahe, Big Bend, and Fort Rendall. The Missouri dams are more than 29 m high and have large reservoirs (between 2300 km^3 and 29 500 km^3) upstream. A series of dams on major rivers engineered rivers such as the Dneiper or Volga into staircases where the tail end of the next downstream reservoir started below the upstream dam. The Columbia in the United States also was modified into a staircase by a number of dams along the river (Magilligan and Nislow 2005).

The second half of the twentieth century saw a proliferation of dams on the major rivers of the developing countries, modelled after the big dams and multipurpose projects of North America and Europe. A series of dams on the Tennessee River which flowed into the Ohio and its tributaries was considered as an illustrative example of an integrated scheme of dams and reservoirs for multipurpose application of water and power from these projects which can be used for improving environmental and social conditions. Several dams and a barrage in the basin of the flood prone Damodar River in eastern India followed such planning. The Tennessee basin used to suffer from floods and extensive soil erosion following uncontrolled logging and strip-mining practices. In 1933, the Tennessee Valley Authority (TVA) was set up in order to generate hydroelectricity in this relatively poor rural part of the United States and enhance its economic development. The scheme grew into an integrated multipurpose project that included flood control, power generation, navigation, and irrigation. This necessitated construction of a number of dams and reservoirs in the basin, treating the entire basin as an integrated functional unit. Apart from building engineering structures, it was deemed necessary to manage the land use in the basin, especially soil conservation. The concept recognised the role of state involvement in order to achieve social welfare. The governments of a number of developing countries were attracted by the concept of a state implementing a large and multipurpose engineering project in order to save people from floods, prevent soil erosion, and provide them with water, power, and employment. The dams became important and desirable. TVA was not as successful as planned, neither were similar projects, but a multipurpose project in a large river basin was seen as a social commitment and a monument to national pride. The Aswan High Dam on the Nile is an excellent example of early state-directed modification on a large river (Box 8.1).

Box 8.1 Aswan High Dam on the Nile River

Several dams were built in the twentieth century across the two main sources of the Nile: the White Nile which originates in Equatorial Africa and the Blue Nile that rises in the Ethiopian Highlands. Several dams such as the Aswan Low Dam, Senner, and others were earlier constructed to provide irrigation for the commercial cotton enterprise in Egypt and also for producing crops for local consumption. For example, the Ghezira scheme was in operation for irrigating the land between the two Niles.

The well-known Aswan Dam (properly called the Aswan High Dam) was completed in 1970 on the Nile with a 600 km long reservoir that stretched from southern Egypt to northern Sudan, covering 5250 km^2 of the valley. The part of this reservoir in Egypt is known as Lake Nasser and the tail end in Sudan as Lake Nubia. The 111 m high and 3.8 km long rock and clay dam is also famous as one of the earliest state-built dams in the developing world on a major river that was supported by overseas financing and assistance, and in the process went through a long and serious external political controversy. It was finally built on Soviet technical assistance and funding of one billion dollars on a low interest (Figure 8.3).

Multifaceted benefits were achieved at Aswan including flood control, power generation, crop irrigation, improvement in navigation, and lake fishing. Its construction, however, displaced a significant number of people from the reservoir area, and also required removal and re-siting of archaeological monuments such as the Abu Simbel temple.

The large reservoir was filled in 1976. Unlike the previous low dams on the Nile, this project can store the annual flood of the river and nearly the entire sediment load. The downstream river, the delta, and the coastline were all deprived of sediment and underwent erosion and geomorphic changes. Only about 15% of the flow at Aswan now reaches the Damietta Branch of the river in the delta. This resulted not only in morphological changes in the channel but also increases in salt solution along much of the lower Nile

(Continued)

Box 8.1 (Continued)

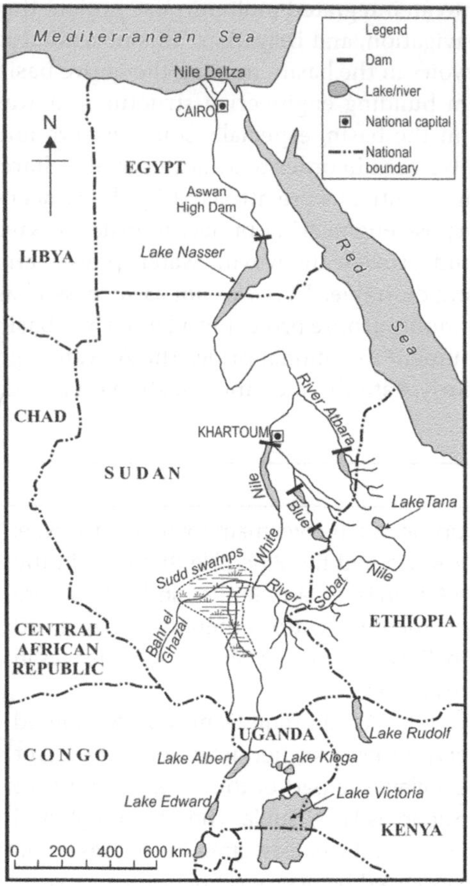

Figure 8.3 Schematic diagram showing the Aswan High Dam in the Nile Basin.

Valley and the delta (Milliman and Farnsworth 2011 and references therein). A number of mostly negative environmental impacts of the Aswan Dam are:

- The water stored in the large reservoir is subjected to high evaporation and seepage loss.
- High sedimentation of flood sand and silt occurs in the reservoir, leading to a reduction in its water storage capacity. Stanley (1996) has compared the sedimentation filling up the upstream end of the lake behind the Aswan Dam with building of a new delta, 200 km x 12 km in area and about 40 m thick. This sediment used to travel down the river prior to the closure of the Aswan Dam. Stanley and Wingerath (1996) estimated the sediment to be about 98% of the pre-dam amount that used to reach the Nile Delta and the Mediterranean. Very little bed load is in the Nile, most of the sediment being carried in suspension (Woodward et al. 2007 and references therein). The mean annual pre-Aswan suspended load has been computed as 120 million tonnes by Milliman and Syvitski (1992) and 125 million tonnes by Shahin (1985). Woodward et al. (2007) referred to an annual variability of 50–228 million tonnes at Abu El-Ata. The Blue Nile and the Atbara contribute about 72% and 25% of the suspended sediment, respectively, from

the Ethiopian Highlands. The Ethiopian sediment arrives seasonally in summer flood, weeks before the flood peak. Most of the sediment ends in the lake, only the finest fraction of the suspended load reaches the end of the reservoir and downstream of the dam (Woodward et al. 2007).

- Loss of Nile sediment resulted in 25–70 cm channel erosion downstream of the dam and erosion of the Nile Delta. Post-dam erosion of the coastline occurs at an annual rate of several metres.
- Saltwater intrudes in the delta area due to a reduction in Nile discharge. Only a fraction of the discharge released from the dam reaches the delta, after loss due to evapotranspiration, groundwater infiltration, and irrigation demands (Stanley 1996).
- The influence of the Nile sediment has been reduced in the eastern Mediterranean, especially concerning marine biology.
- Proliferation of aquatic weeds has occurred in the clear water below the dam.
- Increased salinisation and alkalization of the agricultural fields is a hazard following irrigation.
- Rising groundwater level requires improvement in local drainage; increase in bilharzia.

These are the major negative impacts of the dam, although several other changes can be listed. Clearly certain anthropogenic changes, such as construction of large dams and reservoirs, alter and modify even large rivers and their floodplains and deltas. It is fascinating to compare such effects with that of another large dam, the Three Gorges on the Changjiang (Box 8.2). The effect of the Aswan may be further modified when the Grand Ethiopian Renaissance Dam (GERD) is completed on the Blue Nile.

A number of large rivers, as in the basin of the Missouri, have been dammed and as a result very little variation in sediment or water discharges is experienced downstream. Such rivers become different rivers.

Strangely, rivers discharging water to the sea have been considered wasteful by many political authorities, whereas a large engineering structure impounding a major river is often considered a showpiece of success. Large projects have been completed across the world, under almost all socio-political systems, and the number of new dams inaugurated in a year rose fivefold between 1940 and 1970. The dams also got larger. Nearly all of the current 100 largest dams in the world were constructed after the Second World War, the majority of them in the developing countries (World Resources Institute 1988). According to the International Commission on Large Dams (ICOLD), more than 45 000 large dams exist worldwide, about 35 000 of which have been built after 1950. ICOLD defines a large dam as 17 m high from its foundation to the crest. Dams 5–17 m high and with a reservoir capacity of at least 3 million m^3 of water are also considered as large dams (World Commission of Dams 2000). On average, nearly 300 dams used to be constructed each year, although this number has tailed off, especially in the developed countries. The practice of building large dams on major rivers has been common in the developing countries for some time, such as the 1965 Akosombo Dam in Ghana with a 8500 km^2 reservoir that inundated about 4% of the country (McCully 1996). The loss of sediment from the Volta River has accelerated coastal erosion along the Bight of Benin, east of the river mouth. Other recent examples of very large dams include the 1984 Itaipu Dam at the border of Brazil and Paraguay with enormous power-generating capacity, the 1986 Guri Dam in Venezuela, the 2003 Three Gorges Dam on the Changjiang in China, and the 2010 Xiaowan Dam on the upper Mekong, also in China. Ethiopia is building a very large dam on the Blue Nile, primarily for power generation near its border with Sudan, close to the town of Assosa. This 170 m high and 1.8 km long dam is known as the GERD. When completed, it will have a reservoir capable of storing 63 billion m^3 of water and generating 6000 MW of power, the biggest in Africa.

Anthropogenic activities in drainage basins and on river channels can have significant impacts on the physical environment. Destruction of vegetation cover and intensive farming usually lead to severe sheet and gully erosion and accelerated transfer of sediment to river channels. The large number of impoundments prevent such sediment from travelling downstream and reduce the amount of water in the channel below dams. The slopes, floodplains and channels of even large rivers need to adjust to such changes while the reservoirs fill up with sediment behind dams. The Colorado with many dams and large water extraction systems only manages to reach the sea in flood. It is probably not the only example.

8.5 Evaluating the Impact of Anthropogenic Changes

The impact of anthropogenic changes can be broadly grouped into two classes: (i) changes in water and sediment reaching the river following alterations in land use and land cover in the basin; and (ii) structural alterations in or near the channel, such as dams

Table 8.1 Impact of dams and reservoirs on large rivers.

Accumulation of sediment in reservoirs. Could be controlled by using the dead storage capacity of reservoirs as a sediment trap
Triggering of earthquakes by the weight of the reservoir water
Changes in groundwater level, could lead to slope instability around reservoirs and sedimentation
Reservoir related biological problems: fish kills, algal production, proliferation of aquatic weeds, loss of wildlife
Inundation of forests around the reservoir and changes to vegetation below dams
Decrease of flood peaks in the channel below dams after its construction; other changes vary
Significant decrease in sediment load and concentration in the river for hundreds of kilometres below dam site
Bed degradation and lowering of the longitudinal profile in alluvial rivers but the intensity varies amongst rivers. Most of the degradation happens in the early years after dam construction
Bed material may coarsen in an alluvial river early after dam construction, leading to armouring
Width of an alluvial channel may change below dams
In an alluvial valley flat, riparian vegetation usually increases downstream from dams, possibly due to decreased peak flow and floodplain building
In rock channels, the competence of rivers downstream of a dam is reduced, resulting in discordant tributary junctions and aggradation of debris fans in tributary mouths
Fining of sediment below dams, leading to bed scouring and removal of sand bars
Enhanced erosion in rock channels, resulting in degradation of ecological habitats, changes in flora and fish species; erosion of previous depositional forms in the channel
Reduction of sediment reaching lower floodplains and deltas

and reservoirs that directly affect the river. Both impacts may operate simultaneously, the resultant effect depending on the particular river and its basin (Table 8.1).

8.5.1 Land Use and Land Cover Changes

Alteration of the land cover due to deforestation, farming, and urbanisation has been in operation for thousands of years. By the mid-twentieth century the forests of the world had been reduced to an area between a half and a third of the original cover (Eckholm 1976). The resultant effects on water, erosion, and sediment transfer across local hill slopes and along small rivers have been reported for many parts of the world, e.g. Rapp et al. (1972) for Tanzania; Dunne (1979) for Kenya; Douglas et al. (1992) for Sabah, Malaysia; Lai et al. (1996) for forests of West Malaysia; and Latrubesse and Stevaux (2002) for eastern Brazil. The alterations usually result in accelerated sheetwash, gully development, and slope failures. These processes also tend to operate episodically rather than at a uniform rate and their effectiveness varies from year to year. Considerably less sediment will be released in drier years, but significantly more during years with a number of storms with heavy rainfall. Erosion and sediment transfer will be seasonally conspicuous for many parts of the world during the wet season. White (1995) listed enhanced erosion, short-distance sediment transfers and storage in frequent thundershowers, and impressive long-distance transfers in episodic tropical cyclones in the Philippines. Such changes are well documented for smaller basins and streams but are difficult to determine without caveats for major rivers.

Reliable and long-term data on the impact of basin land use changes on major rivers are difficult to find. Walling and Fang (2003) reviewed the recent trends in suspended sediment loads of rivers worldwide but as they pointed out, the coverage that was available left out huge parts of the world, and although data for suspended load are available for a number of rivers, there are hardly any data for bedload. As bedload for many rivers is less than 10% of the total load, it can be ignored in many cases but this does not hold for all rivers. Bedload for the Brahmaputra, for example, apparently is a high percentage of the total load. Even when only the suspended load is considered, it varies over time due to changes in climate and land cover. Walling and Fang (2003) used the example of the Huanghe to show that if the mean annual load of this river between 1950 and 1970 is considered, the figure is 1.6×10^9 tonnes. The figure for the 1980s halved to about 0.799×10^9 tonnes due to reduced precipitation, increased abstraction of water, and soil management programmes. A large reservoir on the Huanghe, the Sanmenxia, filled up with 50×10^9 tonnes of sediment within three years of reservoir impoundment (McCully 1996). This is an extreme case but loss of sediment and changes in water quality and ecology are generally expected following closure of impoundments on large rivers. Rosieres Dam on the Blue Nile, closed in 1966, provides another example of an eroding basin. The mean annual suspended sediment has been estimated at 54 million m^3. The reservoir lost 21% of its storage capacity in 15 years since closure (Woodward et al. 2007 and references therein). Such figures for sedimentation in reservoirs can be compared with the mean annual soil loss from the Ethiopian Highlands which Hurni (1999) has estimated as 40 t ha^{-1} in the mountains, rising to 300 t ha^{-1} on cultivated slopes.

Alford (1992) investigated the effects of deforestation and extensive slash-and-burn type agriculture in the upper part of the basin of the Chao Phraya River in Thailand over about three decades, the late 1950s to the mid-1980s. This is a 14 000 km^2 basin. There was no significant change to the river. Similarly, Wilks et al. (2001) showed that although forest clearance tends to have a measurable effect on small basins, deforestation and other anthropogenic alterations did not display widespread changes in the 12 100 km^2 Nam Pong catchment in northeast Thailand. Areas identified as forests were reduced from about 80% of the Nam Pong Basin to 30%. Land use is usually not uniform in large river basins which may explain the difficulty of measuring any definite change in water or sediment discharge. Walling and Fang (2003) referred to two other examples of very little change in suspended sediment yield for major rivers: the Ob in Salekhard, Russia, draining a nearly 3 million km^2 basin, and the Upper Changjiang at Yichang, China, draining about 1 million km^2.

The Ebro River of southern Spain is not exactly a major river, but it is well-documented and provides an excellent linkage between changes in land use and sediment in streams. Deforestation and land clearance started under Roman occupation in the first and second centuries CE followed by a rise in agricultural activities in the tenth and eleventh centuries under the Moors. Erosion increased again around 1600 due to the rise in herds of domestic animals. In contrast, dam construction over the Ebro in the twentieth century led to water discharge being halved and sediment discharge dropping to less than 1% of its nineteenth century values (Guillén and Palanques 1997). Rivers need to adjust to such changes, although often a time lag is present.

It may be concluded from the rather limited data available that the effect of land use changes, even though it may result in enhanced erosion, should not be assumed as always significant for the channel of a major river. A very large part of the eroded sediment remains close to the areas undergoing erosion: at breaks of slope; in small tributaries; on floodplains. Only a small proportion reaches the major river and is transported out of the basin. Accelerated erosion and sediment production due to land use changes can be demonstrated to impact small channels but not necessarily large rivers. Milliman and Syvitski (1992) estimated the total mean annual flux of the sediment of the rivers reaching the ocean to be about 20×10^9 tonnes. This probably is the best estimate we have got on a global scale. Hooke (2000) showed that people have moved enough sediment over the last 5000 years to build a mountain range 100 km long, 40 km wide, and 4000 m high.

We may conclude that so far as large rivers are concerned, the effect of basin-scale changes in land use are not necessarily recognisable. Such effects may be significant in major rivers only where the changed volumes of water and sediment are striking enough to modify channel morphology. The usual effects are raising of the channel floor, building of bars, heightening of levees, expansion of floodplains, and reduction of the channel. The channel pattern also may change as a result of excessive sediment reaching the river. The reduction of the water discharge of the Huanghe in the 1970s has been explained partly by the widespread reforestation of the loess plateau through which the river flows and from which a very large amount of fine sediment used to arrive in the channel. Reforestation has reduced sediment loads in gullies by 92% (Li 1992).

The changing morphology of large rivers, however, varies along their course, mainly altering between rock and alluvium. Thus, the effect of sedimentation in the channel would not be the same all along the channel. The Lower Mekong, usually defined as part of the river below the border with China, can be divided into eight reaches (Gupta and Liew 2007). These reaches differ in their forms and functions and are expected to adjust differently to excessive gain or loss of sediment. Temporal and seasonal changes in erosion and sediment transfer have been described for 225 km^2 of steep slopes draining directly to the Mekong River in northern Lao PDR (Gupta and Chen 2002). The annual rainfall there is around 1700 mm with 85–90% falling between May and October. The steep slopes are under forest, but partly cleared for shifting cultivation. A semiquantitative estimate of sediment yield indicates that (i) a seasonal pattern of erosion and sediment transfer and storage occurs every year and (ii) the amount of sediment loss and in transfer is directly related to rainfall. Fresh yellow sand is seen in the Mekong as inset to banks, around rock protrusions in the channel, and as river-mouth fans where tributaries flow into the Mekong. All this sediment presumably originated from enhanced recent erosion following land development (Figure 8.4). A huge amount of sediment apparently is making its way to the South China Sea, although a large part of it may remain stored for a long time in the alluvial valley flat in Cambodia and Vietnam. Sediment plumes, however, are visible off the mouth of the Mekong using satellite imagery, the size of which changes with the season being directly related to discharge (Gupta et al. 2006). Such plumes occur along the coasts of South and Southeast Asia at the mouths of major rivers (Gupta and Krishnan 1994).

A large volume of sediment may impact a major river by building bars, floodplains, and a delta, and locally restricting its channel. However, it is difficult to demonstrate this impact clearly, except in a few extreme cases such as the Huanghe. The impact of direct impoundments on the channel are clearer.

8.5.2 Channel Impoundments

According to the ICOLD, about 45 000 large dams were in operation towards the end of the last century (World Commission of Dams 2000). Large dams on major rivers are still being built in China and various developing countries, not only as engineering achievements but also as symbols of national economic development and pride. Such dams not

Figure 8.4 Fresh sediment against the rocky bank of the Mekong, Lao PDR. Photographer: Avijit Gupta.

only provide water for irrigation and power generation and control flood discharges up to a degree but also act as barriers to the downstream passage of water and sediment, forcing rivers to adjust their morphology and functions (Kondolf 1997). As stated earlier, a World Bank study indicated that sediment trapped every year behind these dams amounts to about 50 km^3 (Mahmood 1987). The cumulative effect over years must be horrendous.

A review of the global impact of sediment retention from river impoundments was carried out by Vörösmarty et al. (2003) using data from 633 large reservoirs, each with a maximum storage capacity of at least 0.5 km^2. In 1950, the global retention of sediment in these reservoirs was 5% of total sediment in transport. By 1968, this figure rose to 15%, and by 1985 to 30%. The amount apparently has stabilised since then. The annual impounded amount has been estimated to total 4–5 Gt (Vörösmarty et al. 2003). It certainly would be a very high figure. In the pre-impoundment stage, this sediment would have built bars and floodplains and reached deltas of major rivers. Thus, impoundments on rivers significantly modify water and sediment fluxes on a global scale plus carbon, various chemical components, and nutrients that ride piggyback on sediment grains. The water quality and ecology of rivers may change with the geomorphology of the rivers. Changes also happen to the social and economic characteristics of the river valley.

Dams and reservoirs impact strikingly on fish. River fisheries and commercial aquaculture are frequently of great importance to people living near big rivers. Ecological interference from dams not only degrades the aquatic environment but it may also create an economic scarcity. Threats to aquatic fisheries occur in many ways: loss of suitable habitat for breeding, spawning, and feeding; fragmentation of the rivers' ecosystem by dam construction; and commercialisation of the catch (Middleton 2017). Impoundments on big rivers and large-scale water utilisation may pollute rivers, redistribute sediment and water discharge, and impact on the aquatic fauna. Such degradation Is often enhanced by the fragmentation of rivers by dams and reservoirs. More examples can be found in Chapters 5 and 11.

Impoundments, unlike basin land use changes, impact directly on major rivers which need to adjust to a high loss of sediment and also to a controlled release of water. Dams and reservoirs thereby alter the form and function of rivers, and such impacts work across the entire range of river sizes. Eight-three per cent of total sediment load is trapped behind the Hoa Bin Dam on the Da River, a large tributary to the Sõng Hóng in Vietnam (Milliman and Farnsworth 2011). Reduction in sediment can be as striking as at Aswan on the Nile. A number of other changes may impact rivers directly. Even major rivers can be changed by the interbasin transfer of water from a wet to a dry basin. Large-scale irrigation also transforms the downstream character of a river. This has been attributed to increased consumption within drainage basins. In arid rivers such as the Indus, Colorado, and Nile, discharge reduction due to irrigation may approach 100% (Milliman and Farnsworth, 2011). The effect of impoundments, however, differs between rivers in alluvium and rivers on rock.

8.6 Effect of Impoundments on Alluvial Rivers

Williams and Wolman (1984) reviewed channel and floodplain changes downstream of 21 dams built on alluvial rivers in the United States. There are other studies but their generalisation is based on very good records on discharge, sediment, morphology of the channel, morphology of the floodplain, and valley flat vegetation for a number of years from the pre-dam to the post-dam conditions. Although the data are from the United States, the listed changes should occur globally to impounded alluvial rivers and the effects appear to be scale-independent. The impact of dams and reservoirs on alluvial rivers, determined by Williams and Wolman, can be summarised as:

1. Commonly, the post-dam flood peaks are reduced, but the impact on other discharge characteristics vary from river to river.
2. A marked decrease is seen in sediment concentration and suspended load for hundreds of kilometres below dams (Figure 8.5).
3. Such changes result in bed erosion and a lowering of the longitudinal profile (Figure 8.6), although the level of degradation varies. Most of the erosion happens

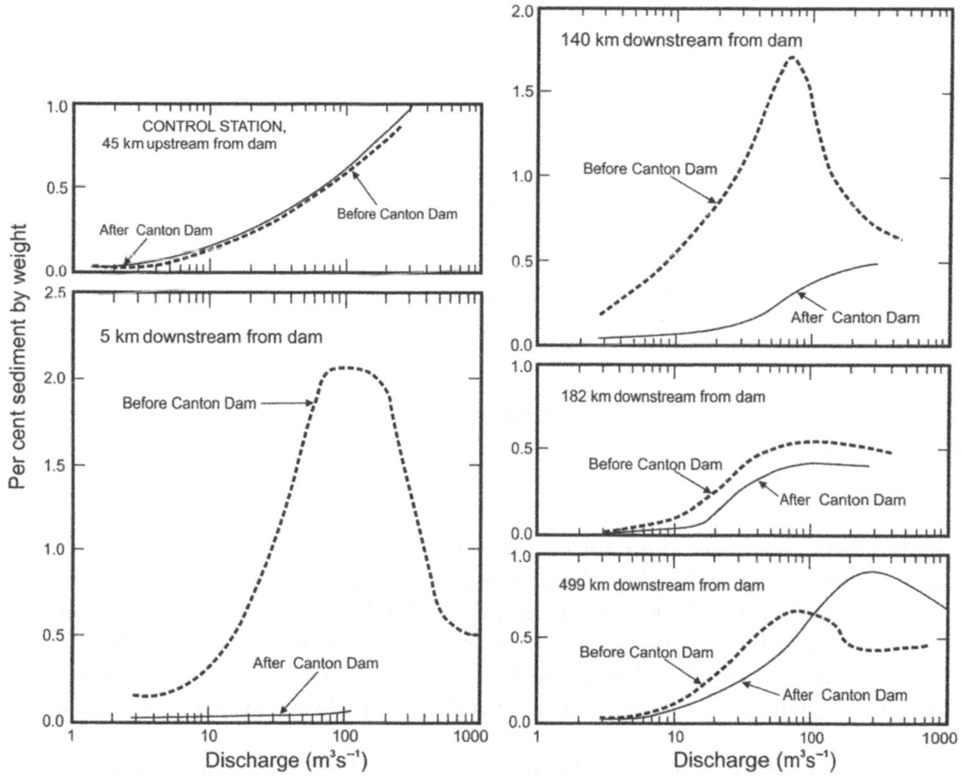

Figure 8.5 Sediment change downstream of dams. Source: Williams and Wolman 1984. Courtesy of USGS.

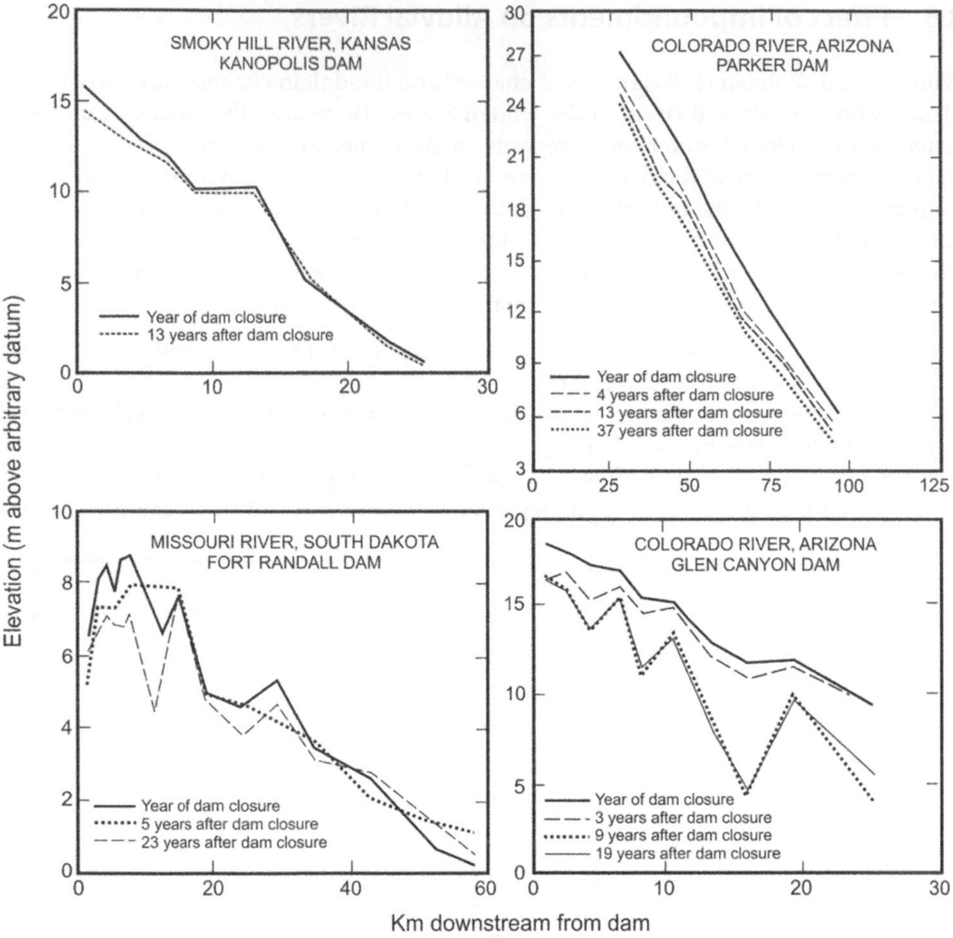

Figure 8.6 Degradation and changes in long profile downstream from dams. Source: Williams and Wolman 1984. Courtesy of USGS.

within two decades of the dam closure. Incidentally, Schumm (2007) stated that the presence of gravel below the channel alluvium reduced bed erosion for the Nile, Missouri, and Mississippi. According to Winkley (1994) gravel stabilised the perimeter and bars of the channel, restricting sediment movement. The gravel may originate from the channel bed of certain rivers or arrive later from the tributaries, but an extensive bed of gravels is not always available for reducing erosion.

4. A coarsening of the bed material may happen after dam closure but this trend may change over time.
5. Width of the channel below the dam can increase, decrease, or remain similar to the pre-dam conditions.
6. The channel pattern below the dam may change from a single-island braided type to a number of smaller multiple channels separated by small islands. The river may

replace its sediment load by eroding the banks and, as a consequence, change the character of the river.

7. Riparian vegetation may increase below the dam, probably because of the decreased peak flow, which may lead to channel narrowing and floodplain formation.

The river downstream of the dam is affected by a more uniform and reduced discharge and reduced sediment. As a result, the river may deepen the channel and erode its bed. The general absence of big floods and proportional rise of finer sediment would allow vegetation to grow, binding the finer material into bigger bars and floodplain. These changes may also result in altered channel patterns below the dam. Williams and Wolman (1984) showed that alluvial rivers they studied may return to the old channel character after a distance of about 500 km. This may happen even over a shorter distance, depending on the size of the river and the level of alteration due to impoundment.

Grant et al. (2003) suggested that the downstream effect of a dam on a river is determined by two dimensionless variables. These are (i) the ratio of sediment supply below the dam to that above the impoundment (S^*) and (ii) the change in frequency of sediment-transporting flow expressed as a fraction (T^*). A high S^* and a low T^* lead to textural shifts at confluences, construction of bars and islands in the channel, general deposition of sediment, and encroachment of vegetation in the channel. A low S^* and a high T^* may lead to scouring erosion of bars and islands, armouring of the channel, and evolution of a degraded narrow channel. A variety of changes are possible in between these two extreme cases, making correct forecasting and management of an impounded river difficult.

8.7 Effect of Impoundments on Rivers in Rock

Rocky gorges are frequently selected as dam sites, but few detailed studies are available on the effect of dams on rivers in rock. The case of the Colorado River below the Glen Canyon Dam, however, is an excellent example (Webb et al. 1999). Glen Canyon Dam is at the head of the Grand Canyon with a large reservoir, Lake Powell, behind it. Before the dam was constructed, the Colorado River used to move a large amount of sediment, including a high proportion of bed load, in flood through the canyon (Figure 8.7). Discharge became uniform after the dam, large floods did not enter the Grand Canyon, and annual variability was significantly reduced. Most of the high sediment load is now deposited in Lake Powell, and very little is carried by the river below the dam. The texture of sediment in the canyon has also become finer. Such changes have reduced the competence of the river below the Glen Canyon Dam, resulting in a series of morphological changes in the channel. These include extension of debris fans at the mouth of the tributaries in the canyon and modification of the channel morphology in places, especially at rapids. Bed scour has occurred and the river bed is armoured with coarse clasts. The river is eroding former sand bars as the earlier sediment-carrying, bar-building floods no longer come down the canyon. All this resulted in degradation of ecological habitats, changes in flora and fish species in the channel, and erosion of tourist campsites on bars and fans in the channel. Dams obviously degrade rivers in both rock and alluvium.

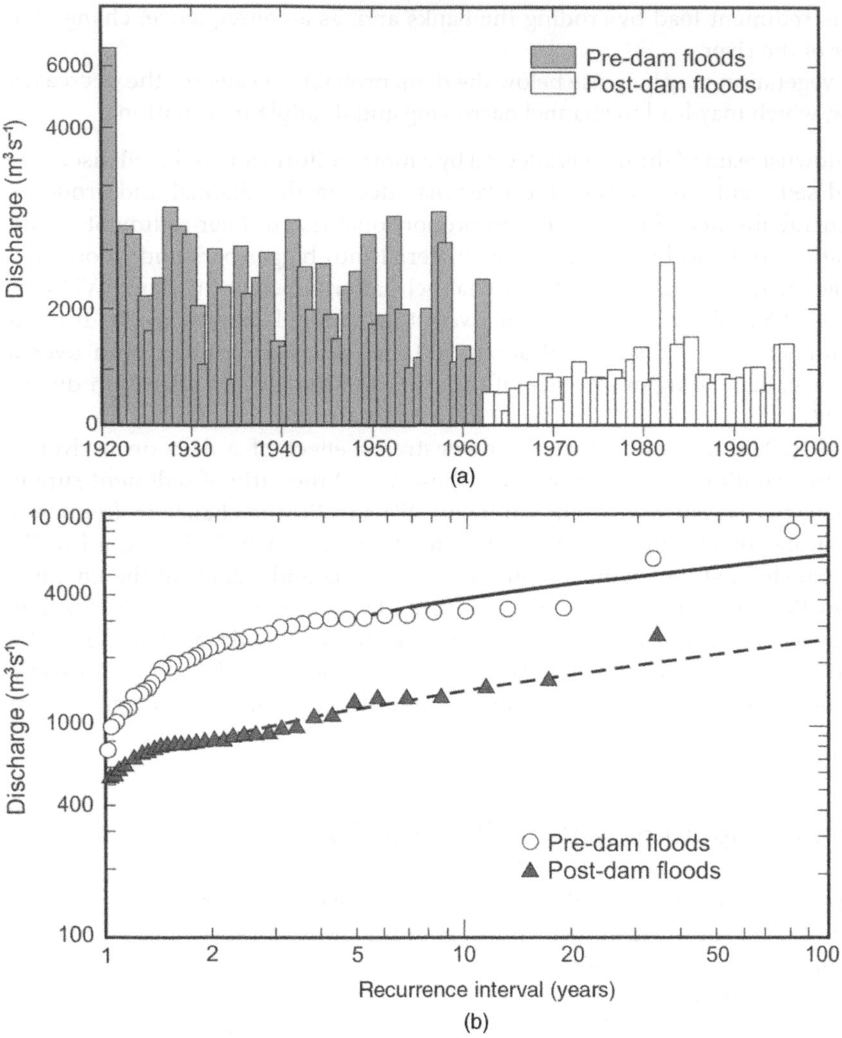

Figure 8.7 Effect of the Glen Canyon Dam on the Colorado River: (a) reduced variability in annual discharge; and (b) reduction in size of large floods entering Grand Canyon. Source: Webb et al. 1999. Courtesy of AGU.

Box 8.2 The Three Gorges Dam on the Changjiang

The Three Gorges Dam on the Changjiang is enormous in design and controversial in evaluation. The 6300 km long Changjiang, the longest river in Asia, drains 1.8 million km^2 from the Qinghai-Tibet Plateau to the East China Sea. The metropolitan region of Shanghai is located near its mouth. Four hundred million people and more than 50 000 dams are found in the Changjiang Basin, making it a highly altered river system (Xu et al. 2007 and references therein). It carries 480 million tonnes of sediment annually, only 5% of which is bed load (Chen et al. 2001).

The upper river flows through a complex variety of high relief landforms, ending in a set of three gorges in Sichuan, about 660 km in length. The dam is located at Sandouping in a wider section of the third gorge, 40 km upstream of the city of Yichang, and about 4500 km from the source of the river. Several large tributaries, namely the Min, Wu, and Jialing, known for contributing floodwater and sediment, join the Changjiang upstream of the gorges. There is an older dam on the Changjiang near the end of the gorges, the Gezhou. Below the gorges the valley flattens, the slope of the river drops, a number of lakes appear, and the river meanders amongst fluvial plains and terraces. According to Xu et al. (2007), sediment in the river used to increase directly with the river length, reaching the maximum at Yichang (which has a sediment station), and then decrease along the middle and lower courses of the river where part of it was lost in floodplain channels and lakes (Figure 8.8).

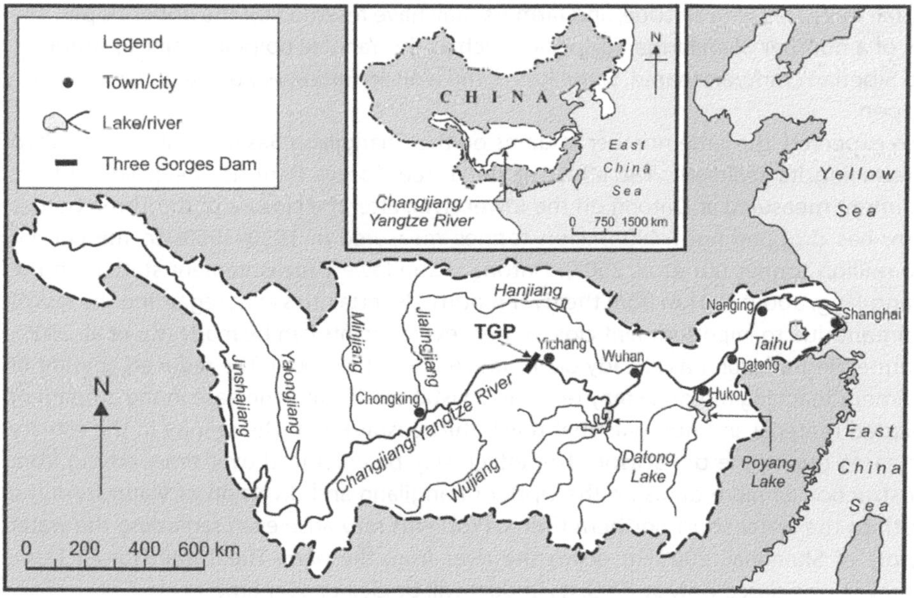

Figure 8.8 The course of the Changjiang and the location of the Three Gorges Dam.

The history of Imperial China includes a series of monumental changes of its river systems. The Three Gorges Dam probably is a continuation of this practice although it was constructed in modern times under a different style of leadership. The project has been on the drawing board since the 1930s but work on the dam did not start until 1993, due to controversies and strategic reasons. It is a concrete gravity dam, 185 m high and 2.3 km long. The project is expected to produce at least 13 x 10^9 W of hydroelectricity, control floods, and improve navigation from the river mouth almost to the city of Chongqing located upstream of the gorges. Irrigation is not a major objective. Hydroelectricity production started in 2003, taking several more years to reach full

(Continued)

Box 8.2 (Continued)

production. The total planned production is 18 200 MW from 26 turbines. The project has the potential to provide China with a large amount of clean energy, replacing burning of huge quantities of fossil fuel and biomass. Several environmental problems, however, may arise.

A very large population was moved from the lower slopes of the gorges to other areas before the closing of the reservoir. Estimates of the total number displaced vary but a figure well in excess of a million is commonly discussed. The level of water in the gorges has inundated a number of urban riverside settlements and industries that supported more than half the displaced population.

The long and narrow reservoir along the gorges stretches for more than 650 km. It may function as a potential sediment trap, especially in the early years after the closure of the dam. The slopes are steep, soils are shallow, a large amount of forest has been destroyed, rainfall increases with altitude, and farmers may have resettled on the upper slopes. The loss of a number of endangered species such as the Yangtze dolphin, Chinese sturgeon, and Siberian crane, are feared. Some loss of the wonderful scenery of the gorges may also happen.

As expected, the large number of dams on the Changjiang Basin has already resulted in reducing its sediment. The effect of the Three Gorges Dam has been added to it. Sediment measured at Datong on the lower river, after the closure of the Three Gorges Dam, has dropped from 480 million tonnes measured in 1950–1960s to the current 140 million tonnes (Xu et al. 2007). Datong is the lowest measurement station on the Changjiang, about 600 km from the river mouth. Uncertainties exist regarding the loss of sediment due to impoundment closure but several points can be made (Xu et al. 2006). Channel degradation has already been noticed (Xu et al. 2007). The reduced amount of sediment reaching the delta may result in coastal erosion and increase in the amount of polluted material and nutrients that reach the coastal water. The salinity in the estuary increases during the dry season. The effect may be enhanced over time, arising from construction of more dams on the Upper Changjiang and diversion of water from the south to the water-short northern China. Problems may arise even regarding the water supply of Shanghai, 2000 km down the river from the dam. The Three Gorges Dam, however, was closed only in 2003. Its impact will be clearer over time.

8.8 Large-scale Transfer of River Water

The physical transfer of water from one river draining a humid basin to another one flowing through an arid zone at some distance apart, usually is an impressive but controversial project. The concept is to bring excess water form a humid zone to an area where water is scarce and supply unreliable. This is a large-scale engineering project which could be both beneficial and environmentally problematical. The quality and quantity of water and sediment discharges, morphology, and ecology of both rivers need to be carefully considered. For example, the Barwon-Darling Unregulated River Water Transfer Project is designed to carry water across arid Australia. Large-scale transfer of river water is also being planned in India but probably the most impressive plans come from China.

In China, demand for water is high in the drier north and a three-way transfer of water from the humid south has been planned as the South–North Water Transfer Project. About 44 x 10^9 m^3 of water is planned to be transferred northwards from the Changjiang and its tributaries. A three-way transfer is planned through the following canal systems:

1. The Eastern Route following the historical Grand Canal.
2. The Central Route from the upper Han River to Beijing and nearby areas.
3. The Western Route from the upper tributaries of the Changjiang to dry areas of the northwest.

This water transfer project obviously would require expensive large-scale advanced engineering such as tunnelling under pre-existing rivers such as the Huanghe, construction of powerful pumping stations and aquaducts for transferring water against general slopes of the region, raising the crest- elevation of existing dams, construction of new large dams and long tunnels, and maintaining the quality of the huge amount of water planned to be delivered. A very large number of people have to resettled.

This is a huge project but all the river basins are located within one country. Water transfer from rivers draining multiple countries may add political problems to these as discussed in Chapter 9. Construction of big dams and large-scale river water transfer are the two projects likely to affect large rivers in the near future.

8.9 Conclusion

The effect of anthropogenic alterations on large rivers vary. Certain large rivers, such as the Congo or Amazon, may still exist in a near natural state. In contrast, some others may have been converted into a staircase of dams and reservoirs as has been done to the Columbia and Volga. A river may lose most of its water to widespread irrigation, as happened to the Amu Darya and Syr Darya leading to the near loss of the inland Aral Sea where these rivers end. A number of impoundments may exist across the river but a very large one may dominate the passage of water and sediment downstream, as for the Aswan High Dam on the Nile.

The form and function of a river may be drastically modified by anthropogenic activities, and the type and scale of modification will depend on the nature of changes brought to the basin and the channel. Many rivers are impacted by the combined effects of marked changes in land use and land cover in the basin and impoundments in the channel. The effects are usually contradictory, and an excellent example is provided by the Missouri River which used to contribute most of the sediment of the Mississippi before a number of large dams were built across it (Meade and Parker 1985). It is obvious that the scale of anthropogenic changes needs to be rather large in order to impact a major river.

To illustrate, the Mississippi River experienced a 30% increase in water discharge due to climatic varations (Hurrell 1997) and a huge and sharp decline in sediment after the construction of large dams on the Missouri. The annual sediment discharge dropped from approximately 700 x 10^6 tonnes in 1950 to about 200 tonnes in the mid-1990s. The state of the Changjiang can be compared over this period. The water discharge in the Changjiang showed a small-scale variation but its sediment discharge has decreased gradually since 1986. Over 50 000 dams were built in the Changjiang Basin between 1950 and 2005. Only about 17% of the sediment of the 1950–1960s was being transferred by the river in 2003–2005 past Yichang below the huge Three Gorges Dam. The decrease in sediment transfer at Datong on the lower river was 38% for the same period. The scale of impact varies along the main river and amongst the main stem and the tributaries (Xu et al. 2007).

Occasionally, people-related catastrophic events contribute huge volumes of water and sediment to a river. A number of hydroelectric dams have been planned and are being constructed in Lao PDR for exporting electricity to other countries such as Thailand with need for power. In July 2018, the Xepian-Xe Nam Noy Dam in the south of the country collapsed while under construction with an estimated 5 billion m^3 of water spreading through the countryside (*The Guardian Weekly* 2018), not only destroying people and villages but also sending a huge amount of water and sediment downslope which ultimately would reach the Mekong Valley.

The effect of dams and reservoirs are usually perceptible in the immediate downstream vicinity of the impoundment. However, dams may also influence river morphology, behaviour, and ecology over a long distance. China is constructing seven dams, collectively known as the Lancang Cascade, on the upper Mekong River. It has been suggested that the effect of these dams would be significant even thousands of kilometres downstream on the lower river, the Tonlé Sap Lake, and the Mekong Delta (Gupta et al. 2006; Kummu and Varis 2007; Kummu et al. 2010). Currently, the management of many large rivers has to take into consideration not only the river's natural geomorphology but also the anthropogenic modification. The state of the river may be in flux, further complicating the situation. Environmental Impact Assessments (EIAs: Chapter 9) are expected to be examined prior to the construction of large projects in order to clarify their problems. Related environmental degradation, however, seems to continue. The common effect of anthropogenic alterations on rivers is fairly well known but changes in the future may bring more complications. New problems may rise from (i) dam withdrawal, i.e. removal of an existing dam from the river bed, and (ii) climate changes caused by anthropogenic global warming. The state of a large river is conditioned by both the physical environment of its basin and anthropogenic alterations imposed on the channel (Milliman et al. 2008).

Questions

1 A number of case studies indicate that basin land use changes and channel impoundments modify small rivers. Are large rivers affected?

2 List large rivers (e.g. the Nile, Mississippi) that have been significantly modified by anthropogenic alterations. Name a few which have not.

3 A number of politicians opined that water is wasted if it reaches the sea. Do you agree?

4 A large volume of sediment is stored behind dams and does not reach the channels downstream. What are the effects of such storage?

5 Explain the impact of big dams on the morphology, behaviour, and ecology of alluvial rivers.

6 Do large rivers of the world appear natural?

7 The course of rivers such as the Irrawaddy or Narmada includes several discrete rock gorges separated by alluvial basins. What could be the effect of building a large dam on the first (most upstream) gorge?

8 Given the sequence of Pleistocene glaciations, Holocene anthropogenic alterations, and progressive global warming, could we expect any river to be in a stable state?

9 The Mekong River may still be considered as nearly in a natural state. Seven dams (the Lancang Cascade) have been built over the rocky gorges of the upper Mekong in China. What changes do you expect downstream? How far do you expect such changes to be found? You may need to read a few descriptions of the Mekong (Gupta and Liew 2007; Gupta et al. 2006; Kummu and Varis 2007).

References

Alford, D. (1992). Streamflow and sediment transport from mountainous watersheds of the Chao Phraya basin, northern Thailand: a reconnaissance study. *Mountain Research and Development* 12: 257–268.

Chen, Z., Li, J., Shen, H., and Zhanghua, W. (2001). Yangtze River of China: historical analysis of discharge variability and sediment flux. *Geomorphology* 41: 77–91.

Douglas, I., Spencer, T., Greer, T. et al. (1992). The impact of selective commercial logging on stream hydrology, chemistry and sediment loads in the Ulu Segama rain forest, Sabah, Malaysia. *Philosophical Transaction, Royal Society of London, B.* 335: 397–406.

Dunne, T. (1979). Sediment yield and land use in tropical catchments. *Journal of Hydrology* 42: 281–300.

Eckholm, E.P. (1976). *Losing Ground: Environmental Stress and World Food Prospects*. New York: W.W. Norton.

Grant, G.E., Schmidt, J.C., and Lewis, S.L. (2003). A geological framework for interpreting downstream effects of dams on river. In: *A Peculiar River: Geology, Geomorphology, and Hydrology of the Deschutes River, Oregon* (eds. J.E. O'Connor and G.E. Grant), 209–225. Washington, DC: American Geophysical Union.

Guillén, J. and Palanques, A. (1997). A historical perspective of the morphological evolution in the lower Ebro River. *Environmental Geology* 30: 174–180.

Gupta, A. and Chen, P. (2002). Sediment movement on steep slopes to the Mekong River: an application of remote sensing. In: *The Structure, Function and Management Implications of Fluvial Sedimentary Systems* (eds. F.J. Dyer, M.C. Thoms and J.M. Olley), 399–406. Wallingford: International Association of Hydrological Sciences.

Gupta, A. and Krishnan, P. (1994). Spatial distribution of sediment discharge to the coastal waters of South and Southeast Asia. In: *Variability in Stream Erosion and Sediment Transport* (eds. L.J. Olive, R.J. Loughran and J.A. Kesby), 457–463. Wallingford: International Association of Hydological Sciences.

Gupta, A. and Liew, S.C. (2007). The Mekong from satellite imagery: a quick look at a large river. *Geomorphology* 85: 259–274.

Gupta, A., Liew, S.C., and Heng, A.W.C. (2006). Sediment storage and transfer in the Mekong: generalisations on a large river. In: *Sediment Dynamics and Hydromorphology of Fluvial Systems* (eds. J.S. Rowan, R.W. Duck and A. Werritty), 450–459. Wallingford: International Association of Hydrological Sciences.

Hooke, R.L.B. (2000). Toward a uniform theory of clastic sediment yield in fluvial systems. *Geological Society of America Bulletin* 112: 1778.

Hurni, H. (1999). Sustainable management of natural resources in African and Asian Mountains. *Ambio* 28: 382–389.

Hurrell, J.W. (1997). Decade trends in the North Atlantic Oscillation: regional temperatures and precipitation. *Science* 269: 676–679.

Hurst, H.E. (1952). *The Nile*. London: Constable.

Knox, J.C. (2002). Agriculture, erosion and sediment yields. In: *The Physical Geography of North America* (ed. A.R. Orme), 482–500. Oxford: Oxford University Press.

Knox, J.C. (2007). The Mississippi River system. In: *Large Rivers: Geomorphology and Management* (ed. A. Gupta), 145–182. Chichester: Wiley.

Kondolf, G.M. (1997). Hungry water: effects of dams and gravel mixing on river channels. *Environmental Management* 21: 533–551.

Kummu, M. and Varis, O. (2007). Sediment-related impacts due to upstream reservoir trapping, the Lower Mekong River. *Geomorphology* 85: 275–293.

Kummu, M., Lu, X.X., and Wang, J.J. (2010). Basin-wide sediment trapping efficiency of emerging reservoirs along the Mekong. *Geomorphology* 119: 181–197.

Lai, F.S., Ahmad, J.S., and Zaki, A.M. (1996). Sediment yields from selected catchments in peninsular Malaysia. In: *Erosion and Sediment Yield: Global and Regional Perspectives* (eds. D.F. Walling and B.W. Webb), 223–231. Wallingford: International Association of Hydrological Sciences.

Latrubesse, E.M. and Stevaux, J.C. (2002). Geomorphology and environmental aspects of the Araguaia fluvial basin, Brazil. *Zeitschrift für Geomorphologie* 129: 109–127.

Li, Z. (1992). *The Effects of Forests in Controlling Gully Erosion*, vol. 209, 429–437. Wallingford: International Association of Hydrological Sciences.

Magilligan, F.J. and Nislow, K.H. (2005). Changes in hydrologic regime by dams. *Geomorphology* 71: 61–78.

Mahmood, K. (1987) Reservoir Sedimentation: Impact, Extent and Mitigation. Technical paper 71. World Bank, Washington, DC.

McCully, P. (1996). *Silenced Rivers: The Ecology and Politics of Large Dams*. London: Zed Books.

Meade, R.H. and Moody, J.A. (2010). Causes for the decline of suspended-sediment discharge in the Mississippi River system, 1940-2007. *Hydrological Processes* 24: 35–49.

Meade, R.H. and Parker, R.S. (1985) Sediment in the rivers of the United States. US Geological Survey Water-Supply Paper 2275, 49-60.

Middleton, C. (2017). Water, rivers and dams. In: *Routledge Handbook of the Environment in Southeat Asia* (ed. P. Hirsch), 204–223. London and New York: Routledge.

Milliman, J.D. and Farnsworth, K.L. (2011). *River Discharge to the Coastal Oceans: A Global Synthesis*. Cambridge: Cambridge University Press.

Milliman, J.D. and Syvitski, J.P.M. (1992). Geomorphic/tectonic control of sediment discharge to the ocean: the importance of small mountainous rivers. *Journal of Geology* 100: 525–544.

Milliman, J.D., Farnsworth, K., Jones, P.D. et al. (2008). Climatic and anthropogenic factors affecting river discharge to the global ocean, 1951-2000. *Global Planetary Change* 62: 187–194.

Nilsson, C., Reidy, C.A., Dynesius, M., and Revenga, C. (2005). Fragmentation and flow regulation of the world's large river systems. *Science* 305: 405–408.

Ramankutty, N. and Foley, J.A. (1999). Estmating historical changes in global land cover: croplands from 1700 to 1992. *Global Biochemical Cycles* 13: 997–1027.

Rapp, A., Murray-Rust, D.H., Christiansson, C., and Berry, L. (1972). Soil erosion and sedimentation in four catchments near Dodoma, Tanzania. *Geografiska Annaler* 54A: 255–318.

Schumm, S.A. (2007). Rivers and humans – unintended consequences. In: *Large Rivers: Geomorphology and Management* (ed. A. Gupta), 517–533. Chichester: Wiley.

Shahin, M. (1985). *Hydrology of the Nile Basin*. Amsterdam: Elsevier.

Stanley, D.J. (1996). Nile Delta; extreme case of sediment entrapment on a coastal plain and consequent coastal land loss. *Marine Geology* 129: 189–195.

Stanley, D.J. and Wingerath, J.G. (1996). Nile sediment dispersal altered by the Aswan High Dam: the kaolinite trace. *Marine Geology* 133: 1–9.

Syvitski, J.P.M., Vörösmarty, C.J., Kettner, A.J., and Green, P. (2003). Impact of humans on the flux of terrestrial sediment to the global coastal ocean. *Science* 308: 376–380.

The Guardian Weekly (2018) Laos, Dam collapse flood region. (27 July), p. 3.

Vörösmarty, C.J., Maybeck, M., and Fekete, B. (2003). Anthropogenic sediment retention: major global impact from registered river impoundments. *Global and Planetary Change* 39: 169–190.

Walling, D.E. and Fang, D. (2003). Recent trends in the suspended loads of the world's rivers. *Global and Planetary Change* 39: 111–126.

Webb, R.H., Wegner, D.L., Andrews, E.D. et al. (1999). Downstream effects of Glen Canyon Dam on the Colorado River in Grand Canyon: a review. In: *The Controlled Flood in Grand Canyon* (eds. R.H. Webb, J.C. Schmidt, G.R. Marzolf and R.A. Valdez), 1–21. Washington, DC: American Geophysical Union.

White, S. (1995). Soil erosion and sediment yield in the Philippines. In: *Sediment and Water Quality in River Catchments* (eds. I.D.L. Foster, A.M. Gurnell and B.W. Webb), 391–406. Chichester: Wiley.

Wilk, J., Anderssson, L., and Plermkamon, V. (2001). Hydrological impacts of forest conversion to agriculture in a large river basin in northeast Thailand. *Hydrological Processes* 15: 2729–2748.

Williams, G.P. and Wolman, M.G. (1984) Downstream effect of dams on alluvial rivers. US Geological Survey Professional Paper 1286, Washington, DC.

Winkley, B.R. (1994). Response of the lower Mississippi River to flood control and navigation improvements. In: *The Variability of Large Alluvial Rivers* (eds. S.A. Schumm and B.R. Winkley), 45–74. New York: American Society of Civil Engineers.

Wohl, E.E. (2002). *Virtual Rivers: Lessons from the Mountain Rivers of the Colorado Front Ranges*. New Haven: Yale University Press.

Woodward, J.C., Macklin, M.G., Krom, M.D., and Williams, M.A.J. (2007). The Nile: evolution, Quaternary river environments and material fluxes. In: *Large Rivers: Geomorphology and Management* (ed. A. Gupta), 261–292. Chichester: Wiley.

World Commission of Dams (2000) Dams and Development: A New Framework for Decision Making. Earthscan, London.

World Resources Institute (1988) World Resources 1988-89. Basic Books, New York.
Xu, K., Milliman, J.D., Yang, Z., and Wang, H.J. (2006). Yangtze sediment decline partly from Three Gorges Dam. *Eos Transactions American Geophysical Union* 87: 185–190.
Xu, K., Milliman, J.D., Yang, Z., and Xu, H. (2007). Climatic and anthropogenic impacts on water and sediment discharges from the Changjiang River (Changjiang), 1950-2005. In: *Large Rivers: Geomorphology and Management* (ed. A. Gupta), 609–626. Chichester: Wiley.

9

Management of Large Rivers

9.1 Introduction

Rivers seldom remain in a natural state. Their water is utilised for irrigation, power generation, navigation, and other purposes. Engineering structures are built for ameliorating floods in valley flats where settlements and farmlands are located. Such physical structures tend to degrade the geomorphic and ecological characteristics of river channels and floodplains. In an ideal scenario, while utilising a river as a resource, its appearance and behaviour should remain as natural as possible. This is difficult, particularly as the balance between utilisation and naturalness should be maintained not only for the present but also for the future, a practice generally referred to as sustainable development (Box 9.1). Furthermore, the resources of the river should be reasonably divided among the inhabitants of its basin. This often becomes problematical, especially for large rivers which cross several sovereign countries, each with different expectations, capacity, and political power for resource utilisation and river maintenance. The Nile or the Mekong is an excellent example. Even when the river flows through a single country, e.g. the Changjiang in China or the Murray-Darling in Australia, expectations vary in different parts of the basin.

Box 9.1 Sustainable Development

The term 'sustainable development' was coined to address the conflict between economic growth and environmental protection. The term was popularised by the report of the Brundtland Commission which was published before the world environment conference of 1992 in Rio de Janeiro (WCED 1988). The term received almost universal endorsement, and was used to guide economic growth without ignoring the environmental responsibility. According to the Brundtland Commission 'humanity has the ability to make development sustainable – to ensure that it meets the needs of the present without compromising the ability of future generations to meet their own needs' (WCED 1988, p. 8). Taking this as a general guideline on development of river landscapes, we should avoid long-term environmental degradation while utilising the resources of the river at present. The river should maintain its form and function reasonably well, in spite of resource utilisation. This may be possible, although it would be difficult to reach the more positive definition of sustainability given by Serageldin who said we should leave 'future generations as many opportunities as, if not more than, we have had ourselves' (Serageldin 1996). Sustainable development should be the general objective in development, including river utilisation.

Introducing Large Rivers, First Edition. Avijit Gupta.

Managing a river is difficult even when good intention, strong legislation, and considerable resource are all present. Campbell (2007) identified the management of major rivers as meeting two major kinds of challenges: technical and political. Both could be difficult, more so in combination. The management of large rivers is discussed in this chapter starting with technical challenges as biophysical management. Rivers are perceived differently by different people depending on their expectation. For example, rivers can be seen as storehouses of resource (water, fish, power) that should be utilised, and as an important part of the physical environment, it should be sustained. The holistic nature of rivers is ignored at times, perceiving them more as a single-dimension phenomenon matching the specialist training of a river manager as an earth scientist, ecologist, economist, or engineer. This creates problems in term of both resource utilisation and river management, as the job is often seen only as an engineering task, and structural control the expected way to proceed. Instead, rivers should be recognised as complex objects.

Development, especially when requiring construction of engineering structures or resource utilisation, should be designed very carefully. The planning stage of a project should not only consider the efficiency of the project but also the possible degradation of the environment. Environmental Impact Assessment (EIA) is a technique often used for this purpose. The basic structure of an EIA is described in Box 9.2. It can be used for river-related projects with appropriate modifications. EIAs are good for planning multifaceted changes and for complex issues such as large rivers.

Box 9.2 Environmental Impact Assessment

We may accept that anthropogenic projects carried out by the state or community for improving existing economic conditions, may lead to degradation of rivers. It is therefore necessary at an early planning stage to identify probable serious degradation from the proposed project, and also the next step, possible project modifications for mitigating an environmental disaster. EIA is one of the techniques used for this purpose. The basic objective and methodology of each EIA appraisal is similar, all studies are flexible, and the original EIA is subject to improvisations and innovations with time. An EIA deals with both the ambient environment and the development project. It should be noted that countries may vary significantly in public participation and openness in preparation of EIAs.

Primarily, an EIA is a study of the effect of a proposed development project on the environment before execution. It is a predictive mechanism that considers both environmental and economic benefits and costs by comparing the reasonable alternatives to the proposed project and evaluating them also for both environmental and economic benefits and costs. Thus, an EIA is an objective evaluation of the environmental problems that may arise with the implementation of a project such as constructing a big dam or large-scale withdrawal of river water. It should provide the public with this information, clearly and completely.

An EIA usually follows a general structure divided into several sections, which may be called steps. This general arrangement is listed below, following Ahmad and Sammy (1985):

1. Preliminary activities and terms of reference (TOR)
   Preliminary activities include defining the project (TOR) and listing the personnel involved, such as for the construction of a dam at a particular location and generation of a targeted amount of power for a given area. Ideally, members of an EIA team should have relevant but different expertise.
2. Scoping (impact identification)
   Scoping is identification of impacts that the project would have on the environment. It is determined at an early stage of the EIA. Either a list of exhaustive changes or a shorter list of significant changes is drawn up following various techniques which are described in standard books on EIA. Scoping has become easier with the advent of the geographical information system and computer technology. In the early days, various matrices constructed by Leopold and his associates were used. These techniques were determined in the late 1960s for evaluating landscape aesthetics (Leopold 1969; Leopold and Marchand 1968) and were probably the most used matrix linking a specific action of the development project with its impact (Leopold et al. 1971). They have also been used in modified forms.
3. Baseline study
   This is the understanding of the environment before project implementation, serving as a datum against which changes from the project could be measured. This type of study requires technical expertise, long-term data (e.g. river discharge), and existing knowledge of the river basin. Such knowledge may not be available in certain countries and external consultants have been employed for river basin studies.
4. Environmental impact evaluation
   This grows out of scoping and baseline study. Primarily it is the quantification of the different aspects of the project on the environment. This is the most difficult, technical, and controversial part of the EIA.
5. Mitigation measures
   Such measures are considered as the post-impact evaluation step. These are proposed measures for reducing the degradational impact of the project.
6. Assessment of alternate measures
   At this stage the proposed project and its alternate versions can be compared, taking into account the possible mitigation measures. The chosen final project could be selected subjectively or technically using the cost–benefit analysis.
7. Preparation of the final document
   Usually, two versions of the EIA are prepared. One is the complete technical draft (reference document) which is used by the technical and management personnel associated with the project. The other is a condensed non-technical version (working document) for communicating with the decision-maker. The writing of the working document is important as often a decision is taken on the project based on the working document by non-technical decision-makers.
8. Decision-making
   This happens after the EIA report has been prepared, probably including discussions on several alternative approaches to the project concerned. A decision-maker, either a person or a committee, has three choices:

(Continued)

Box 9.2 (Continued)

- One of the project alternatives could be accepted.
- The EIA could be returned for further study in specific areas.
- The proposed project could be totally rejected.

9. Monitoring of project implementation and impacts
These activities are carried out while the project is on stream. This is basically to ensure that the project is being carried out following the guidelines and recommendations in the EIA. An inspection could take place following a time lapse after the completion of the project, to determine the accuracy of the EIA predictions.

These are the standard guidelines for treating the EIA as an evolving concept. Like any project, these can be used for river-related work such as construction of dams for power generation or levees for flood control. An EIA may recognise in the project the problem of the upstream-moving native fish in the river, channel incision, or water pollution. As a result, an alternative approach to the original design may be necessary. EIAs should be seen as techniques for consensus-building and integrating environmental with social or economic planning and management. These are open-ended techniques for identifying, exploring, and resolving issues. EIAs require technical and regional knowledge which could be a demanding task for countries with limited human, economic, and technical capacities (Gupta and Asher 1998).

Various techniques have been suggested for maintaining a river without degradation for the present and the future. These are complicated techniques, especially difficult for large rivers where the lack of international cooperation, commitments, and weak political organisations may hinder river management, even if the needed tools and techniques are available. River management requires solving the conflict between aesthetics and water development and needs good science, correct intention, well-defined legislation, and ample resources. Multifactorial knowledge about the river is essential.

The objectives, concept and techniques of river management have changed focus over time. In the early days river management only implied engineering practices that would (i) control the movement of flow (including floods) of a river or (ii) transform the river towards the better utilisation of a specific goal – improved navigation, fishing, irrigation, etc. Snag removal and levee building along the Mississippi in the mid-nineteenth century are examples of this kind of approach. Management used to mean utilisation in easier terms. With passage of time and more intense utilisation of river water, rivers became polluted and less suitable for life, described by Rachel Carson as silent rivers in the middle of the last century (Carson 1965). The term 'river restoration' came in use, implying a return to a condition close to natural. A strong ecological or aesthetical bias existed in early river restoration attempts; river restoration used to be described as habitat creation.

A river, however, can never be completely restored. At best, a river can be partially restored, when its degraded condition is improved to achieve a set of specific objectives such as channel width–depth ratio, water quality measures, or species of fish found in its water. Even when a river appears natural, it may have a history of form change and unusual functions. These are the rivers Wohl (2001) has described as virtual rivers. It is difficult to identify the original naturalness of a river, and thus there may not be an identifiable target for restoration. Furthermore, a restored natural river cannot survive within an altered drainage basin and world environment. This is an impossible concept. In all likelihood, a so-called restored river would change back to a form and function reflecting the altered basin environment. It is rational to use a broader term such as river management instead, suggesting environmental enhancement. The wide variety of natural rivers also demands a range of approaches.

9.2 Biophysical Management

River management can be recognised as an attempt to reach an ideal and stable state that may represent a natural environment. The usual indices to measure the nature of the river refer to aquatic ecology and water quality, rarely the channel form. Examples of such tools are instream flow, channel-maintenance flow, or sediment volume control. One determines a theoretical index reflecting a management objective and attempts to transform the river so that it may reach such an index value. The general term for such discharges is 'environmental flows', several varieties of which are in practice.

In an environmental flow the quantity, quality, and time distribution of a river's discharge are calculated as required indices that should be maintained, so that a healthy riverine ecosystem is preserved even when water is extracted for anthropogenic use. A common index, such as the Biological Oxygen Demand (BOD), is often used. Instream flow, a type of environmental flow, is a discharge confined within the river channel which allows a balance between the amount of water extracted from the river for human use and simultaneous maintenance of the aquatic habitat. Ideally an instream flow should reflect the natural flow as much as possible. The ecological health of a river is preserved when the natural channel discharge regime is approximated and the instream flow is backed by out-of-stream needs.

Channel-maintenance flow is designed mostly to maintain the form of a river. A river needs to maintain water of a certain magnitude at a required frequency to maintain the channel. This is usually associated with bankfull discharge in temperate countries but is more complicated for rivers in the seasonal tropics and arid regions where the seasonal or temporal differences in flow and flood history have to be taken in consideration. Theoretically this is the flow required to maintain the channel and move the sediment. Another measure in use is the environmental flow release (EFR) which is the flow that has to be released from an upstream dam in order to maintain downstream ecosystem integrity and community livelihood (World Commission on Dams 2000). The selection and determination of an environmental flow depend on the river manager's objective and knowledge of the river. River management studies in practice tend to be driven more by ecology and aesthetics than other factors. The habitat suitable index of the US Fish and Wildlife Service or standard guidelines for water quality are examples of tools for managing river functions. Improvement in a river may require changes in both form and function.

Environmental flows could be difficult to determine for large rivers. Difficulties arise because of scale. River management is normally multifunctional, and environmental and application perceptions vary between managers and end-users. Ideal river management may require a combination of geomorphology, ecology, water chemistry, and engineering. River management may be driven also by management/customer expectation rather than science (Haltiner et al. 1996). One example comes from a small but flood prone Californian stream. Two different models were designed for managing the flood in the river, one by the US Corps of Engineers and the other by the residents. The final accepted design was a compromise. There could be more than one way of looking after a river. Managing rivers could be a multifaceted process that requires channel, floodplain, and basin maintenance. Some of the characteristics could be long-term, such as the width–depth ratio of the channel; while others may be transient, such as passage of flood peaks. The volume and pattern of discharge and the amount and type of sediment load need to be managed, so that some desired form of the river (width–depth ratio, distance between pools and riffles in the channel, bar forms, etc.) are maintained.

All rivers require maintenance of their form and function to be as close to natural as possible while they meet the demand for water as a resource. Different types of river hydrology require different adjustments. Flood in certain rivers needs to be recognised as an instrument of channel maintenance. It is necessary in a monsoon river to have enough flow in the dry season to maintain the channel and keep it ecologically sustainable and pollution free. This becomes a problem even in large rivers. Dissolved oxygen in the Ganga, for example, falls to the lowest level in the summer before the rains arrive (Das Gupta 1984).

A changing river requires dynamic management which changes procedure over time. This is known as adaptive management. A current example is the review of the altered ecosystem (channel and floodplain) of the Missouri River and exploration of the prospects of its recovery (National Research Council 2002). The Missouri is a 3970 km long tributary of the Mississippi draining a basin of 1.37 million km^2 with a high natural sediment load from the Rocky Mountains which used to be around 10^8 tonnes annually. More than one hundred years of engineering and river modification has been carried out on the Missouri for navigation, flood control, and cropland expansion. As a result, more than 300 km of river bends were straightened, about a third of the river was canalised, about the same distance inundated by reservoirs behind a number of dams including the seven large ones listed in Chapter 8, and the sediment flow was reduced to a fraction of the former load. These substantial ecosystem changes require repeated adjustments of the downstream channel.

Adaptive management uses interdisciplinary collaboration and enquiry to maintain or restore ecosystem resilience. The approach therefore represents many interest groups. The river ecosystem is monitored to evaluate impacts of management actions. Adaptive management, however, is not a panacea but it encourages a continuing search for improvement because it recognises uncertainty, especially in large systems. Dealing with uncertainty could be useful for large rivers, such as the Ganga or the Changjiang which are utilised and modified or undergo seasonal changes.

A large number of techniques are currently available for determining different environmental flows. However, only a few may be appropriate for covering variable discharge as associated with seasonal streams, flood prone rivers, and modified channels. Very useful computer models are available from the websites of organisations such as the United States Army Corps of Engineers (USACE) Hydrologic Engineering Center.

9.3 Social and Political Management

Social and political management of large river basins may have to deal with regional economic disparities among individual residents, separate communities within a single country, and international sovereign states. Several large rivers, such as the Mississippi, Murray-Darling or the Changjiang, flow through a large single country. Drainage basins of large rivers, however, generally stretch across the sovereign areas of several countries, and their management requires an international agreement on the techniques to be followed and resources to be used. The Danube, Ganga, and Nile are good examples. If the drainage basin covers countries of varying human, technical, and financial capacity, its management becomes problematical. A drainage basin that is shared by several countries may require interfacing between political and value judgements and technical issues (Campbell 2007). The Mekong River Commission (MRC), an international organisation supervising a basin of six countries of varying expectation and capability and run with considerable help from external donor countries and a dedicated centre, illustrates such difficulties (Box 9.3). It is always essential to build a common agreement for basin and river management among the stakeholders but it is not an easy task.

9.3.1 Values and Objectives in River Management

Management of a river with its characteristic features, resources, and regional expectations requires the identification of a set of objectives that need to be achieved. Usually such objectives are stated as the vision of the managing organisation. The vision summarily identifies the objectives, and also may act as a socio-political statement justifying the building of an expensive and demanding organisation. The vision of the MRC is stated as 'An economically prosperous, socially just and environmentally sound Mekong River Basin' (MRC 1998).

Such statements imply three kinds of challenges. First, the policy and procedure of river management must tackle a number of interfaces. Such interfaces usually arise among the objectives to be reached and are based on different subjects such as geomorphology, ecology, economics or engineering. This is a difficult task and even more disciplines could be involved. Each river is different and management needs to stress individual goals, such as fish in the river, flood control, or a combination of dam building and fish preservation. There is no single panacea for all rivers beyond the general guideline of sustainable development.

Secondly, if the river passes through several countries of varying development and technical or economic capacity, then improving the quality of the river, its basin, and the life of the residents to an acceptable level would be another demanding task. Differences in political and military power among the countries are also likely to leave an imprint on the management objectives, irrespective of the principles of international basin arrangements, especially regarding the amount of water available. Even within a single country, geographical variations would demand different treatments, and separate government agencies may approach the river differently with their specialised emphasis. Varying expectations could be difficult to merge.

Thirdly, managing a river is expensive. Its water, sediment and channel pattern have to be studied and monitored regularly. The economic, social, and physical characteristics of the basin need to be understood before management planning or engineering structures could be put in place. All this is costly, time-consuming, and not possible for many developing countries. International basins in developed regions such as Western Europe may not have such a problem but elsewhere a shortage of financing, capacity, and administration is to be expected. Financial and technical assistance should be available from international development banks and donor countries, but such arrangements also come with the different expectations of the funding sources and their appointed personnel. Such demands are added to the existing expectations of the basin countries. Rationality and knowledge may or may not be the primary choice.

These complications are best realised with case studies. Examples of treaties and arrangements for several international basins of large rivers are briefly discussed in the next section. A detailed account of managing the Mekong River that illustrates a number of problems mentioned earlier is given in Box 9.3.

9.3.2 International Basin Arrangements

Successful management of an international river depends to a large extent 'on the preparedness of states to sacrifice some of their own interests to accommodate those of their neighbours' (Campbell 2007, p. 578). Three pertinent principles exist, although there does not seem to be definite international laws applicable in dispute between countries. The principles (Beach et al. 2000; Campbell 2007 and references therein) are:

- principle of prior appropriation
- principle of equitability
- principle of sovereign rights

The principle of prior appropriation confirms that a past user (a sovereign state in this discussion) has the right to continue using the river the same way as before, so long such use would not prevent the future use of the river by any other riverine country or several countries. An upstream country thus cannot prevent or impinge upon the use of the river by a downstream country, even if it has been an earlier utiliser of the river and its resources. So long as water is available, the later users have the right to use the amount of water in excess of the standard use of the upstream stakeholder.

The principle of equitability confirms the rights of all users of the river to a fair share of its resources.

The use of an international river cannot be monopolised by a single user. It suggests overriding of the principle of prior appropriation. A prior downstream utiliser of the river can continue its use provided the river is used equitably by all upstream users.

The principle of sovereign rights is an extreme viewpoint and rarely used. It claims that a sovereign state has the sole right to the river and its use within its territory. It is also known as the Harmon Doctrine.

9.4 The Importance of the Channel, Floodplain, and Drainage Basin

Proper river management implies maintenance of the channel, floodplain, and the catchment area. The pattern of channel flow is important. The flow regime of the river controls not only the form of the channel but also stream biota, and by extension, the relationship between stream biota and the catchment. The two extremes of a discharge distribution, floods and low flows, are relevant for determining not only the channel forms but also channel vegetation and aquatic fauna. The hydrology of the channel is influenced also by the land use of the drainage basin. Time-based changes in discharge affect the life cycle of the fish in the river, especially reproduction and migration. Channel and streamside vegetation and many invertebrates are also sensitive to such changes. Features such as sediment accumulation in bars and pool-riffle arrangement of the channel are determined by regional geomorphology and hydrology. Channel forms in turn determine the habitat spaces of the organic material and their accumulation.

Many large rivers, for example, the Amazon, Mississippi, or Ganga, flow between floodplains for most of their course. Large floods overflow levees or escape through crevasses to inundate the floodplain and in places link up with individual bodies of water in the valley. The floodplain is thus coordinated with flood pulses of the channel, and usually carrying a high concentration of sediment. Thus the water bodies and depressions on the floodplain are linked and replenished with the high flow and sediment of the channel as described for the Amazon (Junk 1997; Dunne et al. 1998). The waterbodies on the floodplain may also drain to the main channel with the falling of the flood, transferring excess sediment and stream and terrestrial biota. These flood pulses not only build up the floodplain with high-flow-derived sediment, but they also transmit additional food and space to the aquatic biota and disperse propagules. The seasonal inundation of the Tonlé Sap Lake during the wet monsoon by the floodwaters of the Mekong reaching the lake by the reversed flow of the Tonlé Sap River (Figure 9.1) is another example which illustrates the importance of flood plains (Campbell 2009a).

The building of dams and extension of levees destroy the existential relationship between the river and its floodplain. Very large floods are usually not stored behind the dam but released downriver, with their high water overtopping or breaking the

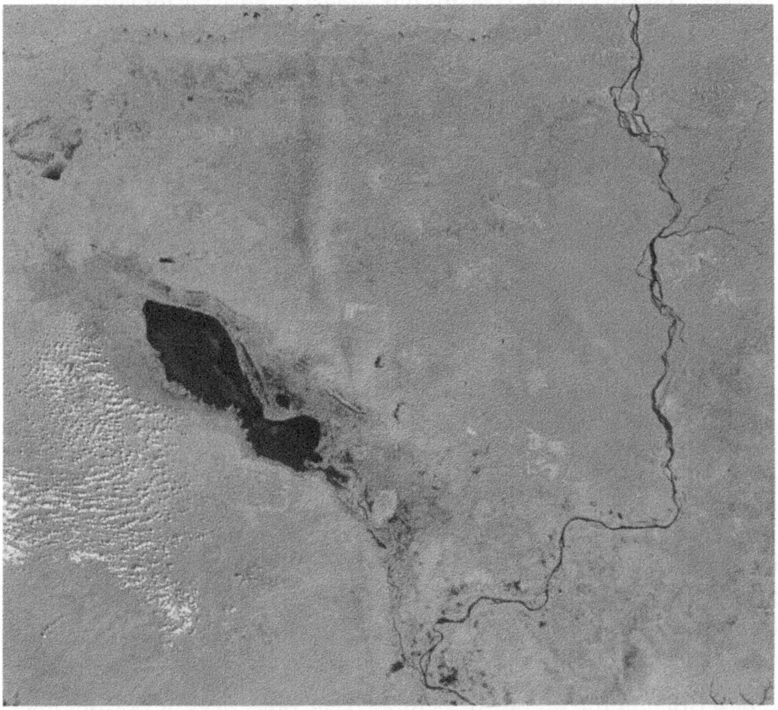

Figure 9.1 Tonlé Sap and the Mekong. Source: NASA worldview application (https://worldview.earthdata.nasa.gov), part of the NASA Earth Observatory System Data and Information System (EOSDIS).

levees. Dams, however, interrupt the natural function of silting, food supply, and seed propagation in small and medium floods. The importance of allowing lesser floods to inundate floodplains is being progressively understood, and the technique should be built into river management manuals.

Finally, management of the river involves its entire drainage basin. The source of almost the entire sediment of a large river is usually located near its headwaters. Meade, while discussing the movement of sediment in the Amazon and Orinoco rivers, described large rivers as 'massive conveyance systems for moving detrital sediment and dissolved matter across transcontinental distances' (Meade 2007, p. 45). The eastern slope of the northern half of the Andean mountain ranges has been the source of sediment for these long rivers. Similarly, most of the sediment of the Brahmaputra comes from the Eastern Himalayas (Singh 2007). Sediment control for large rivers thus may require long-distance control or protection of vulnerable sites following land use changes. Instream ecological processes are thus dependent on catchment processes. Riverside vegetation directly influences biological processes in the headwaters, less so downstream where the valley widens, but as Campbell (2007) has indicated, imports of carbon from upstream, groundwater, and floodplain may become important. The quality of water diminishes due to both point (such as urban or industrial pollution) and areal sources (as from polluted runoff from agricultural and urbanised areas). Pollution from a point source is easier to rectify but diffused pollution is hard to manage as it generally requires extensive controls such as buffer strips next to the river or planned land use. Both are possible for small rural streams but almost impossible for large rivers. River management in many ways therefore requires management of very large areas, covering at least a substantial part of the catchment. As Campbell (2007) pointed out, the techniques require landholder support and participation. This will be a demanding and time-consuming process for large basins.

9.5 Integrated Water Resources Management

Integrated water resources management (IWRM) currently is often considered as the most appropriate technique for river management, covering the entire range from the channel to the basin. It is usually described as a management technique which brings together development and management of land, water, and all related resources. This is done to maximise and equalise economic and social welfare but still not to compromise the sustainability of the physical environment and the vital ecosystems. Three basic objectives are integrated in the description:

- Environmental and ecological sustainability of the river environment.
- Economic efficiency of utilsed water; Campbell (2007) has highlighted the increasing scarcity of water relative to demands.
- Social equity of water, i.e. everyone should have a right to the access of a reasonable share of water of appropriate quality.

These have been described in the literature as three basic objectives or bottom lines in river management (Campbell 2007, 2009b and references therein). It is difficult to incorporate all three in appropriate fashion in the management of a large river. The excellent concept of IWRM is highlighted in textbooks but examples of its application in the real world are limited. It also implies a multifactorial approach, whereas river experts tend to operate in a single field such as engineering or ecology. The these major impediments of IWRM need to be overcome. We should therefore look for alternative techniques involving IWRM at the scale of large basins (Biswas et al. 2005). River management should be an act of sustainable environmental caretaking, careful utilisation of water, and poverty alleviation. We also need to find a way for experts from different water-related disciplines to interface with each other. The increasing acceptance of IWRM reflects the general understanding of the interconnected nature of hydrologic resources. It is a democratic cross-disciplinary approach to management based on the concept that unregulated use of scarce and unsustainable water resources is inappropriate. IWRM is an excellent choice given the complex nature of the water environment of a large river and the wide range of demands of the basin residents. It is a classic approach to the general concept of sustainable river development.

9.6 Techniques for Managing Large River Basins

A small river basin is easy to manage, even when management involves all three features: channel; floodplain; drainage basin. The background information could be collected in a short time, the development project is likely to be restricted in size, and, if the basin falls within the sovereign area of a single country, as small basins usually do, they can be administered solely by the government concerned. In contrast, management of large rivers basins is difficult, especially if the basins stretch across several developing countries. There are likely to be gaps in the knowledge of the area, the required database may not be available to all, and the necessary technical personnel and laboratory and library capacities are frequently insufficient for the entire basin. The project often is dependent on development banks or donor countries for financing and trained personnel. Cooperation among the basin countries may or may not be good, and development agencies and donor countries may bring their own expectations to the project. The demand and capacity for utilising the river's resources may vary and demand for water may even create a prelude to war (Campbell 2007). The ordinary people of the basin may have expectations but usually provide a limited contribution in planning.

It is easier to manage large river basins in developed countries, such as that of the Rhine or Danube, which have available technical expertise and economic prosperity. The oldest agreement is probably on the navigation of the Rhine which was arranged at the Congress of Vienna in 1815, when the political map of Western Europe was reconstructed. Agreements, however, change. The present navigation arrangement for the Rhine was structured in 1963 and so was the international treaty for cleaning up this river. The International Commission for the Protection of the Rhine (ICPR), based at Koblenz, exists as a supervising institution, but most of the work is carried out by various environmental ministers and government officials following national and state legislations. The role of the ICPR is limited to monitoring the agreed implementations and checking the river.

International agreement on the other large river of Europe, the Danube, started probably as far back as 1862 when it was focused on boundary demarcation. The agreement has expanded to include other objectives over time: navigation, water use, floods. The current agreement, the Danube River Protection Convention, came into existence in the 1990s, signed by 11 riparian states.

Negotiations and new objectives become necessary even within the same country. In Australia, the Murray-Darling River Basin is currently supervised by the Murray-Darling Basin Commission (MDBC). The original organisation was the River Murray Commission set up in 1917, mainly to ensure supply of water to the state of South Australia and improve navigation on the river. Over time, its responsibility was extended to controlling salinity in the basin and maintaining water quality. Seventy years after the beginning of the first commission, the MDBC was formed. However, two years of negotiation among the stakeholders (three state governments and the central one) was needed before the MDBC could start (Campbell 2007). Therefore, it is not surprising that it takes a long time to build an international agreement, which usually is more complicated than a national one. The agreement is followed by a research and supervising organisation on an international scale, even when the countries display a wide range of resource, capacity, political arrangement, and geographical location with reference to the basin and the river. This takes time but countries on large rivers, such as the Amazon or Paraná, now operate through international agreements with some success. Two examples, the Nile and Mekong, will be discussed.

The quantity and quality of the water is not always maintained to an expected standard producing extremely negative results. The rivers of the Murray-Darling Basin provide a recent example (2018–2019). The high demand for water for large-scale crop irrigation coupled with a very hot summer reduced the flow of the rivers, resulting in exceptional shortage of water for the local residents, fish kills in huge numbers, and even drying up of major tributaries of the Murray-Darling. The basin rivers are extremely degraded environmentally.

9.7 Administering the Nile

The Nile is a long river, flowing from the highlands and lakes of Central Africa to the Mediterranean Sea. Its drainage basin currently forms part of 10 countries: Rwanda, Burundi, Democratic Republic of Congo, Kenya, Uganda, Tanzania, South Sudan, Sudan, Ethiopia, and Egypt. The river straddles 35° of latitude, flows through a wide range of physical environment, and the countries vary in their knowledge base, capacity, and human resource. They have emerged from a colonial past and currently are independent and sovereign. The early international agreements on the Nile were bilateral arrangements for sharing water in the colonial times. The first basin-wide approach to management occurred in 1967 with the Hydromet project which emphasised hydrological data-building for the Lakes Region in the upper basin. A parallel approach from 1983 to 1992 was the Undugu project on economic development for a group of riverine countries, strongly supported by Egypt although not without disagreements. A bigger river basin organisation was set up in 1993: the Technical Cooperation Committee for the Promotion of Development and Environmental Protection of the Nile Basin (TECCONILE). It focused on technical cooperation, involved several Nile states, and was supported by the World Bank, United Nations Development Programme, Food and Agriculture Organization, and Canada.

In 1999, TECCONILE was replaced by the Nile Basin Initiative (NBI) which ultimately expanded to include all 10 Nile countries listed above. It has been described as an intergovernmental partnership to develop the river in a cooperative manner, share socio-economic benefits, and promote regional peace and security. The vision of the NBI is 'to achieve sustainable socio-economic development through equitable utilization of, and benefit from, the common Nile Basin water resources' (www.nilebasin.org). Unlike the previous attempts, the NBI is characterised by a shared vision and includes all Nile countries. The highest decision maker on the action and policy of the NBI is the Nile Council of Ministers which includes the ministers in charge of water in each NBI member state. They are assisted by the Nile Technical Advisory Committee, comprising 20 senior government officials, two from each of the member states. It is administered by a general secretariat, based in Entebbe, and much of the work is carried out using donor support. It is interesting to note that the early stages of the NBI focused on establishment, confidence building, and institutional strengthening before the project was scheduled to deliver. Emphasis is given to basin monitoring, enhancing the Nile knowledge base and sharing such knowledge with the member countries, increasing technical capacity, and having an interactive information system. Three areas are given priority in resource development: power; agriculture and regional trade; and river basin management and development. The idea of management for this long river appears to be based on concepts of sustainable development, poverty alleviation, and an IWRM type of approach. The management of the Mekong River (Box 9.3) is also multifaceted.

Box 9.3 Problems and Management of the Mekong River Basin

The Mekong Basin illustrates a number of issues relevant for managing a large river in developing countries which have a wide range of capacity, funding, and political power. Shortly after the Second World War the United Nations established the Economic Committee for Asia and the Far East (ECAFE) in Bangkok to improve development in Asia. It was suggested that an international committee should be established to collect and exchange information on the Mekong Basin and also coordinate planned projects for development.

The Committee for Coordination of Investigations of the Lower Mekong Basin (Mekong Committee) was established in 1957, mainly to harness the resources of the river. For example, a series of dams were planned on the Mekong and its main tributaries. This was an international approach with the cooperation of four riverine countries: Thailand, Laos, Cambodia, and Vietnam. Mainland China was then not in the United Nations and Myanmar did not join as a member. However, the recommendations of the Committee were not followed for a number of reasons including the war in Vietnam that eventually spread to Laos and Cambodia. A modified organisation, the Interim Mekong Committee operated from 1977 to 1995, with only three members. Cambodia was not part of the Interim Committee. However, studies on the basin continued along with water and sediment collection data for the basin. As a result, a time series of discharge data is now available for a number of stations on the Mekong and its major tributaries, which is very useful for understanding the hydrology of the river.

In 1995, with Cambodia joining the other three countries, the current MRC was formed. The other two Mekong countries, China and Myanmar, participate with observer status. As stated earlier, the vision of the Commission encompasses sustainable environment, economic development, and social justice, all three for the part of the river and basin

(Continued)

Box 9.3 (Continued)

below the Chinese border. The Agreement of the Commission includes a range of topics listed as articles:

- Areas of cooperation.
- Projects, programs, and planning.
- Protection of the environment and ecological balance.
- Sovereign equality and territorial integrity.
- Reasonable and equitable use of the water.
- Maintenance of the mainstream flows.
- Prevention and cessation of harmful effect.
- State responsibility for damages.
- Freedom of navigation.
- Emergency situations.

(summarised from Campbell 2009a).

A ministerial council, meeting annually and representing the four member countries, sits at the top of the administrative structure of the Commission. A joint committee of bureaucrats, placed below the ministerial council, meets three times a year. A secretariat, headed by a Chief Executive Office with four directors from the riparian member countries, is entrusted with providing logistics and technical support. The secretariat has moved among the major centres of the river countries. It has been in Bangkok, Phnom Penh, and Vientiane. The MRC is designed to target both basin management and economic development. It is supported by various development banks and donor countries, and external countries have also provided senior administrative and technical personnel. Given the nature of its structure, the Commission is bound to have a wide range of expectations and expertise (Campbell 2009a).

The Mekong is a 4880 km long river that drains a 795 000 km^2 basin which includes parts of the six countries to the South China Sea (see Chapter 10). A disproportionately high amount of water arrives to the main river from (i) the steep northern hills of Lao PDR and the northern Annamite Chain and (ii) the southern Annamite slopes via the large tributaries of Kong, San, and Srepok. Before dams were built on the Mekong in Yunnan (the Lancang Cascade), about 18% of the annual discharge used to arrive from the upper basin. The upper basin was also the source of about half of the Mekong's sediment.

The river runs on rock though narrow valleys in mountainous regions for about 3000 km. The next 1000 km is on both rock and alluvium in a narrow valley flat. This river runs freely on alluvium in a wide lowland only for the last 600 km, ending in a large delta. The river can be divided into eight units (Figure 10.1), according to its varying physiography (Gupta and Liew 2007). Each of these units needs to be treated differently for river management. The flow of the river is strongly seasonal with occasional floods. The use of land use varies across the basin, but in general it is forested or being farmed. The population density is low, except in several urban centres along the river. The physical environment and the socio-economic conditions vary across the basin.

The Mekong is a river seasonally affected by flood pulses. Such flood pulses inundating the floodplain create an ecological linkage between the river and the floodplain. This controls the aquatic life and rice cultivation. The impact would be noticeable mostly in the lower river, especially in the environment of the Tonlé Sap Lake. The flood pulses determine the upper limit of the river water controlling the distribution, movement, and life cycle of aquatic fauna and riverside vegetation. The lowering of water level in the dry season also acts as control. The seasonal movement of water across the floodplains where the land is low and wide allows widespread and intensive cultivation of rice. As people in the Lower Mekong Basin live on rice, fish, and aquatic vegetation, periodic inundation of the valley flat is crucial for their existence. Construction of a dam on a river like the Mekong therefore has the potential to be especially harmful by (i) lowering the wet season flow and reducing channel movement, (ii) keeping the low season flow higher than usual, and (iii) preventing the movement of fish along the river.

Concern for the environment is growing due to certain types of changes, especially the building of a series of dams in China over the Upper Mekong. Controversies continue regarding the utilisation of the water for power generation and the channel for navigation. These events carry the likelihood of environmental disasters impacting the river and its people. The performance of the MRC is crucial following a number of changes and proposed future developments in the basin. The major areas of possible environmental impact can be identified as:

- Proposed and constructed dams and navigation projects on the river.
- Increased erosion and accelerated sediment production on the basin slopes.
- Degradation of the aquatic life.
- Degradation of crop cultivation.

A series of seven dams and reservoirs, known as the Lancang Cascade, are being constructed on the Upper Mekong in Yunnan, China. These include the Xiaowan Dam which is 300 m high with an active storage of 990 million m^3, the second largest in China after the Three Gorges Dam. There may be considerable benefits from irrigation, power generation, and navigation but not for the downstream countries. The impacts on the other hand could be substantial. It appears likely that only limited environmental assessments have been determined and the MRC or any other organisation has very little power over planning and development of large structures on the river. The Lancang Cascade is built by the most upstream and powerful country in the basin.

Currently, the list of possible environmental and socio-economic problems in the Mekong Basin includes:

- Changes to natural flow regimes of the Mekong due to dams constructed or planned in China, Thailand, and Cambodia. A large number of dams are planned and being constructed in Lao DPR, Cambodia, and Vietnam. Export of hydroelectricity to Thailand is financially beneficial for Lao PDR. Widespread physical, ecological, and social changes should be expected.
- Changes in the channel due to the dams of the Lancang Cascade: channel scour, incision, bed armouring, and stripping of loose sediment in the rock-cut channel.
- Erosion at the base of the islands, many of which are farmed and several of which are settled.
- Accelerated transfer of sediment to the lowlands and the delta on a decadal scale, and accumulation of sediment in the Tonlé Sap Lake altering its ecological function.
- Deterioration of aquatic life and wetland vegetation
- Prevention of inundation of floodplains due to changes in flow regime and construction of levees and roads on embankments.

(Continued)

Box 9.3 (Continued)

- Controversial suggestions to improve navigation with bigger boats
- Pressure on fishing, the main supply of protein to the basin population. Changes in life cycle and migration pattern of the fish may happen due to alteration in flood pulse related inundation and barriers engineered across river channels.
- Post-sediment erosion of the delta face, especially with sea-level rise.

It is not clear whether the MRC can solve these problems. Not only information on the detailed geography of the basin is needed along with the necessary technical capacity but also the understanding of the holistic nature of the basin. A shared vision and extensive coordination among the countries is not always forthcoming. For example, dams have been planned and discussed for northern Cambodia which may cause direct interference to migration and spawning of the aquatic fauna in channels and on floodplains. This would be impact directly on many people of the Lower Mekong Basin who have been described as rural farmers with fishing as the second occupation.

National and political desires may override careful planning. An international organisation may build a database and carefully constructed plans, including an IWRM type of management procedure but success often depends on the expectation and willingness of individual countries. The fishery of the Lower Mekong around Tonlé Sap depends on cooperative actions of all the Mekong countries, including upland China. The country which reaps the maximum benefit from a project, as from dam construction, would not solely bear its physical, ecological, and social costs. Such costs undoubtedly will also be borne by other countries, certainly the downstream ones, which may or may not have any share in the benefits.

9.8 Conclusion

Successful management of river basins requires understanding and organisation of both biophysical and socio-political environments. Such tasks may be difficult but not impossible. However, sovereignty of individual countries in a transboundary basin may create conflicting demands unless the countries concerned are willing to recognise their shared right to the water. International organisations generally have limited success in coordinating such arrangements. Biswas (2008) has described the activities of most international and bilateral organisations as becoming progressively more risk-averse and politically correct. Large river basins are hard to manage because a number of countries, often with varying demands and capacity, may be involved and environmental conditions change across a large basin. However, success has been achieved in developed countries, and occasionally even in major basins shared by developing nations (Varis et al. 2008). Controversies, however, are more common, as the water of the Nile and the unilateral building of the Lancang Cascade on the Upper Mekong show.

Currently, a number of large river basins are overseen by dedicated river basin organisations. Such organisations have a political superstructure with ministers and senior bureaucrats of the countries involved. A secretariat studies the basin and collects a widespread dataset. The vision of the organisation these days tends to assimilate sustainable environment, social equity, and economic development. Such objectives are wonderful but very difficult to achieve, especially if the interests of the countries differ among themselves and with those of the donor organisations and countries who provide funds, and of the external technical workers. Expectations generally differ.

Basin management at least may succeed in acquiring and understanding relevant datasets, realising varying expectations across the basin, and confidence-building in the vision of the organisation. It is hoped that properly managed large international basins with a multifaceted environment and variable demands would be common in the future. The alternative could be a disastrous conflict.

One of the recent examples is the problem with the drying up of the Murray-Darling Basin in the summer of 2018–2019. The 1.07 million km^2 Murray-Darling Basin drains the inland slopes of the south-eastern highland of Australia. The rivers have been regulated with a number of dams and weirs and water transfer between the sub-basins. A massive increase in agricultural production and expectations resulted from such huge efforts. The rivers, however, have dried up in the 2018–2019 summer due to increased demand for water and enhanced drought conditions. This has been disastrous for the river, its ecology, and the people who depend on it. To illustrate, there have been three major fish kills. Meeting the demand for water of the river and its management can be difficult in a degraded environment. We may see more of such cases as the current climate change continues (Chapter 12).

Questions

1 What should be the ideal vision of a managed large international river?

2 How would you check the quality of the biophysical environment of a large river?

3 What is adaptive management? Why is it necessary?

4 What does social and political management of a river primarily deal with?

5 Is there a single technique for river management? Justify your answer.

6 What principles do international basin arrangements depend on? Do they work?

7 It is said that proper river management requires maintenance of all three of the channel, floodplain, and basin. Why?

8 What is IWRM? Why has it been described as a good idea seldom practiced?

9 Describe the general structure of an international river basin management organisation.

10 Familiarise yourself with the Mekong River. Explain the concerns about the Lancang Cascade.

References

Ahmad, Y.J. and Sammy, G.K. (1985). *Guidelines to Environmental Impact Assessment*. London: Hodder and Stoughton.

Beach, H.L., Hammer, J., Hewitt, J.J. et al. (2000). *Transboundary Freshwater Dispute Resolution. Theory, Practice and Annotated References*. Tokyo: United Nations University Press.

Biswas, A.K. (2008). Management of transboundary waters: an overview. In: *Management of Transboundary Rivers and Lakes* (eds. O. Varis, C. Tortajada and A.K. Biswas), 1–20. Heidelberg: Springer.

Biswas, A.K., Varis, O., and Tortajada, C. (2005). *Integrated Water Resources Management in South and Southeast Asia*. New Delhi: Oxford University Press.

Campbell, I.C. (2007). The management of large river: technical and political challenges. In: *Large Rivers: Geomorphology and Management* (ed. A. Gupta), 571–585. Chichester: Wiley.

Campbell, I.C. (2009a). *The Mekong: Biophysical Environment of an International River Basin*. Amsterdam: Academic Press.

Campbell, I.C. (2009b). Development scenarios and Mekong River flows. In: *The Mekong: Biophysical Environment of an International River Basin* (ed. I.C. Campbell), 389–402. Amsterdam: Academic Press.

Carson, R. (1965). *Silent Spring*. Penguin Books.

Das Gupta, S.P. (1984). *The Ganga Basin (Part II)*. New Delhi: Central Board for the Prevention and Control of Water Pollution.

Dunne, T., Mertes, L.A.K., Meade, R.H. et al. (1998). Exchanges of sediment between the flood plain and channel of the Amazon River in Brazil. *Geological Society of America Bulletin* 110 (4): 450–467.

Gupta, A. and Asher, M.G. (1998). *Environment and the Developing World: Principles, Policies and Management*. Chichester: Wiley.

Gupta, A. and Liew, S.C. (2007). The Mekong from satellite imagery: a quick look at a large river. *Geomorphology* 85: 259–274.

Haltiner, J.P., Kondolf, G.M., and Williams, P.B. (1996). Restoration approaches in California. In: *River Channel Restoration: Guiding Principles for Sustainable Projects* (eds. A. Brooks and F.D. Shields Jr.), 291–329. Wiley.

Junk, W.J. (ed.) (1997). *The Central Amazon Floodplain. Ecology of a Pulsating System*. Berlin: Springer.

Leopold, L.B. (1969) Quantitative comparison of some aesthetic factors among rivers. US Geological Survey Circular 620.

Leopold, L.B. and Marchand, M.O. (1968). On the quantitative inventory of the riverscape. *Water Resources Research* 4: 709–717.

Leopold, L.B., Clark, F.E., Hanshaw, B.B. and Balsey, J.R. (1971) A procedure for evaluating environmental impact. US Geological Survey Circular 645.

Meade, R.H. (2007). Transcontinental moving and storage: the Orinoco and Amazon rivers transfer the Andes to the Atlantic. In: *Large Rivers: Geomorphology and Management* (ed. A. Gupta), 45–63. Chichester: Wiley.

Mekong River Commission (1998) Water Utilization Program Preparation Project: Final Report. Bangkok: Mekong River Commission, MKG/R.98.013.

National Research Council (2002). *The Missouri River Ecosystem: Exploring the Prospects for Recovery*. Washington, DC: National Academy Press.

Serageldin, I. (1996). Sustainability as opportunity and problem of social capital. *The Brown Journal of World Affairs* 3 (2): 187–203.

Singh, S.K. (2007). Erosion and weathering in the Brahmaputra River system. In: *Large Rivers: Geomorphology and Management* (ed. A. Gupta), 373–383. Chichester: Wiley.

Varis, O., Tortajada, C., and Biswas, A.K. (eds.) (2008). *Management of Transboundary Rivers and Lakes*. Heidelberg: Springer.

Wohl, E.E. (2001). *Virtual Rivers: Lessons from the Mountain Rivers of the Colorado Front Range*. New Haven: Yale University Press.

World Commission on Dams (2000). *Dams and Development: A New Framework for Decision-Making*. London: Earthscan.

World Commission on Environment and Development (1988). *Our Common Future*. Oxford: Oxford University Press.

10

The Mekong: A Case Study on Morphology and Management

10.1 Introduction

The Mekong is a large river (Table 10.1) with a complex physiography and evolutionary history. This is a large flood pulse river with a tight hydrological and ecological linkage between the channel and the floodplain. The Mekong existed in a near-natural state until only decades ago but current changes in basin land use and construction of channel impoundments have significantly modified the river. Its international basin comprises parts of six sovereign states: China, Myanmar, Thailand, Lao PDR, Cambodia, and Vietnam. An international organisation, the Mekong River Commission (MRC), has the responsibility to plan and manage the channel and the basin with collaboration from the six states and several donor agencies such as the Asian Development Bank. The Mekong is an example of a large river undergoing anthropogenic modification driven by the often conflicting interests of multiple users.

Rising approximately at 5000 m in eastern Tibet, the river flows for 4880 km in a northwest–southeast pan-shaped basin to discharge into the South China Sea. For the first 3000 km, or about 60% of its course, the river flows on rock though narrow mountainous valleys. The next 1000 km is on both rock and alluvium, although the channel usually stays confined within a narrow valley flat. For the last 600 km, about 12% of its course, this huge river runs freely on alluvium in a 500 km wide lowland, ending in a large delta.

The valley has been settled for a long time. Archaeological excavations at Oc Eo on the western delta revealed roman coins and trading objects that originated in West Asia and the Indian subcontinent and passed through a first century CE seaport. Chinese records described a state called Funan in the delta between the second and fourth centuries. Subsequent records described a state called Chenla in the present Cambodia and southern Lao PDR. The Khmer Empire lasted for nearly a thousand years and built large settlements and the fabulous Angkor group of temples between the ninth and fifteenth centuries. It was a civilisation renowned for water management. The empire collapsed under Siamese attack from the west, and the capital was moved to Phnom Penh on the west bank of the Mekong. Subsequently, the basin was governed by several small states under the hegemony of three big regional powers: Burma, China, and Siam. Europeans traders and travellers started to arrive, and ultimately France occupied most of the basin as part of French Indochina. However, the river on rock was a barrier to inland transport, and a French expedition travelled up the river to China only in the 1860s. The present day states came into being after the Second World War.

Introducing Large Rivers, First Edition. Avijit Gupta.

Table 10.1 Measurements on the Mekong.

Characteristics	Measurements	World rank
Area of the basin	795 000 km^2	21
Length of the river	4880 km	12
Mean annual discharge at mouth	475 x 10^9 m^3	9
Mean discharge	15 000 m^3 s^{-1}	8
Average annual suspended sediment	160 x 10^6 t	10

Source: Data from Meade 1996 and Mekong River Commission, various dates.

In 1957, the Mekong Committee was convened under the aegis of the United Nations to plan the development of the river in a multistate basin. The general expectation of the Committee of constructing a series of dams in the basin was prevented by the Vietnam War. A number of hydrological stations, however, were started and now provide a fairly long-term record for several points on the Mekong. In 1995, at the end of the war that devastated the region, the organisation was restructured as the MRC with four member states: Thailand, Lao PDR, Cambodia, and Vietnam. Myanmar operated under observer status, and China was not a member of the United Nations then and hence not part of the Commission. The first bridge on the Mekong was constructed only in 1994.

A study of the Mekong River and its basin should explore its source-to-sea physical characteristics; examine the different expectations of a number of stakeholders; and discuss the potential of adapting management techniques which, as described by the MRC, are politically correct, socially just, and environmentally sound. In Chapters 8 and 9, the generalities in management and environmental degradation in a large river were discussed. In this chapter, the Mekong River is discussed as an example.

10.2 Physical Characteristics of the Mekong Basin

10.2.1 Geology and Landforms

The complex geology of the Mekong Basin is only patchily known (Figure 10.1). The basin is part of the Eurasian Plate, an old continental plate regionally deformed by its collision with the Indian Plate in the Neogene. The valley probably opened up due to the extrusion tectonics associated with the collision of the Indian Plate with the Eurasian which led to the formation of the Himalaya Mountains (Tapponnier et al. 1982, 1986; Peltzer and Tapponnier 1988). In China, the several kilometre wide upper basin is eroded into Palaeozoic and Mesozoic granitic and sedimentary rocks. From near the Chinese border to near Vientiane in Lao PDR, the lithology of the basin includes granitic rocks, folded Palaeozoic sedimentary and metamorphic rocks, Mesozoic sedimentary deposits, and local volcanic exposures. Basic and ultrabasic rocks, associated with ancient suture zones, also occur in the area. Near Vientiane, a deposit of Quaternary alluvium extends across the river over the older rocks. The valley of the mountain-girt Mekong widens upstream of Vientiane where it flows through this alluvium inlier.

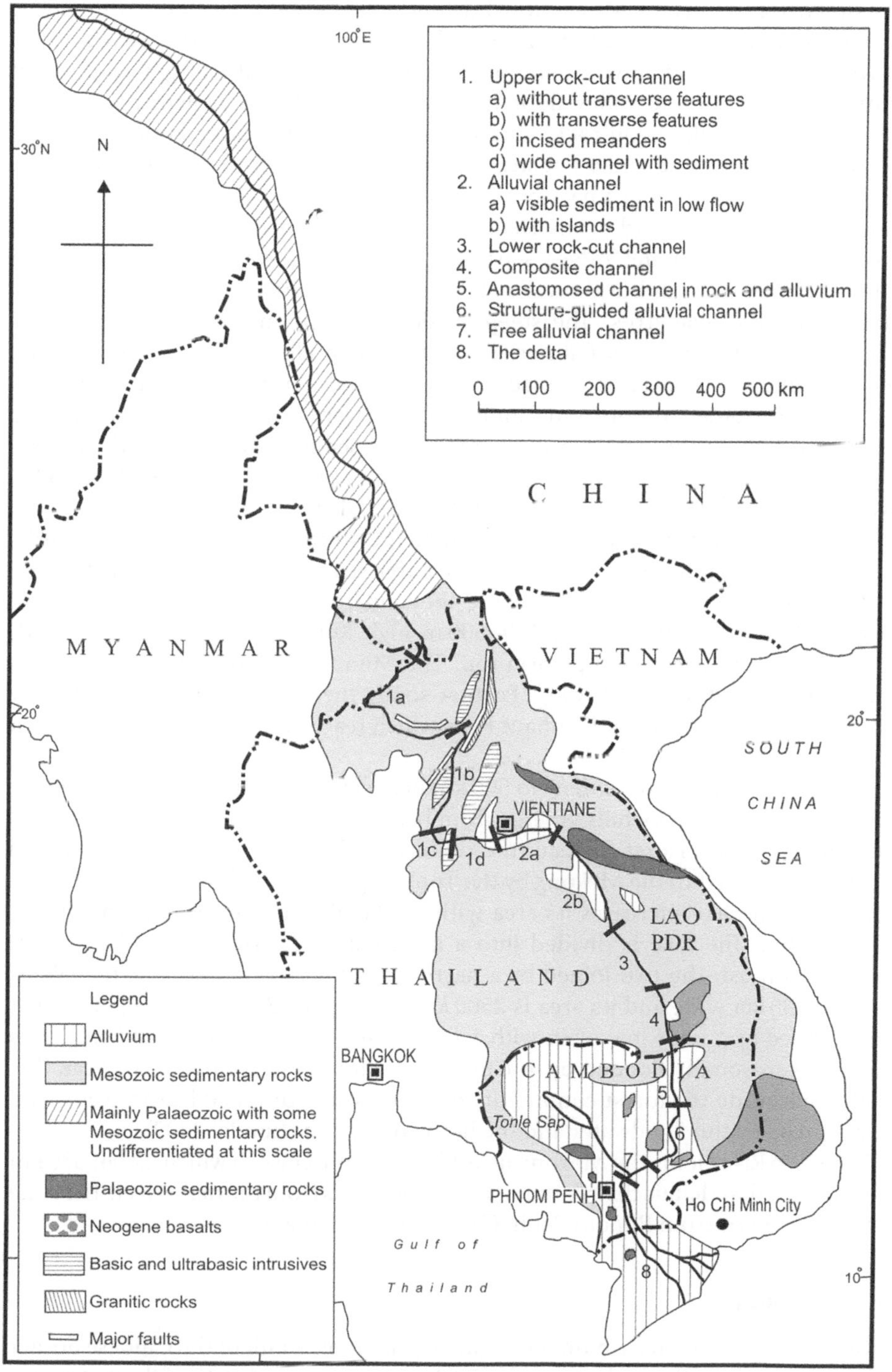

100°E
30°N
N
1. Upper rock-cut channel
a) without transverse features
b) with transverse features
c) incised meanders
d) wide channel with sediment
2. Alluvial channel
a) visible sediment in low flow
b) with islands
3. Lower rock-cut channel
4. Composite channel
5. Anastomosed channel in rock and alluvium
6. Structure-guided alluvial channel
7. Free alluvial channel
8. The delta
0 100 200 300 400 500 km
CHINA
MYANMAR
VIETNAM
20°
20°
SOUTH CHINA SEA
1a
1b
1c
1d
2a
2b
VIENTIANE
LAO PDR
THAILAND
3
4
5
6
7
8
CAMBODIA
BANGKOK
Tonle Sap
PHNOM PENH
Ho Chi Minh City
Gulf of Thailand
10°
Legend
Alluvium
Mesozoic sedimentary rocks
Mainly Palaeozoic with some Mesozoic sedimentary rocks. Undifferentiated at this scale
Palaeozoic sedimentary rocks
Neogene basalts
Basic and ultrabasic intrusives
Granitic rocks
Major faults

Figure 10.1 General geology and river units in the Mekong Basin. Source: Gupta 2007.

Further downstream, mainly Mesozoic sandstones and evaporites emerge from below a thin layer of alluvium to form numerous rock ribs in the Mekong. The narrow valley of the Mekong tends to follow structural lineations and several half-grabens. About 1500 km from the sea, the valley floor is already below 200 m. Downstream of the alluvium inlier, the valley is on rock -- narrow steep, and alternating between sharply demarcated hill ranges and wider flat stretches where tributaries meet the Mekong.

The Upper Mekong flows in a narrow valley with steep side slopes in Yunnan, China. In mountainous Lao PDR and north Thailand, the basin is an assemblage of dissected, narrow-crested, near-parallel ridges separated by deep valleys. The peaks rise to a maximum of 1500–2000 m but the valley bottoms drop to 1000 m within a horizontal distance of several kilometres. The steep hillslopes locally measure more than 25°, and are marked by numerous failures. Dissected steep cliffs of limestone overlook the Mekong and its tributaries in places. Volcanic rocks are exposed near the Lao PDR–Cambodia border. In northern Cambodia, the river flows over an alluvium cover of variable thickness with local exposures of Triassic sedimentary rocks and Neogene basalts. Such rocks also occur in the highlands to the east. From southern Cambodia, the river flows on alluvium in its lowermost reach and delta.

The Annamite Chain of mountains represent the steep eastern divide with peaks rising to 2000 m along almost the entire length. Large streams (Don, Kong, San, Srepok) erode this mountain and build fans where they emerge onto the plains. The gentle western divide is different. It is the 200–500 m high Korat Plateau of Thailand, with a few ridges running across the plateau top. The Mun River and its tributaries drain this area, flowing east to the Mekong. Further south, the western divide is marked by small but steep Cardamom and Elephant Hills which reach the sea at the northwestern corner of the Mekong Delta.

The Mekong plain abruptly widens to 500 km between the divides only about 600 km from the sea. Isolated rocky hills emerge from below this wide alluvial cover. Towards the west, south of the Korat Plateau, the drainage collects in the lake of Tonlé Sap (Figure 9.1), connected to the Mekong by the Tonlé Sap River, the flow of which reverses seasonally and the lake increases its area with the floodwaters of the Mekong during the rainy season. The lake is divided into a large northwestern basin and a smaller one to the southeast, the two joined by a narrow strait. In the dry season, the lake is 120 km long, 35 km wide and its area is 2500 km^2 (Campbell 2009a). It is a permanent lake surrounded by a wide low plain with a zone of stunted forest and rice fields. This community is seasonally flooded by 4–6 m of water for eight months of the year. The seasonality has made the forest community deciduous, and dropped leaves, fruits, and seeds are spread by the floodwater and the herbivorous fish (Rundel 2009).

The Tonlé Sap River joins the Mekong near Phnom Penh, close to which, about 330 km from the sea, the first deltaic distributary, the Bassac, splits from the main river. The large delta stretches south from Phnom Penh, Cambodia to Vietnam.

10.2.2 Hydrology

The MRC currently operates a number of gauging stations, south of the Chinese border including 33 gauges on the Mekong itself. Discharge is measured at 10 of these stations. The length of record could be quite long for several of these stations. For example, the daily discharge record for the Mekong at Vientiane goes back to January 1913 (MRC, various dates). The average annual precipitation that falls over the basin south of China, has been calculated to be 1672 mm (MRC 1997). It is less than 1000 mm in China in the north and over the Korat Plateau in the west. In contrast, annual rainfall is high on the eastern divide (2000–4000 mm). The rainfall is strongly seasonal and arrives with the southwestern monsoon, 85–90% falling between June and October.

The seasonal pattern of discharge of the Mekong matches the pattern of rainfall. The river rises a little in May following the arrival of summer snowmelt on the Tibetan Plateau, but 80% of the annual discharge occurs between June and November (Figure 10.2). A single month may have 20–30% of the annual discharge. Large floods occur later in the wet season and tail off very slowly. Large floods may inundate thousands of square kilometres across the lower basin (Figure 10.3). The floods of the Mekong tend to occur late in the wet season, often triggered by the arrival of tropical storms over the Annamite Mountains when the Mekong is already high.

The tributaries are extremely seasonal, with steep short-term discharge peaks in the wet season and a prolonged low stage at other times. A disproportionately high amount of water arrives from (i) the steep northern hills of Lao PDR and the northern Annamite Chain via the Nam Ngum and Nam Theun systems and (ii) the southern Annamite slopes via the large tributaries of Kong, San, and Srepok (MRC, various dates). Before dams were built on the Upper Mekong in Yunnan (China), 18.2% of the annual discharge used to arrive at Chiang Saen (Thailand). The percentage used to rise a little in summer in the pre-dam days (MRC, various dates).

10.2.3 Land Use

Intensification and changing of land use across the basin and modifications of the river started to happen only in the last few years. Traditionally, slopes in the highlands are

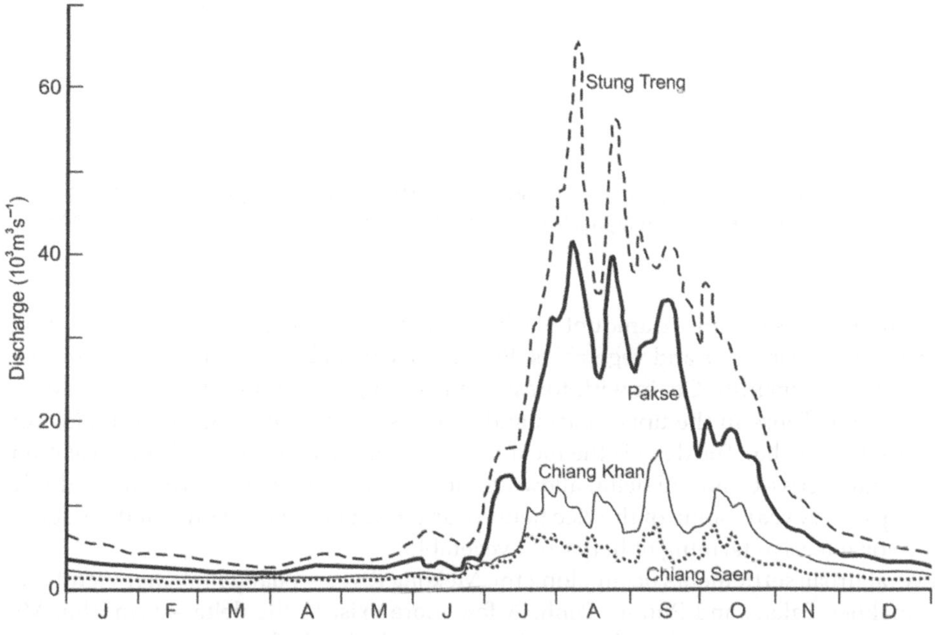

Figure 10.2 Annual hydrographs (1997) of the Mekong at Chang Saen, Thailand (14% of the basin area), Chiang Khan, Thailand (36%), Pakse, Lao PDR (69%), and Stung Treng, Cambodia (80%). Source: Gupta 2007.

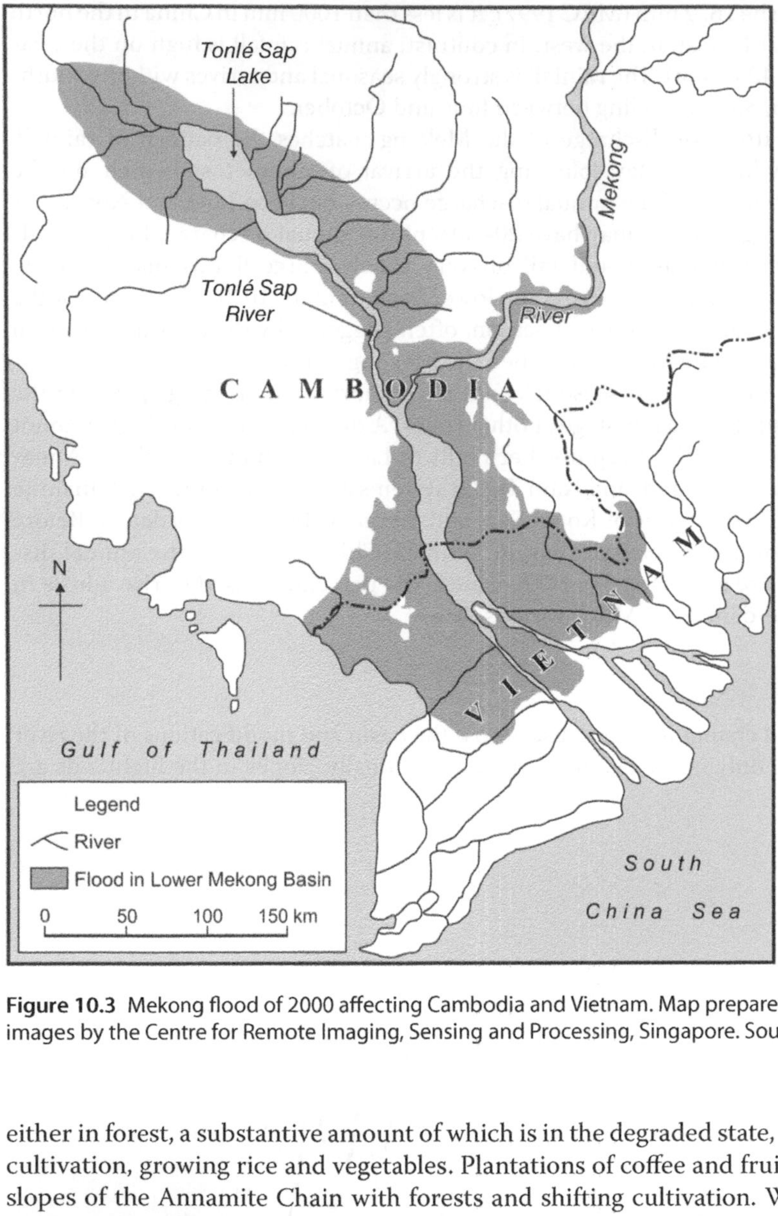

Figure 10.3 Mekong flood of 2000 affecting Cambodia and Vietnam. Map prepared from SPOT 2 images by the Centre for Remote Imaging, Sensing and Processing, Singapore. Source: Gupta 2007.

either in forest, a substantive amount of which is in the degraded state, or under shifting cultivation, growing rice and vegetables. Plantations of coffee and fruit crops share the slopes of the Annamite Chain with forests and shifting cultivation. Wet rice grows in narrow valley floors in the upper basin and is widespread in the large alluvial plain of southern Cambodia. The delta is the most fertile region. A series of canals for irrigation and drainage criss-cross the delta, allowing the common practice of growing multiple rice crops every year. Some of the rice fields near the sea have been converted for aquaculture or intensive farming of herbs and vegetables.

Several urban settlements occur along the Mekong River – Luang Prabang, Vientiane, Savannakhet, Pakse, and Phnom Penh. A few more exist in the delta – Can Tho, My Tho, and Ca Mau. A considerable disparity exists in the level of economic development across the basin.

In 1957, as mentioned earlier, the Committee for Coordination of Investigation of the Lower Mekong Basin was set up under the auspices of the United Nations Economic Commission for Asia and the Far East with four member states, Cambodia, Lao PDR, Thailand and Vietnam. In 1995, at the end of the conflict in Indochina, a new instrument was suggested: The Agreement on the Cooperation for the Sustainable Development of the Mekong Region, also known as the MRC. The agreed vision of the Commission was described as 'An economically prosperous, socially just and environmentally sound Mekong River Basin' (MRC 1998). The other two states that occupy part of the basin, China and Myanmar, have been in dialogue with the original four riverine states.

The Agreement highlighted the need for a timely and appropriately planned development of water resources in the basin. It recognised the potential for conflict and committed the countries concerned to maintain the quantity and quality of flow in the Mekong River (MRC 1998). The Commission also acts as a collector of information and maintains a database. The development of the Mekong Basin thus involves (i) planning by a central agency in a structural fashion and (ii) the expectations of stakeholders. The two factors have not always matched.

10.3 The Mekong: Source to Sea

10.3.1 The Upper Mekong in China

The Mekong begins in rock in Tibet where the river is called Dza Chu (River of Rocks) and in Yunnan it is known as Lancang Jiang (Turbulent River). Both names aptly describe the Upper Mekong. In China, it is generally an extremely steep structure-guided river on rock (average gradient of nearly 0.002), usually flowing fast through rugged, deep gorges with straight sections joined together by short, sharp bends and local offset channels. Short tributaries join the river at a very high angle, draining steep, small sub-basins indicating slope failures. A braided pattern has developed where the channel and valley of the river widen but the river channel commonly is steep, straight, and laden with coarse sediment.

10.3.2 The Lower Mekong South of China

Based on satellite imagery, topographical maps, charts of the MRC's hydrologic atlas, water data from the yearbooks of the MRC (MRC, various dates), and field visits, Gupta and Liew (2007) prepared a detailed description of the river and divided the river south of China into eight physiographic units (Table 10.2). The following is a summary account of the river from south of the Chinese border to its mouth, details are given in Gupta and Liew (2007).

Unit 1. South of the China border, the Mekong runs on rock for 900 km nearly up to Vientiane. For the first 500 km, almost up to Luang Prabang, the Mekong is a rocky conduit consisting of nearly straight reaches joined by sharp bends. In the narrow rocky valley, the river has very little chance to shift its course. A number of the straight reaches coincide with faults. The channel is either trapezoidal in cross-section, or with a deep 100 m wide inner channel bounded by rock benches. Scour pools and rock protrusions occur on the benches and also the channel floor. The local relief inside

Table 10.2 The Mekong: summary of river units, south of China border.

Unit	Bed material	Bank material	Average slope	Length (km)	Width (m)	Low flow depth (m)	Seasonal stage change (m)
1	Rock	Rock or alluvium on rock	0.0003	910	200–2000	<5–10	10–20
2	Alluvium	Alluvium	0.0001–0.00006	500	800–2000	3–5	12–14
3	Rock	Alluvium on rock	0.0002	200	400–2000	variable	20
4	Rock	Alluvium	0.00006	150	750–5000	Variable	~15
5	Rock	Alluvium on rock	0.0005	200	15 000	8	9
6	Alluvium	Alluvium	0.00005	225	3000	~5	14–18
7	Alluvium	Alluvium	0.000005	50			
8	Alluvium	Alluvium	0.000005	330			

All measurements are approximate. Variable implies difficulty in averaging because of too many scour holes. For Unit 7, the width is several kilometres and increasing. Unit 8 consists of several tidal deltaic distributaries. Seasonal stage changes for Unit 7 and Unit 8 are difficult to measure accurately.
Source: Gupta and Liew 2007.

the channel controls sediment accumulation, and sediment becomes visible in low flow as bars on rock benches. The river continues on rock with the inner banks in silt for the next 250 km. A profusion of tens of metre high cross-channel rock ribs, isolated transverse piles of rocks, and lines of rapids distinguish this reach. In places the channel flows through a rocky scabland (Figure. 2.4) (Gupta 2004).

The Mekong changes stage between 15 m and 30 m or more between the wet and dry seasons, leading to a wide difference in shear stress, unit stream power, channel erosion, and sediment transport between seasons (Figure 10.4). Sand and gravel are carried at high flow and accumulate at the falling stage around cross-channel rock protrusions to form bars or fill depressions between the high spots on the channel floor and rock shoulders. The sediment also forms insets against vertical rocky banks as massive episodic beds or is plastered over the rock in small quantity. The competence of the Mekong in flood is demonstrated by boulders resting on sand bars, exposed at low flow.

Further downstream, the Mekong, in rock, sharply changes from a south-flowing river to an east-flowing one for no apparent reason, and within 30 km crosses several low hills in a set of near-symmetrical meanders but no significant bars. Beyond these meanders the Mekong is still on rock but flowing in 10–20 km straight reaches separated by sharp bends. The reach identified as Unit 1 ends cutting through a cuesta with increased gradient. Beyond the cuesta, the channel of Unit 2 is full of sediment forming rock-cored mid-channel bars, bank insets, and shallow channel fills.

Unit 2. The river runs on alluvium for the next 500 km past Vientiane as a wide, shallow, meandering channel filled with sediment. A series of low sandy cross-channel bars appear in the dry season. The river is still on alluvium in the lower part of Unit 2 but probably structure-guided, as it changes into 50–60 km straight courses joined by

Figure 10.4 Rock ribs and flood markers in the Mekong channel upstream of Luang Prabang. Source: Photographer Avijit Gupta. From Gupta 2007.

sharp bends with a very low gradient and fewer bars which are big lozenge-shaped depositional forms, skewed to one side of the channel. The Mekong turns towards the south near the Annamite Mountains in the east and flows parallel to the ranges on rocks again in Unit 3.

Unit 3. In the first part of this unit, large lozenge-shaped islands up to 5 km long occur in a 1.5–3 km wide river on rock. The gradient of the river steepens in the second part, islands disappear, the channel narrows to hundreds of metres, and cross-channel rock exposures and an inner channel mark the Mekong. In one extreme case, a scour pool at a bend measures 1250 m long, 150 m across, and up to 60 m deep. At the end of this unit, the Mun, one of the major tributaries, comes in from the west draining the Korat Plateau, building a small fan at the confluence with the Mekong.

Unit 4. The Mekong continues to flow over rocks but with high alluvial banks. The gradient is gentle and scour pools replace the continuous inner channel of the previous unit.

Unit 5. The next unit, also on rock, is a puzzling 200 km stretch of the river where the river alternates between nearly straight and anastomosed channels. At the beginning of the unit, the Mekong flows for 50 km over a belt of Mesozoic basalt, resulting in a series of rock-cored low islands and multiple channels, a reach locally known as the '4000 Islands'. Downstream of the 4000 Islands, the river flows over a zone of rapids and waterfalls (including the Phapheng Falls) to the large alluvial plain of Cambodia where the Mekong is joined by the combined water and sediment discharge of the Kong, San and Srepok rivers draining the Annamite Mountains to the east joining the Mekong near Stung Treng. The Mekong here displays two or three secondary channels flowing parallel to the main flow with long alluvial islands separating them.

Unit 6. In spite of flowing across a wide alluvial plain, the Mekong is still rock controlled with limited mobility. The next 225 km consists of four nearly 50 km long straight reaches joined by right-angled bends (Figure 9.1).

Unit 7. A meandering Mekong moves freely across a wide floodplain. Old channels and scroll marks are common on both overbank and top of mid-channel islands. Overbank flooding occurs in the rainy season and inundation of backswamps beyond the levee of the Mekong is common. During high flow of the wet season, flow reversal in the Tonlé Sap River, as described earlier, fills the lake (Figure 9.1).

Unit 8. The delta starts near Phnom Penh where the first distributary, the Bassac, leaves the Mekong 330 km from the sea. The delta is commonly defined as the triangular piece of land between Phnom Penh to the north, the mouth of the Saigon River to the east, and the southwestern cape of the Ca Mau Peninsula to the west (Ta et al. 2002). The Mekong has built one of the largest deltas in the world. The subaerial delta can be divided into an upper and a lower part, associated with subaerial and marine agencies, respectively. Channels, levees and backswamps form the upper delta, whereas the lower part is controlled by waves and tides as indicated by low sand ridges separated by linear depressions. Subsurface cores indicate that the Mekong Delta was formed in the last 6000–7000 years, extending the shoreline for about 200 km (Nguyen et al. 2001; Ta et al. 2002). The drainage in the delta has been modified by a set of canals constructed over the last 300 years. In Vietnam, more than 80% of the dry-season flows are used for irrigation in the delta (Asian Development Bank 2012).

10.4 Erosion, Sediment Storage and Sediment Transfer in the Mekong

The Mekong is a seasonal river with high water and floods in the wet season. A disproportionately high amount of water arrives from (i) the steep northern hills of Lao PDR and the northern Annamite Chain via the Nam Ngum and Nam Theun systems and (ii) the southern Annamite slopes via the large tributaries of Kong, San, and Srepok (MRC, various dates). For most of its length, a wide floodplain is absent, and sediment is stored in the channel and locally in the narrow valley flat and then moves downstream during high flows. The main sources of sediment include (i) material from mass movements on steep valley slopes in China and northern Lao PDR that directly reach the channel; (ii) sediment contributed by short tributaries draining steep side slopes that tend to deposit small steep fans of gravel and sand at the confluence with the Mekong; (iii) sediment contributed by larger rivers (the Nam Ou from the northern Lao hills, the Mun from the Korat Highlands, and the Kong, San, and Srepok draining the Annamite Mountains); and (iv) sediment from accelerated erosion on slopes associate with current anthropogenic activities such as deforestation, construction, and shifting agriculture. This sediment for most of its course is stored primarily within its channel as (i) insets against banks, (ii) bars and islands mainly formed by piling up against rock ribs and rock protrusions in the channel, (iii) a mobile belt on the bed, and (iv) plastered on the rock shoulders of the inner channel. It is likely that the coarse sediment is loosely accumulated over a rocky channel bed and banks. For thousands of kilometres, sedimentation is controlled not so much by channel geometry but by channel relief.

Figure 10.5 Sand deposition on boulders by the Mekong after a large flood. Photographer Avijit Gupta.

Most of it is confined within the channel forming a quasi-mobile belt of sand and gravel resting on or against bedrock (Figure 10.5).

Narrow lateral shifts of the river start in Unit 6, the shifts are due to welding of elongated islands to banks. As the river becomes more mobile downstream, significant material is stored in floodplains and depressions from Unit 7, where flood basins may stay submerged during the months of the wet season. Modifications of the channel processes are likely to be common. In the Mekong, sediment transfer is seasonal and episodic. Rapid sediment transfer happens in the channel during high flows when sediment stored around rock barriers are entrained and moved downstream. Given the number of rock protrusions in the channel, bed load is likely to travel in stages. Cataracts and waterfalls with downstream pools may act also as sediment traps in the Mekong. Lateral flux is limited until Unit 6 is reached, the Mekong is free to move, and overbank flooding is expected in the wet season particularly from large storms. The finer sediment travelling from upstream may therefore be deposited in the delta and the Tonlé Sap Lake. In the past, the Mekong did not carry huge amounts of sediment unlike the Ganga or Yangtze. The river was almost at the regional average when sediment yield was plotted against drainage area for the South and Southeast Asian rivers (Gupta and Krishnan 1994). Walling (2009) stated that the overall impression is that of relative stability of the sediment loads transported by the river over the last 40 years. We do not have older records of Mekong's sediment but the stability may have continued for an even longer time, as no significant change seemed to have happened in the basin over this period. The stable situation is likely to change as discussed below. Walling (2008) has reviewed the available literature on the sediment of the Mekong, including unpublished sources.

To sum, the Mekong is a seasonal river with a sediment load essentially of sand and gravel loosely resting on a rock bed for about 4000 km of the upper course and fine material which is deposited overbank in flood and by channel shifting from about 600 km from its mouth. For most of its course, the river flows through a narrow valley; it only becomes mobile about 300 km above its delta. Most of the water and sediment enter the channel in specific areas as listed above.

10.5 Management of the Mekong and its Basin

The MRC has the responsibility of collecting the riverine data and planning the Mekong's management. The expectations of the six sovereign states that comprise the basin, however, vary and may give rise to conflict in management. The capacity and power of the states are also different. For example, China, the strongest state, overlooks the upper part of the catchment where it has built a series of dams (the Lancang Cascade). It is also interested in making the river navigable, by removing rock barriers from the channel. Both processes my give rise to conflicts in expectation from the users of the lower river. We can use the Mekong as a specific example of a large river with problems associated with management which arise from various environmental modifications of the river. Such conflicting issues in management have been discussed in general in Chapters 8 and 9. The performance of the MRC as an international organisation responsible for planning of the river and the basin in a holistic fashion and on a sustainable basis with collective consent from the riparian states is crucial but difficult. This is reflected in certain changes, such as the currently improved river and road transport in the lower basin. Deliberations and conflicts, however, may continue amongst the users, donor agencies, and this international organisation.

Three major areas of possible environmental impact at present or in the near future can be identified:

- Proposed and constructed dams and navigation projects in the river.
- Increased erosion and accelerated sediment production on the basin slopes.
- Degradation of the aquatic life.

Other problems undoubtedly exist or will emerge in the future but these three should be considered as serious barriers to maintain the Mekong as economically prosperous, socially just, and environmentally sustainable.

10.5.1 Impoundments on the Mekong

Dams modify rivers as described in Chapter 8. Such changes may happen in the Mekong but given the nature of the river, implications that are accelerated may be expected. It has been suggested that the effect of the dams on the Mekong will extend downstream, even up to the delta, where water and sediment discharges, soil rejuvenation, nutrient flow, and fish productivity will be reduced, threatening local agriculture and fisheries (Douglas 2005). The Mekong Basin countries plan to increase dry season irrigation from 1.2 to 1.8 million hectares by 2030 (MRC 2011).

Dams and reservoirs were proposed earlier by the older Mekong Committee but never executed. Several dams, however, were constructed on various tributaries of the

Mekong in the 1960s, primarily in Thailand. In 1968, the 150 MW Nam Ngum Dam in Lao PDR started to export hydroelectricity to Thailand. At present, several dams on various tributaries of the Mekong are in operation in Lao PDR, Thailand, Cambodia, and Vietnam. A set of seven dams planned and built by China, the Lancang Cascade, however, has created a considerable amount of discussion and controversy. The dams vary in size and include the big Xiaowan (4200 MW) which is 300 m high and has an active storage of 990 million m^3. It is the second largest dam in China, surpassed only by the Three Gorges Dam.

The effect of the Lancang Cascade should be significant. The controlled release of water from the Lancang Cascade reservoirs may reduce the seasonality in discharge affecting the ecosystem of the Mekong. About 18% of the annual flow comes from the Mekong in China. The available data on sediment are not as extensive or detailed as for water discharge but it is assumed, in general, about 50% of the sediment of the Mekong, mostly coarse-grained, comes from China. Kummu and Varis (2007) concluded, using data from various sources, that the measured annual sediment flux has dropped both at Chiang Sean, Thailand, about 660 km down the river, and also at Pakse in southern Lao PDR. The change in Pakse, however, could be due to other factors as it is a long way downstream. The wet season flooding of the Mekong is caused by heavy rainfall over the mountains of north Lao PDR and the Annamite Chain, well downstream of the Lancang Cascade dams. In combination, however, large floods, shortage of fresh coarse upstream sediment, and sand and gravel sitting loosely on rocky channel beds may lead to stripping of most of the sediment resting in the Mekong's channel within several years, effectively transforming the Mekong to a rock-cut canal, up to the zone of waterfalls near the Lao PDR–Cambodia border.

Steep tributaries of the Mekong are perceived in both riverine countries, Lao PDR and Cambodia, as carrying high potential for electricity generation. A number of dams and reservoirs have been planned on the main river and its tributaries, draining steep hills. The Lower Se San 2 Dam, for example, has been constructed extremely rapidly, in about a year, in the basin of the monsoon-fed Kong, San and Pok rivers which drain the steep western slopes of the Annamite Mountains to the Mekong River. Monsoon rain, steep rocky slopes, and shallow soils of the area require careful planning and management of impoundments of this type. Even a quick assessment highlights the potential loss of vegetation cover, suspected blocking of fish migration, and the high possibility of slope erosion and accumulation of sediment in the reservoir.

The stripped sediment may be deposited further downstream in the next flood when water has to be released from similar reservoirs. It has been suggested that part of the sediment will travel into the Tonlé Sap Lake to partially fill it leading to negative consequences for local ecology and economic well-being. The lake and the Mekong Lowland are the primary sources of rice and fish in the country. A description of the lake and its environmental conditions and probable future conditions is generalised in a number of papers (Kummu and Sarkkula 2008; Penny 2006). Flooding in the wet season and low water level in the dry season are transboundary phenomena and strongly related to the amount of discharge along the Mekong.

Campbell et al. (2009) stated that the Tonlé Sap Lake is progressively coming under environmental pressure with changing conditions of the Mekong Basin. They have indicated that these problems may arise from within the lake and the floodplain, within the lake catchment, and within the Mekong Basin, even globally. Two specific problems affect the lake and the floodplain: clearing of local vegetation, and degradation of fishing resources. Population of the lake basin has increased rapidly simultaneously with the degradation of natural resources. The town of Siem Reap has grown rapidly as a tourist base for the Angkor temples, polluting the Siem Reap River and ultimately the lake where the river drains. The dams and reservoirs of the Mekong Basin, particularly the Lancang Cascade and the proposed impoundment at Sambor rapids are disastrous for the lake altering its hydrology and depositing sediment.

It has also been suggested that the delta face may undergo rapid erosion following a reduced sediment supply combined with a rise in sea level and storminess associated with global warming. All these conjectures, however, require detailed investigation but building any engineering structure on the Mekong should require careful and detailed pre-construction environmental impact analysis and post-construction monitoring.

Ta et al. (2002) have demonstrated that sediment transport to the Mekong Delta has not changed substantially over the last 3000 years. Based on the rate of delta progression from borehole data, they calculated an annual rate of $144 \pm 36 \times 10^6$ tonnes which is very close to 160×10^6 tonnes for the current Mekong as determined by Meade (1996). This is different from the accelerated sediment yield measured for other regional large rivers due to significant anthropogenic impact on their drainage basins (Saito et al. 2001). A lowering of sediment transfer to the delta face and a rise in sea level therefore would be a disastrous combination for this fertile and populated part of the Mekong Basin.

Campbell (2009b) summarised various MRC models and other earlier reviews on the hydrological consequences of changes due to impoundments. The conclusions broadly are similar. The dams of Lancang Cascade would impact the discharge of the Mekong the entire length downstream to the delta. Their effect would be more conspicuous on the dry season flows which would increase. The consequences are harder to determine. For example, water for navigation and irrigation (especially for rice) should increase and reduce the potential for saline incursion in the delta. The dams are not expected to significantly reduce flooding downstream. However, the impact on the fishery could be expected to be high. The impact on inundation of floodplains and the Tonlé Sap Lake are complicated and difficult to determine. The general opinion is of a major decline in the fishery of the Mekong (Campbell 2009b).

10.5.2 Anthropogenic Modification of Erosion and Sedimentation on Slopes

The population of the Mekong Basin is essentially rural and poor. Parts of the basin are affected by deforestation and shifting cultivation. Deforestation has been extensive during the long war in Indochina. Deforestation currently continues on an individual and organised basis, and a significant part of the forest is probably secondary. Currently, vegetation is being removed along small valleys that follow geological lineations, have a local relief of several metres, and the slopes exceed 25°. Proper management of steep slopes is required. Shifting cultivation does not necessarily produce a large amount of sediment while the population density is low (Gupta and Chen 2002); a rise in density, however, erodes slopes. The current development of urban centres and growth of settlements along the river are producing sediment and lowering water quality.

10.5.3 Degradation of the Aquatic Life

The Mekong River system, especially the wetlands that include the Tonlé Sap and the delta, is a very important source for fish and shrimps for the local population. Reliable estimates are difficult to obtain as most of the catch is locally caught and consumed. Campbell (2007) mentioned annual estimates of 2 million tonnes of wild fish and 400 000 tonnes of catch in Cambodia. According to the MRC, the annual value of wild fish catch is about US$ 1 billion. The tributaries and wetlands of Cambodia are important breeding grounds for many species. The Mekong Fish Database, prepared by the MRC, names over 900 species. Fish are caught in the main river, the mature individuals consumed, and the fingerlings cultured in ponds (MRC 1988). Prawn culture has become important in the wetland and the delta. The Tonlé Sap Lake is not only important in Cambodia for growing rice and supplies of fish, the two main food resources of Cambodia, but it is also significant for biodiversity conservation. The Mekong is seasonal, and the extensive fisheries and agricultural production of the lowlands are driven by expected flood pulses.

Degradation of aquatic life and its effect on people are both significant. It is anticipated from excessive sedimentation in the river from local sources, pollution from urban discharges and fertilisers, and the engineering control of the river upstream that allows saline water to move upstream into the system. These factors undoubtedly affect the aquatic species, especially the endangered ones which include the giant freshwater catfish (*Pangasianodon gigas*) of the Mekong and the freshwater dolphins found below the Phapheng Falls. All the huge fishes in the river are now rare: the giant barb (*Catlopcarpio siamensis*), a type of Golden Carp (*Probarbus jullieni*), and others. A high diversity of gastropod snails occurs in the Mekong, especially near the confluence of the Mun with the Mekong. There is also a high bird diversity (Campbell 2009a). During the dry season, a large number of species of fish takes shelter in the elongated deep pools of the Mekong described earlier. Water stage fluctuates considerably through the year in the Mekong. As a result, fish in the river migrate (white fish) to utilise different habitat in the different seasons. White fish migrate from floodplains into channels and also along the river, provided barriers are absent. Certain species (known as black fish), however, reside in swamps, marshes and other depressions or use them as low season refuges. A middle group, known as grey fish, moves into the floodplain to breed and in search of food during high water but uses the river channels during the dry season. Impoundments prevent the migratory fish from reaching the suitable habitat for different stages in their life cycle and ameliorating floods that are necessary for the fish (Valbo-Jorgensen et al. 2009). The proposed channel impoundments therefore would have a negative effect on the main source of protein for most of the inhabitants of the basin.

The Tonlé Sap area (Figure 9.1) is threatened with excessive sedimentation and loss of aquatic species in the future. Degradation of fishing is from various causes: a loss of suitable habitat for spawning, breeding and feeding; commercial fishing replacing the rural catch; and fragmentation of the Mekong River system from impoundments. If the proposed cascade of up to 11 large dams on the Mekong (Middleton 2017) materialises, fragmentation on the Mekong could be disastrous. Apart from the Lancang Cascade and several other dams proposed on the Mekong in China, 12 hydropower dams have been proposed in Lao PDR and Cambodia (Carew-Reid 2017). Other dams on tributaries have also been proposed. The MRC prepared a Strategic Environmental Assessment (SEA) to review this high number of proposed impoundments against the entire plan of power generation on the river. The SEA recommended a 10-year postponement of constructing dams on the mainstream until their effects are better understood and their management becomes possible by the regional states. The SEA has been accepted by Cambodia and Vietnam, but Lao PDR has decided to go ahead with two large dams at Xayaburi and Don Sahong in order to generate hydroelectricity for the region. Carew-Reid (2017) has summarised the development of water resources and power potential of the Mekong River to take place against a web of political, historical, and economic ties amongst the Mekong countries, a web of increasing complexity and financial interests. Degradation of the aquatic environment is to be expected. He has identified the most significant problem for the SEA is the lack of any overarching plan for regulatory and management authority for the river in Lao PDR and Cambodia which could integrate the individual projects.

10.6 Conclusion

This case study on the Mekong illustrates the complexity of the physiography of a large river, which requires varying adjustments to anthropogenic modifications in different parts of the river. For example, a steep river on rock carrying coarse sediment will adjust differently below impoundments than a river on alluvium. If there are multiple stakeholders of the river basin with varying capacity and expectations, managing the river becomes difficult for an overarching agency. If assistance is received from external donor agencies, their expectations could also be different. Such possible conflicts exist in the case of the Mekong and other rivers. The effect of anthropogenic modification significantly affects the pattern of water discharge, sediment storage and transfer, and also channel physiography. Often, as in the case of the Mekong, it is not the country that achieves most from impoundments and other changes that also pays most in terms of environmental degradation. This may lead to a future conflict. Other changes could be disastrous. The effect of multiple dams may affect not only the physical characteristics of the channel but also basin ecology, and thereby the quality of life of river dwellers. The degradation of fishery and rice fields of the Mekong are intricately tied with the construction of large impoundments on the main river and the tributaries. Large-scale modifications should be planned only after proper impact assessments but that does not always happen and the necessary data are often not collected. National politics and economic expectations become more important than the environmental quality of the river and the basin. Furthermore, in the near future large rivers may be affected by anthropogenic alterations at the same time as they have to cope with a rising sea level and a changing hydrology, both due to global warming. In the case of the Mekong this could be disastrous.

Questions

1 Where do you expect the sediment of a natural Mekong to accumulate? Why?

2 Where are the current major impoundments on the Mekong located? What effect may they have on the sediment in the channel?

3 Discuss the future changes of Tonlé Sap Lake. Why is this lake important?

4 List the stakeholders of the Mekong. Discuss the possible areas of conflict amongst them.

5 Is it possible for the Mekong River Commission to establish an environmentally prosperous, socially just and environmentally sustainable Mekong? Discuss in detail.

References

Asian Development Bank (2012). *Greater Mekong Subregion Atlas of the Environment*. Manila: Asian Development Bank.

Campbell, I.C. (2007). The management of large rivers: technical and political challenges. In: *Large Rivers: Geomorphology and Management* (ed. A. Gupta), 571–585. Chichester: Wiley.

Campbell, I.C. (2009a). *The Mekong: Biophysical Environment of an International River Basin*. Amsterdam: Academic Press.

Campbell, I.C. (2009b). Development sceneries and Mekong River flows. In: *The Mekong: Biophysical Environment of an International River Basin*, 389–402. Amsterdam: Academic Press.

Campbell, I.C., Say, S., and Beardall, J. (2009). Tonle Sap Lake, the heart of the lower Mekong. In: *The Mekong: Biophysical Environment of an International River Basin* (ed. I.C. Campbell), 251–272. Amsterdam: Academic Press.

Carew-Reid, J. (2017). The Mekong: strategic environmental assessment of mainstream hydropower development in an international river basin. In: *Routledge Handbook of the Environment in Southeast Asia* (ed. P. Hirsch), 334–355. London: Routledge.

Douglas, I. (2005). The Mekong River Basin. In: *The Physical Geography of Southeast Asia* (ed. A. Gupta), 193–218. Oxford: Oxford University Press.

Gupta, A. (2004). The Mekong River: morphology, evolution and palaeoenvironment. *Journal of the Geological Society of India* 64: 525–533.

Gupta, A. (2007). The Mekong River: morphology, evolution, management. In: *Large Rivers: Geomorphology and Management* (ed. A. Gupta), 435–455. Chichester: Wiley.

Gupta, A. and Chen, P. (2002). Sediment movement on steep slopes in the Mekong Basin: an application of remote sensing. In: *The Structure, Function and Management Implications of Fluvial Sedimentary Systems* (eds. F.J. Dyer, M.C. Thom and J.M. Olley), 399–406. Wallingford: International Association of Hydrological Sciences.

Gupta, A. and Krishnan, P. (1994). Spatial distribution of sediment discharge in the coastal water of south and Southeast Asia. In: *Variability in Stream Erosion and Sediment Transport* (eds. L.J. Olive, R.J. Loughnan and J.A. Kesby), 457–463. Wallingford: International Association of Hydrological Sciences.

Gupta, A. and Liew, S.C. (2007). The Mekong from satellite imagery: a quick look at a large river. *Geomorphology* 85: 259–274.

Kummu, M. and Sarkkula, J. (2008). Impact of the Mekong River flow alteration on the Tonle Sap flood pulse. *Ambio* 37 (3): 185–192.

Kummu, M. and Varis, O. (2007). Sediment-related impacts due to upstream reservoir trapping, the Lower Mekong River. *Geomorphology* 85: 275–291.

Meade, R.H. (1996). River-sediment inputs to major deltas. In: *Sea-Level Rise and Coastal Subsidence: Causes, Consequences and Strategies* (eds. J.D. Milliman and B.U. Huq), 63–85. Kluwer Academic: Dordrecht.

Mekong River Commission (various dates (various dates, published annually)). *Lower Mekong Hydrologic Yearbook*. Bangkok, Phnom Penh and Vientiane: Mekong River Commission.

Mekong River Commission (1998) River Utilization Program Preparation Project: Final Report. Bangkok: Mekong River Commission, MKG/R.98.013.

Mekong River Commission (2011). *Integrated Water Resources Management-Based Basin Development Strategy*. Vientiane: Mekong River Commission.

Middleton, C. (2017). Water, rivers and dams. In: *Routledge Handbook of the Environment in Southeast Asia* (ed. P. Hirsch), 204–223. London: Routledge.

Nguyen, V.I., Ta, T.K.O., and Tatheishi, M. (2001). Late Holocene depositional environments and coastal evolution of the Mekong River Delta, southern Vietnam. *Journal of Asian Earth Science* 18: 427–439.

Peltzer, G. and Tapponnier, P. (1988). Formation and evolution of strike-slip faults, rifts and basins during the India-Asia Collision: an experimental approach. *Journal of Geophysical Research* 93: 15085–15117.

Penny, D. (2006). The Holocene history and development of the Tonle sap, Cambodia. *Quaternary Science Reviews* 25: 310–322.

Rundel, P.W. (2009). Vegetation in the Mekong. In: *The Mekong: Biophysical Environment of an International River Basin* (ed. I.C. Campbell), 143–160. Amsterdam: Academic Press.

Saito, T., Yang, Z., and Hori, K. (2001). The Huanghe (Yellow River) and Changjiang (Yangtze River) deltas; a review on their characteristics, evolution and sediment discharge during the Holocene. *Geomorphology* 41: 219–231.

Ta, T.K.O., Nguyen, V.I., Tatheishi, M. et al. (2002). Holocene delta evolution and sediment discharge of the Mekong River, southern Vietnam. *Quaternary Science Reviews* 21: 1807–1819.

Tapponnier, P., Peltzer, G., Le Dain, A.Y. et al. (1982). Propagating extrusion tectonics in Asia, new insights from simple experiments with plasticene. *Geology* 10: 611–616.

Tapponnier, P., Peltzer, G., and Armijo, R. (1986). On the mechanics of the collision between Indian and Asia. In: *Collision Tectonics* (eds. M.P. Coward and A.C. Ries), 115–157. London: Geological Society of London.

Valbo-Jorgensen, J., Coates, D., and Hortle, K. (2009). Fish diversity in the Mekong River basin. In: *The Mekong: Biophysical Environment of an International River Basin* (ed. I.C. Campbell), 161–190. Amsterdam: Academic Press.

Walling, D.E. (2008). The changing sediment load of the Mekong River. *Ambio* 37 (3): 150–157.

Walling, D.E. (2009). The sediment load of the Mekong River. In: *The Mekong: Biophysical Environment of an International River Basin* (ed. I.C. Campbell), 113–142. Amsterdam: Academic Press.

11

Large Arctic Rivers

Olav Slaymaker

Department of Geography, The University of British Columbia, Vancouver, Canada

11.1 Introduction

The rivers that flow through the arctic region have a set of special characteristics. The pattern of their annual discharge, the kind of sediment they transfer, their nutrient and contaminant loads, and their progressive morphological changes following global warming tend to distinguish themselves. The term 'large arctic rivers' may be a misnomer as all the rivers considered in this chapter also flow through a wide range of subarctic and even temperate biomes before debouching into the Arctic Ocean. They all flow to the Arctic Ocean or, as in the case of Yukon River, influence the composition of the water, sediments, dissolved solids, nutrients, and pollutants in the Arctic Ocean via marine transport through the Bering Strait. Large arctic rivers behave in much the same way as large rivers anywhere else but the severe thermal climate introduces distinctive effects associated with the extreme hydrological regimen and the seasonal occurrence of ice. This is the most important feature of the hydrological cycle in northern latitudes (Church 1977).

Large arctic rivers, however, continue to flow throughout the winter; baseflow being derived from unfrozen gravels along the channel, from deep springs or from lakes. Because of the large water retaining capacity of muskeg and healthy tundra vegetation, flood flows are greatly attenuated in watersheds dominated by these surface cover types. Lowland tundra with many standing water bodies exhibits a similar regime.

The Arctic Ocean is the shallowest and smallest of all the world's oceans and therefore the largest rivers that drain into the Arctic Ocean have a relatively large impact on that ocean (McClelland et al. 2012). The Arctic Ocean receives c. 11% of global river discharge while accounting for only c. 1% of global ocean volume. (Lammers et al. 2001). For example, the five largest arctic rivers discharge 2055 km^3 of fresh water annually, which is 40% of mean annual discharge of all freshwater river inputs to the Arctic Ocean (Shiklomanov et al. 2000). In comparison, only about 1000–1500 km^3 of atmospheric freshwater is contributed annually by precipitation minus evaporation over the Arctic Ocean (Walsh 2005). This means that freshwater added to the Arctic Ocean is made up of four parts river water and one part precipitation.

Marked seasonal variations in continental runoff are a defining feature around the pan-arctic domain. Most notably, runoff is relatively low during winter months, and

Introducing Large Rivers, First Edition. Avijit Gupta.

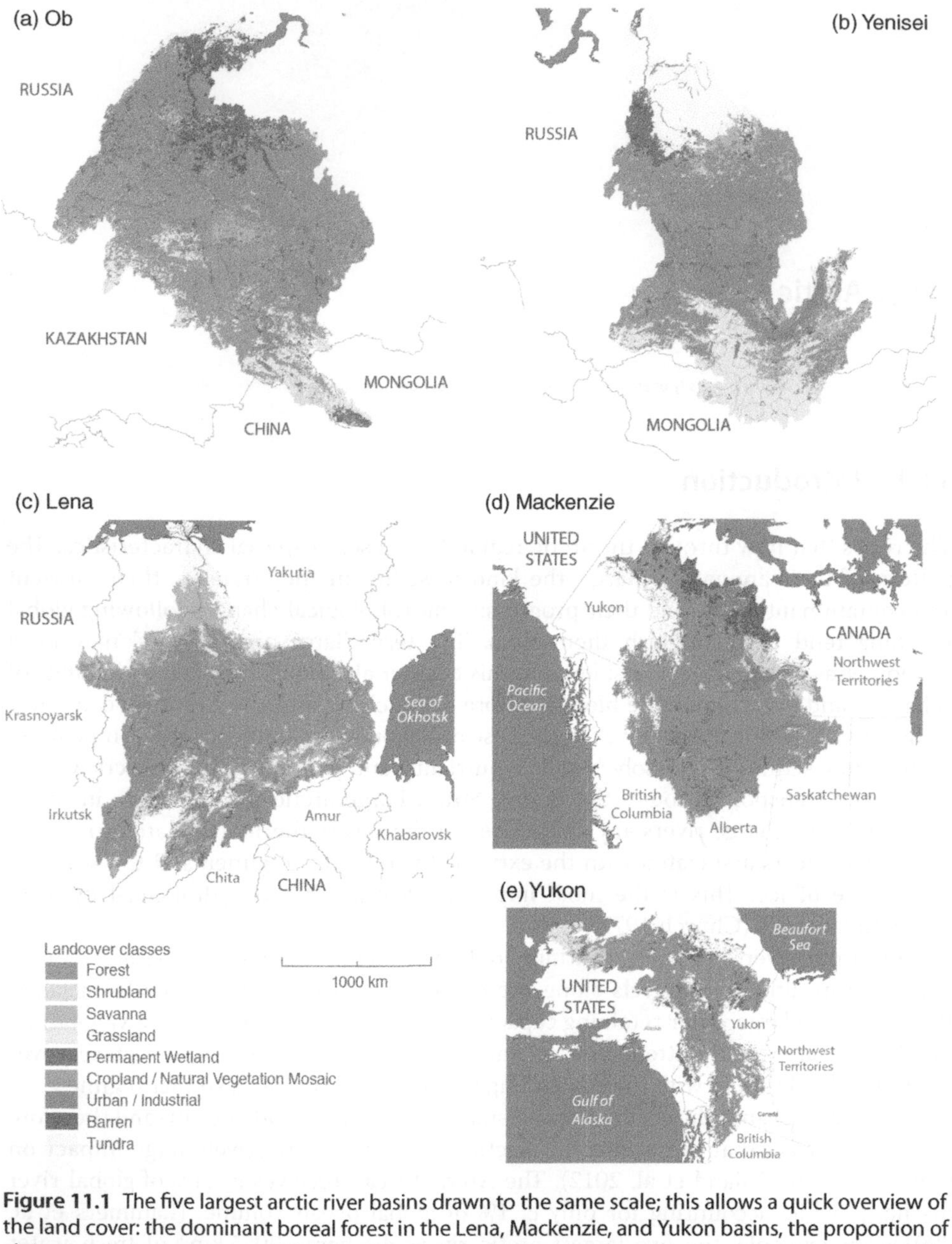

Figure 11.1 The five largest arctic river basins drawn to the same scale; this allows a quick overview of the land cover: the dominant boreal forest in the Lena, Mackenzie, and Yukon basins, the proportion of alpine tundra in the Lena and MacKenzie basins, the high proportion of permanent wetland in the Ob, Yenisei, and Mackenzie basins, the extensive agriculturally modified Ob basin, and the grassland for grazing in the Yenisei basins. Source: Adapted from Water Resources Institute 2003.

peaks with snowmelt in the spring. Rain induced floods occur in the smaller rivers during the spring and autumn but rarely do the largest rivers respond to rainfall events. The winter–spring transition in runoff occurs in April to June in the Kara Sea (Yenisei and Ob), Laptev Sea (Lena), Chukchi Sea (Yukon) and Beaufort Sea (Mackenzie) regions.

With clear signals of climate change occurring in the Arctic and greater changes anticipated in the future, there is good reason to focus on this region. In particular, determining fluxes of water, sediment, dissolved solids, nutrients, and pollutants is vital for understanding potential changes in local ecosystems and broader environmental feedbacks associated with storage and release of fresh water, sediment, and carbon. But perhaps the single most distinctive feature of the largest arctic rivers is that their drainage basins extend over many physiographic, hydroclimatic and ecological zones. The first half of this chapter therefore focuses on the spatial variability of the contributing drainage basins before addressing the regimes, behaviours and impacts of the large rivers themselves.

11.1.1 The Five Largest Arctic River Basins

The largest rivers, in order of basin magnitude, are the Ob, Yenisei, Lena, Mackenzie, and Yukon (Figure 11.1).

Among these five basins, Ob-Irtysh is the longest river and is the fourth largest river basin in the world (Table 11.1). These five basins are among the 22 largest river basins on the planet.

The Ob, Yenisei, and Yukon are international basins; the Mackenzie and Lena basins are confined within a single country (Table 11.1). A common characteristic of these basins is that they also straddle several biomes, eventually flowing into the Arctic Ocean through the Kara (Ob and Yenisei), the Laptev (Lena), the Beaufort (Mackenzie) and the Bering(Yukon) seas.

11.1.2 Climate Change in the Five Large Arctic Basins

The Arctic Climate Impact Assessment (ACIA 2005) identified similar trajectories of climate warming during the 1950–2000 period for all five large arctic basins. Average annual temperatures have increased by 1–3 °C, with most of the warming occurring during winter, when average temperatures increased 3–5 °C. The largest warming occurred inland in the Lena basin in Siberia in areas where a reduced period of snow

Table 11.1 Global rank and countries encompassed by each of the five large arctic river basins.

River and basin	Length (km)	Basin area (10^6 km^2)	Global rank (length)	Global rank (basin)	Countries
Ob	5400	3.00	5	4	Russia, Kazakhstan, China, Mongolia
Yenisei	4100	2.60	11	7	Russia, Mongolia
Lena	4400	2.50	8	8	Russia
Mackenzie	4200	1.80	10	11	Canada
Yukon	3700	0.83	15	22	USA, Canada

Source: Data from Milliman and Farnsworth 2011.

cover amplified the warming. Thawing of permafrost is already presenting problems, and coastal erosion on the shores of the Bering and Beaufort seas has accelerated.

Environmental changes which are envisaged over the next 50 years should be borne in mind for these vast river basins (McCarthy and Martello 2005). Model simulations of future climate suggest continuation of the warming trend which, if the warming continues, will have a major impact on river flows. Ice roads will be usable for shorter periods. The discontinuous permafrost zone is expected to move northwards and thermokarst development is seen as a potential natural hazard. Projected increase in winter precipitation may lead to increase in mean annual discharge and a shift in timing of peak flows to earlier in the spring. Greater flows of nutrients and sediments can be expected to provide both positive and negative effects. Increased absolute water resources are likely to produce greater risk of flooding. Extension of boreal forest northwards is predicted and fire frequency and intensity are expected to increase.

11.1.3 River Basin Zones

Schumm (1977) presented an idealised fluvial system model in the context of the study of small river basins (Figure 11.2). The model identifies three zones: (i) sediment production; (ii) sediment transfer; and (iii) sediment deposition zones. Here we suggest that this small basin model is applicable to all river basins, however large they may be. However, the following considerations, inter alia, add complexity. The largest arctic river basins have complex physiography, Quaternary history, hydroclimatic and vegetation zones, and different degrees of modification by human agency within a single basin. For example, the Siberian basins are heavily regulated by dams and reservoirs. The Mackenzie basin is regulated by its very large natural lakes but the Yukon basin effectively is unregulated. Every dam, reservoir, and large lake becomes a sediment deposition zone (zone 3) at some intermediate point within the basin and a new sediment production zone (zone 1) is initiated downstream. Because of the extensive lowlands in the northern reaches of the Siberian rivers and their high level of regulation, the rate of deposition of sediment in their coastal zone (zone 3) is an order of magnitude lower than that of the Mackenzie and Yukon basins (see Box 11.1).

Box 11.1 The Idealised Fluvial System Model (Schumm 1977)

This well-known idealised fluvial system model was developed in the context of the study of small river basins in temperate climate environments (Figure 11.2). The model identifies three zones: (i) sediment production; (ii) sediment transfer; and (iii) sediment deposition zones. Zone 1 is the area of greatest interest to watershed scientists, hydrologists, and geomorphologists working on the evolution of drainage systems. Zone 2 is of major concern to the hydraulic and river control engineer, to stratigraphers and sedimentologists and to geomorphologists interested in river channel morphology. Zone 3 is primarily of concern to the geologist, geomorphologist, and the coastal engineer. The internal structure, stratigraphy, and morphology of alluvial fans, alluvial plains and deltas are of particular geomorphic interest (Schumm 1977). In the most general sense, this small basin model is equally applicable to the largest river basins. The only difference is that the largest river basins have multiple sediment producing, sediment transporting and sediment deposition zones rather than just one of each.

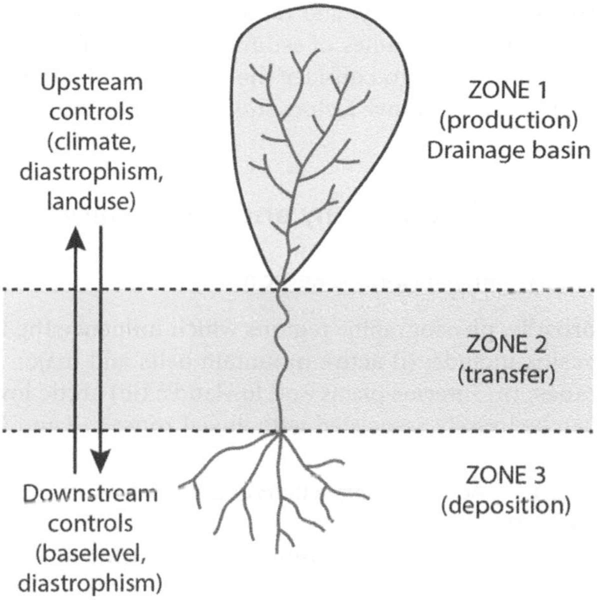

Figure 11.2 Ideal fluvial system model (after Schumm 1977). Every fluvial system at every spatial scale contains each of the zones 1, 2, and 3.

More than 60% of the Lena, Mackenzie, and Yukon basins is covered with taiga or boreal forest (Figure 11.1). The Ob and Yenisei basins have a more diverse land cover because they extend further south and people there have modified the natural forest cover for agricultural and urban purposes. These two flow from undisturbed boreal forest to arable land, and in doing so, they are partially transformed from sediment transporters (zone 2) to sediment producer (zone 1) because the disturbed land surface makes sediment more available to river and gully erosion.

The deltas of the Mackenzie and Lena rivers are a focus of much economic and environmental conflict and are in danger of being transformed from sediment deposition zones (zone 3) to sediment producers (zone 1). Perhaps most distinctively, the Mackenzie and Yukon basins are relatively pristine large basins poised on the verge of large-scale economic development. The Russian basins have already gone through early economic development with extensive damming of rivers and exploitation of their water resources.

The high-latitude location of the northward-flowing arctic rivers implies a number of characteristic geomorphic features. The presence of permafrost, whether continuous, discontinuous, or scattered, impacts the evolution of the fluvial zones 1, 2 and 3. For example, these large arctic basins remain frozen at their lower (northern) ends for six months of the year because of polar climate; they also remain frozen for long periods of the year at their upper ends because of the high-altitude climate. The city of Barnaul, located at the upper (southern) end of the Ob system, is frozen for almost as long a period as Salekhard, at the arctic mouth of River Ob (Wohl 2011). The presence of river and lake ice simultaneously in zones 1 and 3 can drastically reduce the length and transfer capacity of zone 2 and produces unique hydrologic phenomena during break-up and freeze-up periods. The comparatively recent post-glacial rearrangement of arctic river drainage over the last 10 000 years further complicates the picture. Every basin includes an exceptional volume of storage in large and small lakes and poorly drained muskegs.

The size, complexity, and remoteness of these arctic basins make the conducting of source-to-sink studies of sediment and dissolved solids fluxes extremely challenging. When we come to consider the behaviour of large arctic rivers in response to climate change, all the above factors should be taken into account (Ashmore and Church 2001).

11.2 Physiography and Quaternary Legacy

11.2.1 Physiographic Regions

Broadly, physiographic regions which influence the behaviour of the five largest arctic basins include: (i) active mountain belts and major mountain belts with accreted terranes; (ii) interior plains and lowlands; (iii) arctic lowlands. These physiographic units can be loosely associated with fluvial zones 1, 2 and 3 in Schumm's model.

11.2.1.1 Active Mountain Belts and Major Mountain Belts with Accreted Terranes (Zone 1)

The most spectacular mountain belts with active tectonic histories and extensive ice fields are found in the Canadian and Alaskan Cordillera and in the Central Asian Orogenic Belt. The Altai and Sayan mountains, form the southern boundary of the Lena, Yenisei and Ob basins (Figure 11.3); the Wrangell-St Elias Mountains and

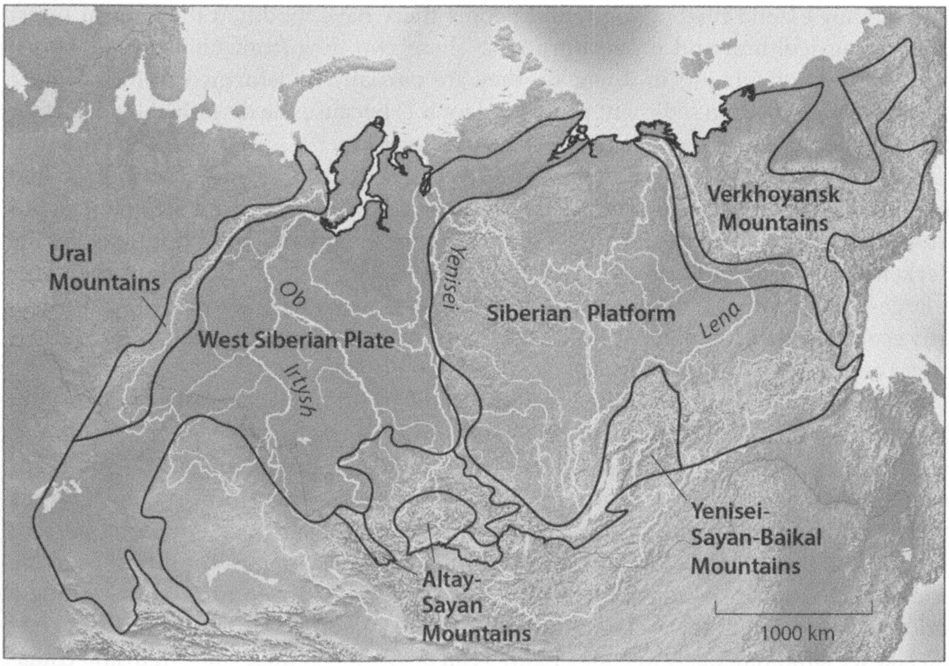

Figure 11.3 Major tectonic structure of the Lena, Yenisei, and Ob basins (after Koronovsky 2002). The Siberian platform forms a craton that is encircled by the younger folded mountain systems and the gently dipping sedimentary strata overlying the West Siberian Plate. Black lines are the boundaries of the tectonic structure, and grey lines demarcate the watersheds of the three rivers.

the Alaska Range mark the southern boundary of the Yukon basin; and the Rocky Mountains form the western boundary of the Mackenzie basin (Figure 11.4). Lesser mountains in the interior of the Cordillera and Alaska form a mosaic of geologic terranes which control the lineaments of the landscape and much of the shapes of the river basins (Koronovsky 2002).

11.2.1.2 Interior Plains, Lowlands, and Plateaux (Zone 2)

The eastern half of the Mackenzie basin and the northern halves of the Lena, Yenisei, and Ob basins are underlain by Precambrian rocks. They are known locally as the Canadian Shield, and the Siberian Platform. The distinctive feature of the Canadian Shield occupied by the Mackenzie basin is the presence of three huge lakes: Athabasca, Great Slave, and Great Bear, with surface areas of 7.8×10^3, 28.6×10^3, and 31.3×10^3 km^2, respectively. The Yenisei basin includes Lake Baikal which, although similar in surface area to Great Bear Lake, is so deep that it contains c. 20% of all terrestrial surface water (23×10^3 km^3). The lake is situated in one of a series of rift zones, which has been active since the Tertiary. Each of these mega-lakes serves as an intermediate sediment deposition site or a local zone 3 in Schumm's terminology.

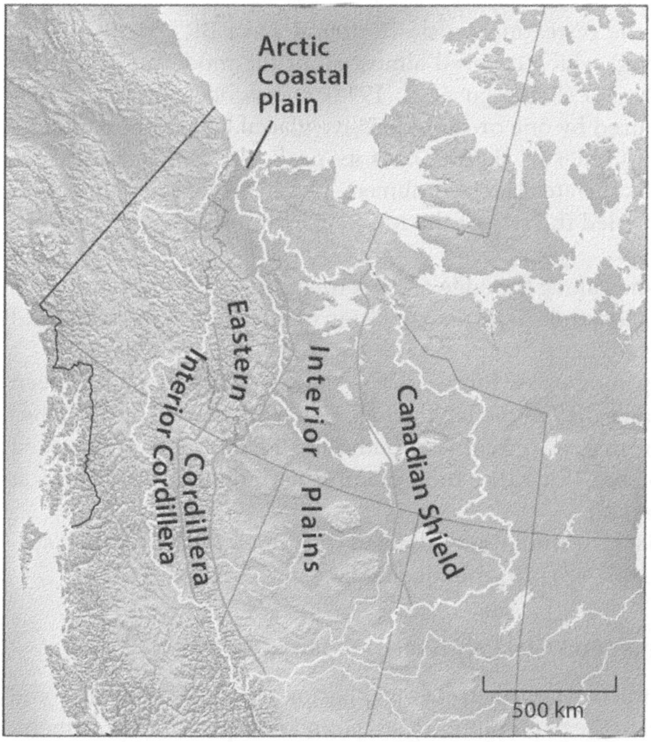

Figure 11.4 Physiographic subdivisions of the Mackenzie basin (after Bostock 2014). The Canadian Shield is a craton that is bounded to the west by younger folded mountains of the Cordillera and the gently dipping sedimentary strata of the Interior Plains.

11.2.1.3 Arctic Lowlands (Zone 3)

Low relief and continuous permafrost zones fringe the Arctic Ocean from the Bering Strait in Alaska to the East Siberian Sea in Russia. The most dramatic physiographic features are the massive deltas of the Lena (30×10^3 km^2), Yukon (23×10^3 km^2) and Mackenzie (13×10^3 km^2) rivers. These deltas prograde continuously into the Laptev, Bering and Beaufort seas, respectively.

11.2.2 Ice Sheets and Their Influence on Drainage Rearrangement

Almost all of the Mackenzie basin and the majority of the Yukon basin were overridden by ice. In comparison, only the lowermost and uppermost parts of the Ob and Yenisei basins were overridden, and the Lena basin experienced limited ice sheet cover, a fact which encouraged the development of the thickest permafrost on Earth. Nevertheless, the pre-glacial drainage of all five basins was significantly rearranged during the Pleistocene by the interaction of the varying trajectories of the ice sheets with the pre-existing topography.

The last continental glaciation redirected much of Mackenzie River drainage away from the Hudson Bay to the Arctic Ocean. The Mackenzie basin is itself entirely post-glacial in age (Duk-Rodkin and Lemmen 2000) and its landscapes can be aptly described as paraglacial or dominated by 'non-glacial processes conditioned by glaciation' (Church and Ryder 1972).

Lakes were impounded by the retreating ice sheet around 13 ka BP, culminating in Lake McConnell (1100 km long) which covered almost all of the modern Great Bear, Great Slave and Athabasca basins (Lemmen et al. 1994). It has been proposed that the Mackenzie valley was scoured by one or more massive glacial lake outburst floods unleashed from Lake Agassiz (Teller et al. 2005). At its peak, Agassiz had a greater volume than all present-day freshwater lakes combined. This hydrological connection would have approximately doubled the size of the Mackenzie drainage basin for about 400 years (Figure 11.5).

Meltwater channels established along the edge of the retreating ice sheet became parts of Mackenzie River tributaries. Post-glacial stream incision of up to 500 m has produced spectacular canyons on the Peel Plateau.

In Siberia, two contrasting models of Eurasian Ice Sheet deglaciation have been described (Grosswald and Hughes 1995; Velichko et al. 1997). Grosswald and Hughes suggest meltwater drainage from east to west with both Ob and Yenisei rivers blocked from flowing north. Velichko et al. think that the Ob and Yenisei flowed northwards at all times during the last glaciation.

A huge West Siberian glacial lake would have drained the upper Yenisei and Ob systems through the Aral and Caspian seas to the Mediterranean. (cf. Baker 2008).

11.2.3 Intense Mass Movement on Glacially Over-steepened Slopes

In general, the landscapes of the five largest arctic river basins are geologically new and are in the process of transition to the contemporary fluvial and non-glacial environment. The Mackenzie and Yukon basins are essentially mountain basins and the Siberian basins are more characteristically plateau and lowland basins.

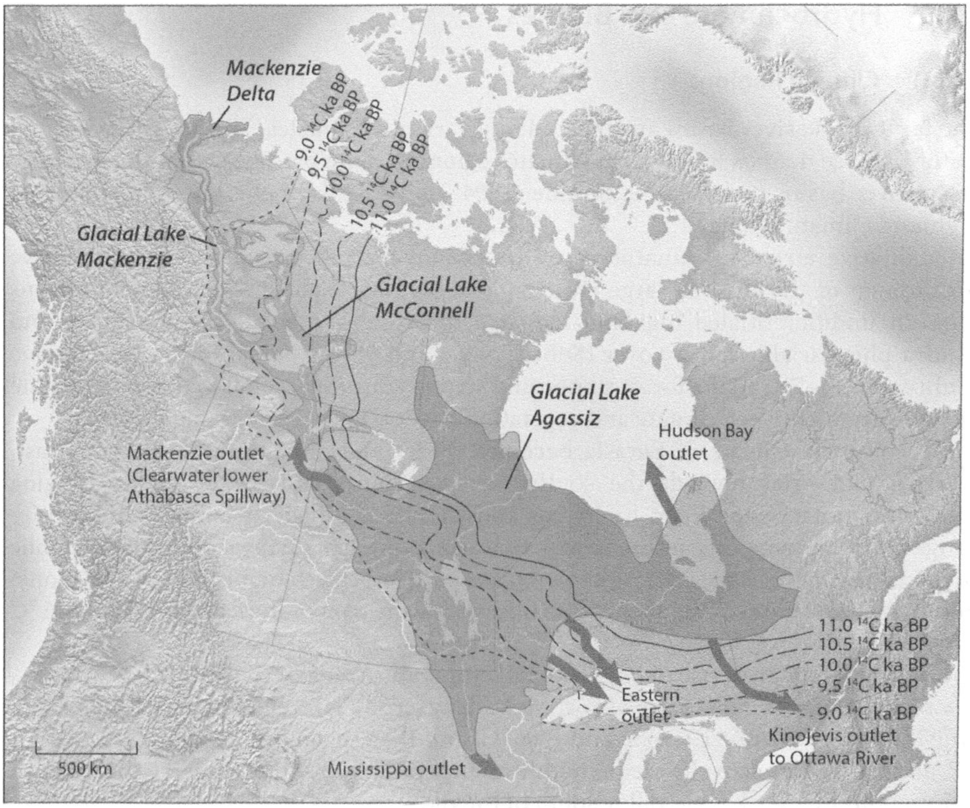

Figure 11.5 Maximum extent of glacial lakes along the margin of the retreating Laurentide Ice Sheet (after Smith 1994; Teller et al. 2005). Numbered lines indicate ice margins during deglaciation in thousands of radiocarbon years from Dyke (2004).

Landslides which occur on slopes that have been 'debuttressed' following the removal of glacial ice continue to provide evidence of the landscape instability caused by Quaternary glaciation. Between 60 °N and the delta, the Mackenzie Valley and adjacent mountains contain more than 3000 active landslides over an area of 250 000 km^2 (Aylsworth et al. 2000). Many of these landslides are related to the contemporary degradation of permafrost and ground ice; others are influenced by permafrost control on ground water movement. Any rise in ground temperature and degradation of permafrost increases the frequency of landslides. The most vulnerable sites include ice-rich, fine grained sediments on slopes near water bodies or in fire scars, and frozen, coarse-grained sediments overlying clay or clayey till in steep riverbanks. In the Mackenzie basin, the major impact of these landslides will be the disruption or destruction of pipelines, roads, and bridges. River navigation can also be temporarily affected in the event of a landslide damming a river. Increased landsliding can also increase sedimentation in streams and may lead to destruction of spawning beds.

11.3 Hydroclimate and Biomes

11.3.1 Climate Regions

The large arctic river basins are affected by a variety of climates. The Thornthwaite soil water balance defines five distinct climate regions that are present within the five large arctic river basins (Strahler and Strahler 1994; Figure 11.6a,b) plus an undifferentiated highland climate (Table 11.2).

Qualitative air mass climatology (Shahgedanova 2002) provides a more sensitive breakdown of the Russian large river regions defining eight distinct climate regions plus an undifferentiated highland climate. The tundra zone (Subregions 1–3) and tundra plus variable forest cover (Subregions 4 and 9); the boreal forest (Taiga) zone (Subregions 5–7) and the forest steppe and steppe zone (Subregion 8) demonstrate the close relation between climate and vegetation cover (Figure 11.6c; Table 11.3).

Basically, the climate of Eurasia becomes colder and drier towards the northeast, warmer and drier towards the southwest, and cooler and wetter in the fringing mountain ranges. The Lena basin has consistently the highest annual temperature range from extreme cold in the winter to warm summers. Oymyakon in the Republic of Sakha in Siberia is the coldest known place in the northern hemisphere: −68 °C on 6 February 1933. From December through February the temperature drops below −46 °C every night.

Continentality gives rise to large annual temperature ranges. In the Mackenzie basin, the mean January temperature of Fort McMurray (56°39′N), Yellowknife (61°46′N), and Inuvik (68°18′N) are −19, −25, and −28 °C, and their mean July temperatures are 17, 17, and 14 °C, respectively. Mean monthly temperatures in summer tend to be uniform throughout the Mackenzie valley south of Inuvik.

11.3.2 Biomes

The Lena, Yukon and Mackenzie basins remain essentially boreal forest landscapes, with widespread wetlands. Only the Yenisei and especially the Ob basins have significantly transformed their boreal forests into agricultural and urban land.

Taiga is the circumpolar boreal forest that rings the northern hemisphere, mostly north of the 50th parallel. It stretches from the most easterly part of the Mackenzie basin to the mouth of the Yukon and from the western Ob basin to the eastern Lena basin (Figure 11.7).

The area is dominated by coniferous forests, particularly spruce (North America) and pine and larch (Siberia), interspersed with vast wetlands, mostly bogs and fens. The Lena (1.6 million km^2 or 65% of the basin), Mackenzie (1.2 million km^2, 66%) and Yukon (0.5 million km^2, 64%) basins consist of more taiga or boreal biome than the Ob and the Yenisei which have only 34 and 40% taiga, respectively. The riverbanks of the Lena, Mackenzie, and Yukon are lined with sparse vegetation like dwarf birch and willows, as well as numerous peat bogs. In North America, the boreal forest is composed largely of spruce (*Picea glauca and Picea mariana*) whereas in Siberia it is composed predominantly of pine (*Pinus silvestris)* and tamarack (*Larix dahurica*). Towards the south the tundra vegetation transitions to black spruce, aspen and poplar forest. A broad swath of boreal forest covers the plains. The boreal forest consists of black and white spruce

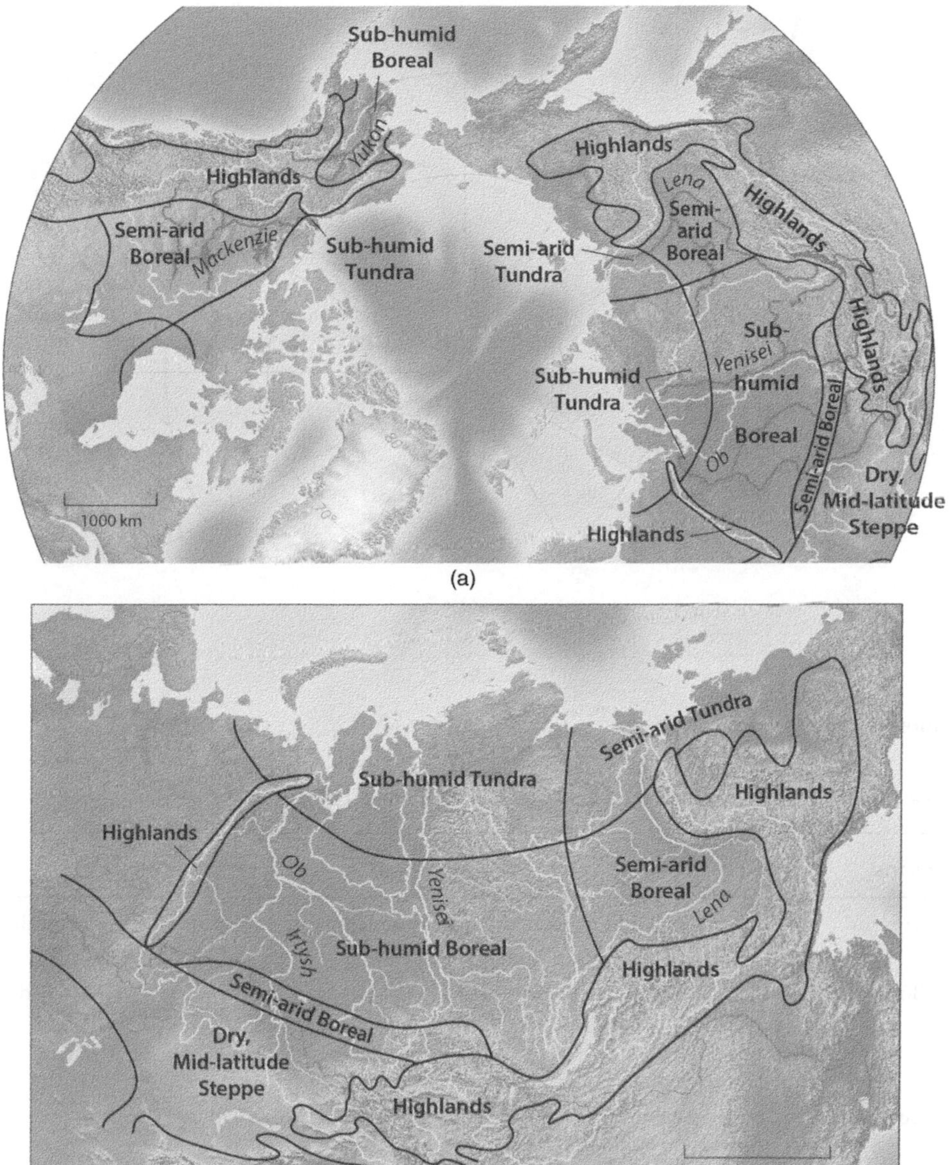

Figure 11.6 (a) Climate regions in the five large river basins based on Thornthwaite's soil water balance as described in Strahler and Strahler (1994). (b) Climate regions in the three largest Russian basins based on Strahler and Strahler (1994). (c) Climate zones and provinces in the three largest Russian basins based on Alisov (1956). See Table 11.3.

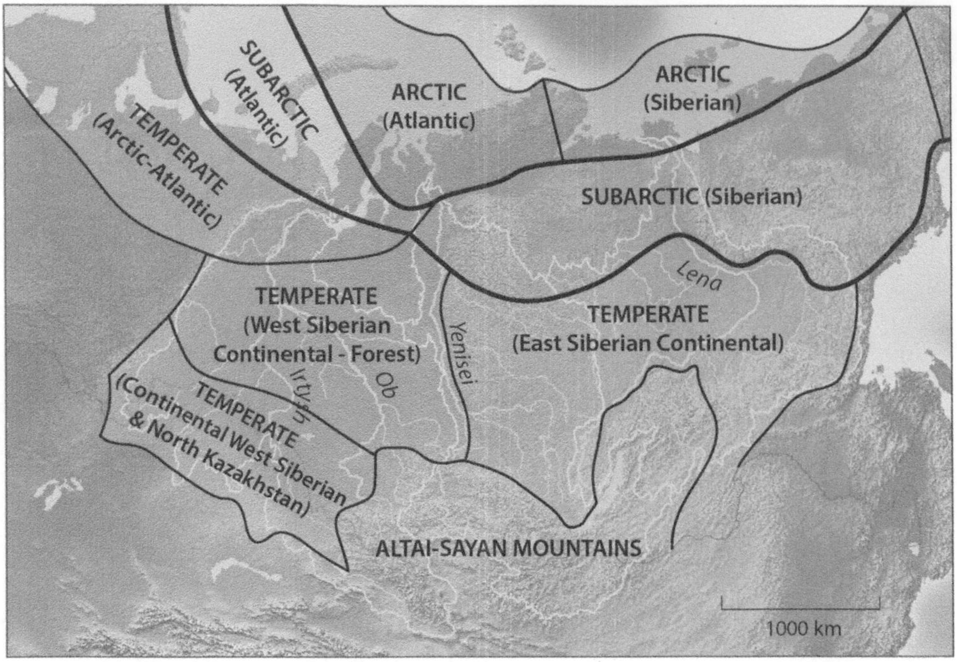

(c)

Figure 11.6 *(Continued)*

Table 11.2 Climate regions based on soil water balance (after Strahler and Strahler 1994).

Regions	Subregions
A. Tundra	1. Semi-arid tundra: found in Lena basin only, very cold, sparse forest and tundra
	2. Sub-humid tundra: found especially in Yenisei, Ob, and Mackenzie basins
B. Boreal	3. Semi-arid boreal: found especially in Mackenzie and Lena basins, relatively dry, boreal, and mixed forests
	4. Sub-humid boreal: found especially in Yukon, Ob, and Yenisei, relatively humid, cool summers, boreal forest
C. Temperate	5. Steppe: found in Ob basin only, relatively warm, mid-latitude steppe. Agriculture and urban land use
D. Highland	6. Found in all basins

mixed with aspen and poplar (*Populus tremuloides; Populus balsamifera*), pines (*Pinus banksiana; Pinus contorta*), birch (Betula spp.), balsam fir (*Abies balsamiea*), and with larch (*Larix laricina*) in poorly drained areas. The subarctic open woodland is dominated by black or white spruce with a ground cover of lichen and low shrubs.

The boreal zone is sparse in climatic and hydrometric data (Lammers et al. 2001) though there are sufficient streamflow gauging stations to provide representative data from across the circumpolar ring of boreal forests. The hydroclimate determines the

Table 11.3 Climate regions of the Russian large arctic rivers based on qualitative air mass climatology (Alisov 1956).

Regions	Subregions	Large river basins
A. Arctic	1. Atlantic	Ob and Yenisei estuaries
	2. Siberian	Lena Delta
B. Subarctic	3. Atlantic	Ob and Yenisei lowlands
	4. Siberian	Yenisei and Lena plateaux
C. Temperate	5. Arctic-Atlantic	Ob northwestern
	6. W. Siberian continental	Ob and western Yenisei
	7. E. Siberian continental	Lena and eastern Yenisei
	8. Continental W. Siberian	Ob forest steppe-steppe
D. Highlands	9. Altai-Sayan Mountains	Undifferentiated

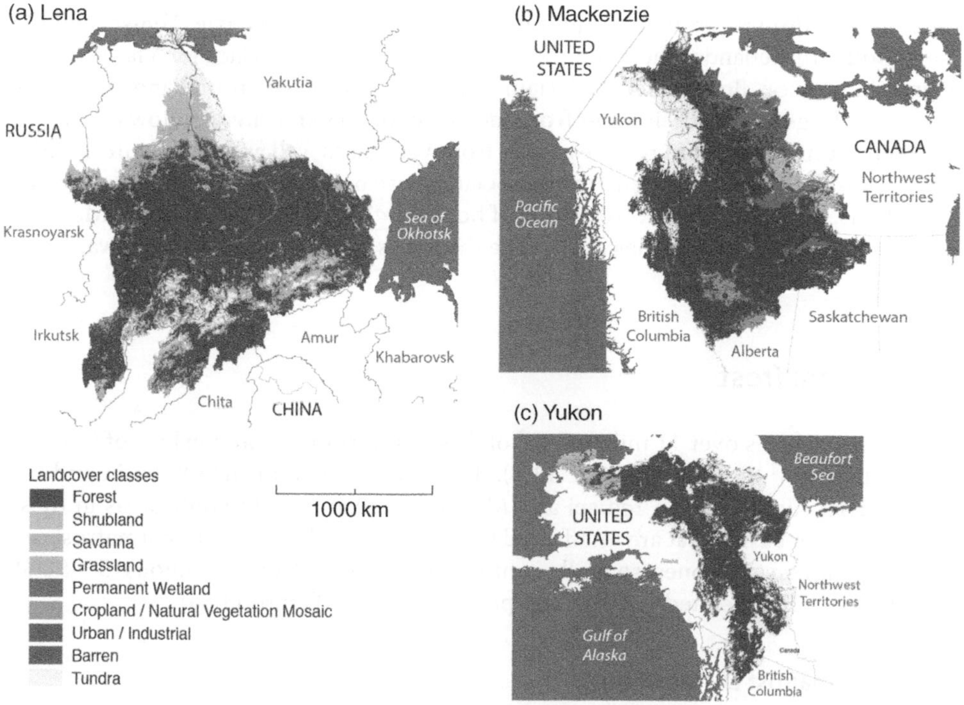

Figure 11.7 The boreal forest (taiga) biome in (a) Lena, (b) Mackenzie and (c) Yukon basins: essential boreal forest landscapes. Source: Adapted from Water Resources Institute 2003.

streamflow regime. The boreal climate is a result of complex interplay among such factors as large seasonal contrasts in solar input, a wide spectrum of disturbances in the mid and high latitude air streams and variable physiography within and adjacent to the boreal zone. Within the boreal zone there is also considerable variety in hydroclimate due to topographic influences, proximity to large water bodies, and position of large-scale atmospheric features such as fronts and storm tracks. Furthermore, the boreal climate exhibits large temporal variability as a function of interactions of the north Pacific flow with the Cordillera, such that winter temperature response in the western Canadian boreal zone is unusually variable. Enhanced response to large-scale conditions has induced exceptional winter warming of northwestern North America and west and central Siberia and cooling over eastern Canada during the last few decades.

The Ob and the Yenisei basins incorporate seven natural vegetation regions of the world in addition to boreal forest or taiga (Zlotin 2002). Alpine tundra, semi-desert scrub, steppe, forest-steppe, mid-latitude deciduous forest, forest-tundra, and arctic tundra. Semi-desert scrub and steppe biomes, within and immediately to the northwest of the central Asian orogenic belt, provide a higher variety in the Ob and Yenisei basins than in the others. The natural biomes have been transformed into arable land (about 37% of their land cover) and urban land surfaces in the Ob basin giving it a distinctive appearance.

11.3.3 Wetlands

Vast wetlands dominated by bog and fen characterise the Mackenzie, Yenisei and Ob basins. Both are wetlands where the water table is close to the surface for a large part of the growing season. Bog is often underlain by perennially frozen ground and ground ice is common. Bogs receive their water from meteoric sources and lateral inflow is insignificant. Fen is unfrozen. A fen receives water from vertical as well as lateral sources and is richer in minerals. Large peatlands are associated with broad, featureless glacilacustrine or till plains. There are about 30 000 km^2 of bog and c10 000 km^2 of fen in the Mackenzie basin (Aylsworth and Kettles 2000). The thickness of the peat increases towards the south.

11.4 Permafrost

Permafrost occupies over 11 million km^2 of Russian territory, 5 million km^2 of Canada, and 1 million km^2 of Alaska (Figure 11.8). The southern half of the Ob basin and the southwestern corners of the Yenisei and Mackenzie basins are the only areas in these large arctic river basins that are unaffected by permafrost. The area with permafrost can be divided into several zones: spatially continuous (90–100%), discontinuous (50–90%), sporadic (10–50%), isolated (<10%), and permafrost-free (Figure 11.8a).

11.4.1 Permafrost Distribution

The extent of permafrost increases from the permafrost-free area in the southwest of the Ob, Yenisei and Mackenzie basins, through isolated patches of permafrost in the mountains of all five basins (Tumel 2002; Woo 2012). This is rapidly followed in the plains towards the north and northeast, and with increasing elevation in the mountains by a sporadic permafrost zone and then a discontinuous permafrost zone. Continuous permafrost is widespread in the Lena and northern half of the Yukon basins but is confined to a relatively narrow coastal strip of the Ob, Yenisei and Mackenzie basins (Figure 11.8a).

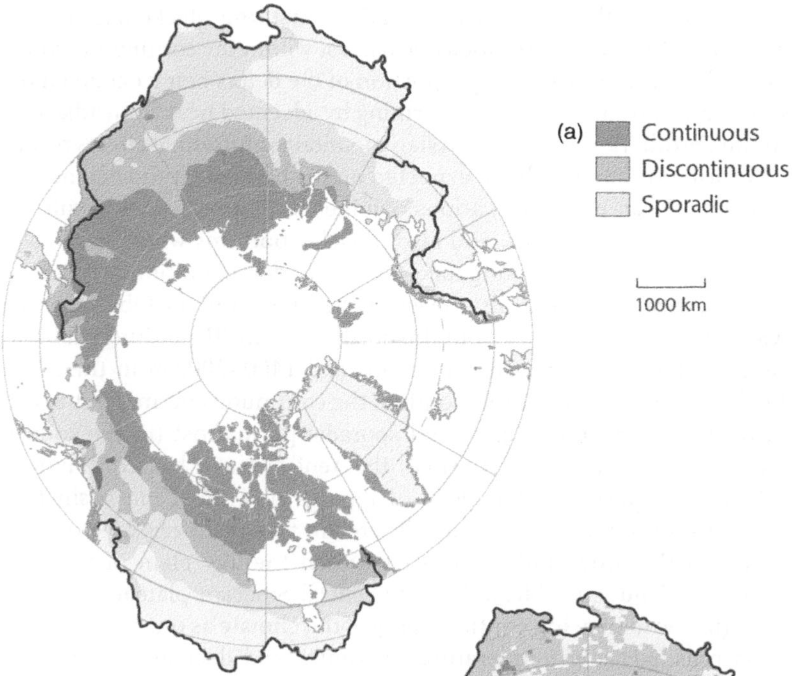

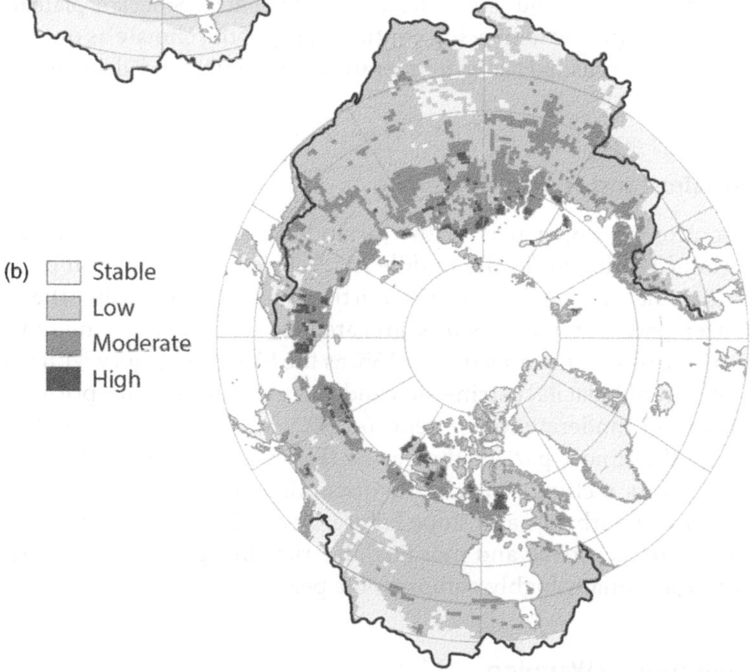

Figure 11.8 (a) Zonation of permafrost in the northern hemisphere under the climatic scenario predicted by a specific kind of Global Circulation Model developed by the Max Planck Institute for Meteorology to predict future climates, ECHAMI-a GCM (after Cubasch et al. 1992), where continental permafrost would be reduced by 12–22% under a doubling of CO_2 concentration in the atmosphere. (b) Hazard potential associated with degradation of permafrost under ECHAMI climate change scenario. Classification is based on a thaw settlement index calculated as the product of existing ground ice content (after Brown et al. 1997) and predicted increase in the depth of thaw (after Anisimov and Fitzharris 2001).

There are some exceptions to this broad generalisation. First, the Mackenzie, Yukon and Lena deltas are areas of local discontinuous permafrost within the continuous zone. The two main reasons for this are (i) a large proportion of the deltas being occupied by lakes with subjacent taliks, and (ii) the ground warming by advected heat from adjacent rivers. Secondly, in the Cordilleran and central Asian mountains altitudinal effects result in more extensive permafrost than would be expected on the basis of latitude alone.

These effects occur both at upper elevations (alpine permafrost) as a result of normal adiabatic temperature lapse rates, and also in valley bottoms where very cold air is ponded and severe radiative cooling takes place under conditions of stable temperature inversion in winter. In the Urals, the western boundary of the Ob basin, the permafrost limit varies from 400 to 800 m from the north to south. The altitudinal limit of permafrost is 600–1200 m in the Sayan Mountains and 1400–2000 m in the Altai Mountains. The Ob basin lies almost entirely south of the continuous permafrost zone. Discontinuous (in the north and north east) and sporadic permafrost (in the south) occupy as little as 20% of the basin. By contrast, the Yenisei basin has continuous, discontinuous and sporadic permafrost in 80% of its area and the Lena basin has 100% continuous and discontinuous permafrost. There are two main reasons for this distributional pattern: (i) the orographic contrast between the low plain of western Siberia (<250 m) and the denuded tableland of the central Siberian plateau (average elevation 600 m), and (ii) the increasing continentality of the climate as one moves from west to east, with increasingly low temperatures in winter and decreasing amounts of snowfall.

11.4.2 Permafrost and Surficial Materials

The ice content of the upper part of the ground ranges from negligible in rock and dry, coarse grained sediment in the middle Mackenzie basin to high in peatlands and fine-grained deposits, particularly in the northern part of the basin. The ice usually occurs as frozen pore water, lenses of segregated ice typically a few millimetres thick or ice veins. The ice content varies from >15% as the highest category with very limited extent to peat-covered glacilacustrine silts and clays. The most widespread category of ground ice is that of moderate ice content (6–15%). Massive icy beds are found exclusively in the northernmost part of the basins (Figure 11.8b). Ground ice is generally within a degree of its melting point in the discontinuous permafrost zone and is susceptible to surface disturbance and slope failure. Geological materials most sensitive to thaw settlement are lacustrine and morainal, ice-rich, fine grained sediments and also the ice-rich and potentially highly compressible peatland.

11.4.3 Contemporary Warming

In the latter half of the twentieth century, mean annual temperatures for most recording stations on average have risen about 1–2 °C (Figure 11.9). Warming is most notable in the winter season and snow depth decreases by the end of February (Derksen and Brown 2012). Permafrost and ground ice conditions are changing with an overall reduction in the extent of permafrost and a melting of ground ice, with a general northward shift in the southern limits of permafrost. On a site-by-site basis, the effects of climate

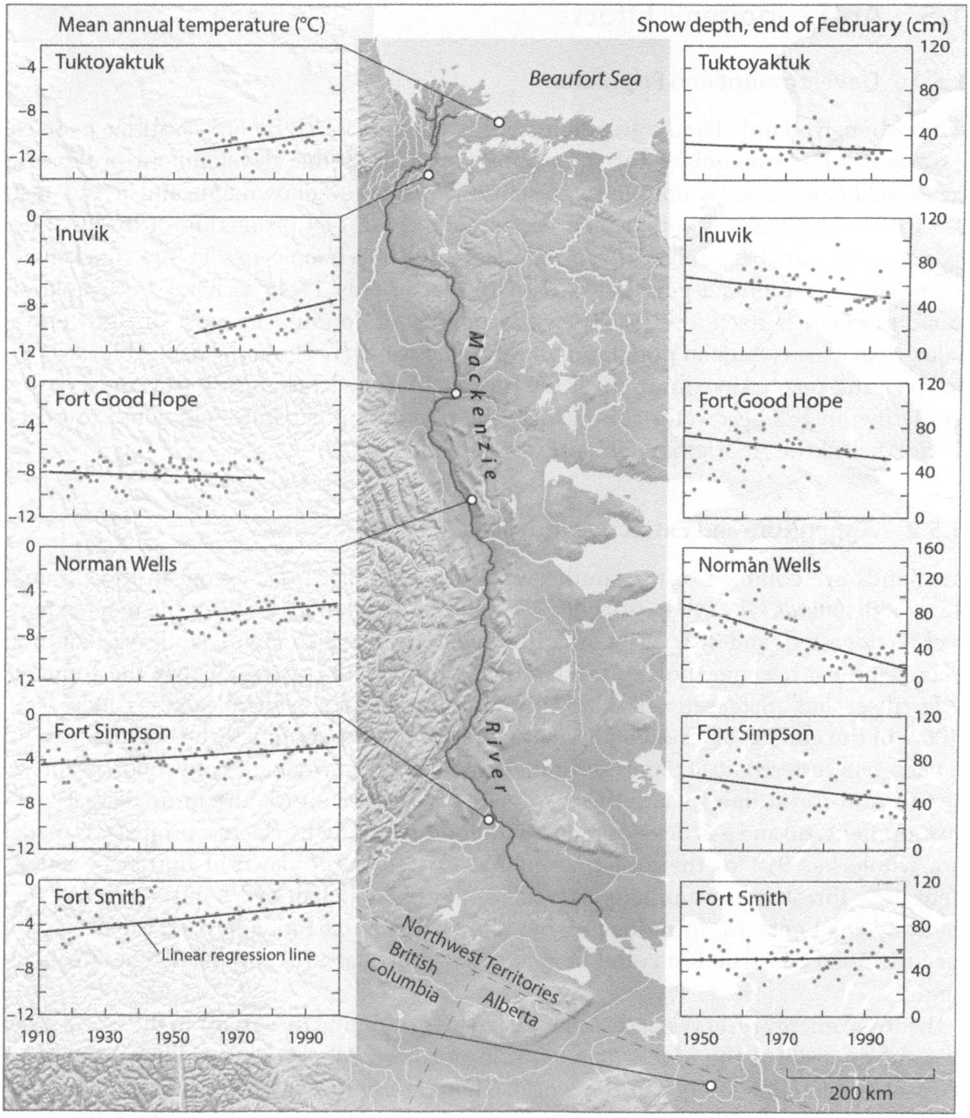

Figure 11.9 The six longest Environment Canada climate records available for mean temperature and snow depth at the end of February for stations in the Mackenzie valley. Linear egressions suggest an overall warming and decrease in snow depth for the intervals of record (after Dyke 2000).

warming can be assessed qualitatively. The low thermal conductivity and diffusivity of earth materials, coupled with the high latent heat required to melt ice, lead to a time lag between warming of the air and ground surface and permafrost thawing. For deep permafrost this time lag may be of the order of thousands of years. The warming trend of the past decades has had serious impact on the discontinuous and sporadic permafrost zone. Thermokarst development complicates road and pipeline corridor maintenance, especially around the southernmost limits of permafrost degradation.

11.5 Anthropogenic Effects

11.5.1 Development and Population

Comprehensive land claims agreements are being demanded by First Nations peoples in Canada as a prerequisite for their approval of economic development proposals. The Mackenzie basin is unique in its natural resource endowment, and it is a test case for the seriousness with which economic growth and protection of the pristine environment can be resolved. Approximately 400 000 people live in the Mackenzie basin. This represents a population density that is only 1% of that for the whole of Canada, which is itself a relatively lightly populated country. In most of the world's major river basins, human population tends to concentrate along the navigable parts of the river and reach a maximum near the mouth. In these large arctic basins most of the people live in the headwater sub-basins. The population becomes less dense towards the north and the Aboriginal population approaches 100%.

11.5.2 Agriculture and Extractive Industry

Farmlands are confined to the southern fringe of the lowlands, except in the Ob and Mackenzie basins. The Ob basin, with 30 million inhabitants, has been the most extensively deforested, and 37% of its area has been converted to arable land. Agricultural practices have become more intensive over the past 40 years such that the tonnage of fertiliser has increased 500% and the amount spent on chemicals has increased 1400% in the context of a stable population. In contemporary times, the boreal forest in Canada has suffered little deforestation, defined as the permanent conversion of forest area to non-forest due to activities associated with agriculture, urban or recreational development, oil and gas development, and flooding for hydroelectric projects. Canada as a whole has 91% of the forest cover that existed at the dawn of European settlement. The forest sector annually harvests approximately half of 1% of the region. The resulting road network from logging has effects that persist long beyond the period of harvest. Road construction could be one of the most harmful and persistent effects of logging.

The fossil energy reserves of these arctic basins are among the largest in the world. The oil sand deposits of the Mackenzie basin in northern Alberta are the largest deposits of recoverable fossil energy in the world. There are numerous natural gas, oil and oil sands production facilities as well as coal mines within the basin. Access to natural gas reserves in the Mackenzie Delta is in the environmental assessment stage. There are uranium, gold, emerald, diamond, jade/nephrite and tungsten mines within the Mackenzie basin. The Lena, Yenisei and Ob rivers have been dammed as has the Peace River in the Mackenzie basin. Exploration for mineral resources, the need for ever more energy, and the highly controversial oil sands development highlights three of the flash points for environmental concerns.

11.5.3 Urbanisation: The Case of Siberia

Six cities (Novosibirsk, Omsk, Barnaul, Novokuznetsk, Tomsk, and Kemerovo) with more than half a million inhabitants occur in the Ob basin and two (Krasnoyarsk and

Irkutsk) in the Yenisei. In the Lena basin, the city of Yakutsk, with a population of 300 000 is the primary conurbation. In the North American arctic basins, however, primary settlements are one order of magnitude smaller than their Russian counterparts. In the Yukon basin Fairbanks (30 000) and in the Mackenzie basin Yellowknife and Fort St John (20 000 each) are major settlements. The linking of Siberia with Moscow by road in the eighteenth century occurred 200 years earlier than the construction of the Alcan Highway during the Second World War. Consequently the urbanisation process in the Russian Arctic is considerably more advanced than in the Alaskan and Canadian Arctic. One of the benefits of this stunted urban growth in North America has been the opportunity to protect large areas of the pristine environment of the Alaskan and Canadian North. Massive economic and land use changes have occurred in the Ob and Yenisei basins and these processes of landscape transformation have begun in the Lena basin. While the impact of such changes is scarcely discernible at the large river basin scale, small basins and individual rivers within these regions have been seriously impacted. The important and yet unanswered question is whether any of the large arctic river basins is approaching a tipping point beyond which the environmental damage will be irreparable.

11.6 Discharge of Large Arctic Rivers

11.6.1 Problems in Discharge Measurement

Several problems occur when attempting to establish the accuracy of discharge estimates in northern rivers. Uncertainties in discharge may exist during periods of extreme flows, especially immediately after spring break up. Complicated backwater and unsafe conditions are created by downstream ice jams in river channels. Errors are largely controlled by three factors: (i) the lack of measurements made in May; (ii) errors from using rating curves; and (iii) the cross-sectional representativeness of the measuring sites. There are also large problems associated with the physical remoteness of the hydrometric stations (Figure 11.10) and their spatial representativeness.

11.6.2 Water Fluxes

Most of the freshwater that enters the polar seas is supplied directly or indirectly by rivers that drain boreal regions. Large annual contributions (Table 11.4) come from the Yenisei, Lena, Ob, and Mackenzie rivers, all of which flow directly into the Arctic Ocean. The Yukon flows to the Bering Sea but much of its water is subsequently conveyed by ocean currents to the Arctic Ocean (Figure 11.11).

The Siberian basins (Ob, Yenisei, and Lena) experience mostly light to modest precipitation which occurs as snow from November to March over most of the region. In the western foothills of the Urals precipitation increases to over 1000 mm yr^{-1} and similar orographic effects are associated with the Altai and Sayan mountains in the south. There are pronounced orographic effects resulting in high precipitation on the windward side of the North American Cordillera, especially the Alaska Ranges from which the Yukon River derives substantial runoff but both the Mackenzie and Yukon basins experience low precipitation on the leeward side of the mountains.

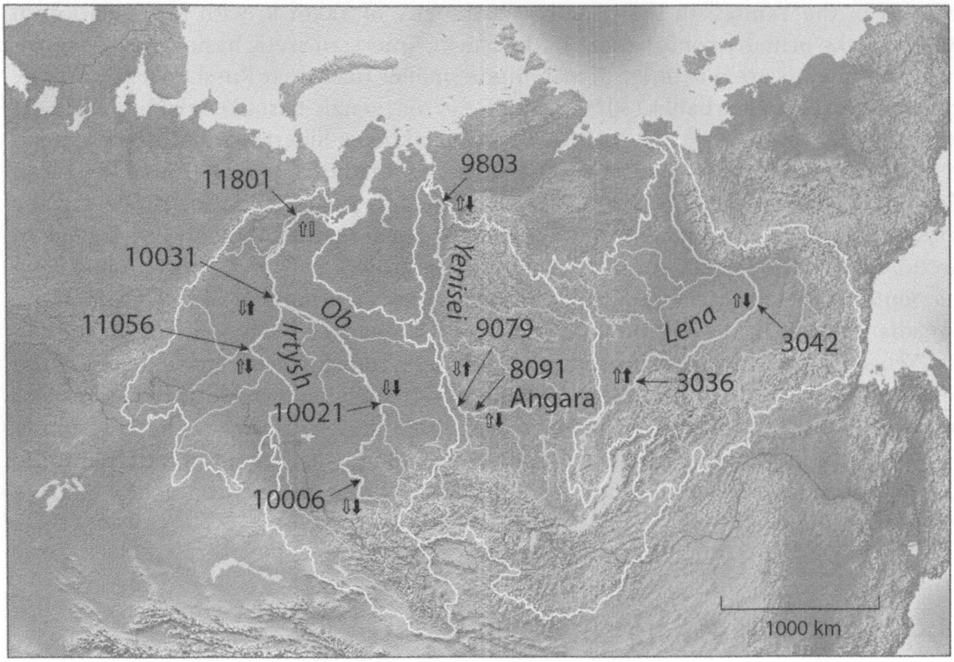

Figure 11.10 Hydrometric stations on Ob-Irtysh, Yenisei-Angara and Lena river systems (after Bobrovitskaya et al. 2003). Black arrows show changes in sediment discharge; open arrows show changes in water discharge. Stations are at Solyanka (3036), Tabaga (3042), Tatarka (8091), Yeniseisk (9079), Igarka (9803), Barnaul (10006), Kolpashevo (10021), Belogorie (10031), Tobolsk (11056), and Salekhard (11801).

Table 11.4 Global rank of the five largest arctic rivers by discharge.

River	Global rank	Mean discharge ($m^3\ s^{-1}$)	Mean annual discharge (km^3)
Yenisei	4	19 600	620
Lena	8	16 530	520
Ob	12	12 760	390
Mackenzie	21	9910	310
Yukon	30	6430	210

Source: Data from Milliman and Farnsworth 2011.

Owing to the large latitudinal extent of all basins, the arrival of spring may be as early as March in the south of the Ob basin and is postponed for at least three months in the northern parts of Siberia. A rise of air temperature above freezing point leads to a period of snowmelt that normally lasts for several weeks, followed by breakup of the river ice cover. Formation and disintegration of lake ice lag behind river ice due to the considerable heat capacity of large water bodies. In summer, many lakes and wetlands supply much moisture for evaporation which gives rise to convectional precipitation.

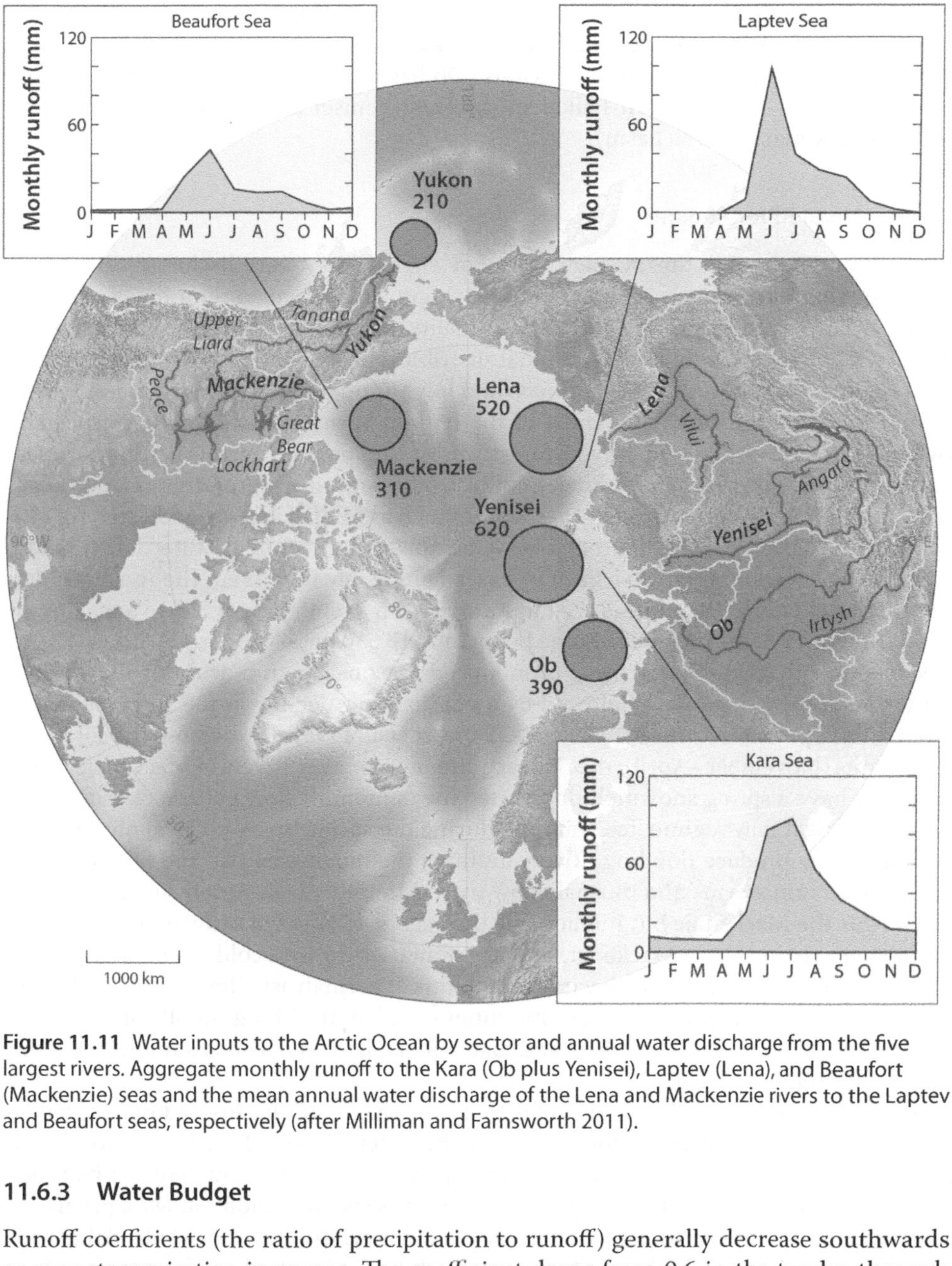

Figure 11.11 Water inputs to the Arctic Ocean by sector and annual water discharge from the five largest rivers. Aggregate monthly runoff to the Kara (Ob plus Yenisei), Laptev (Lena), and Beaufort (Mackenzie) seas and the mean annual water discharge of the Lena and Mackenzie rivers to the Laptev and Beaufort seas, respectively (after Milliman and Farnsworth 2011).

11.6.3 Water Budget

Runoff coefficients (the ratio of precipitation to runoff) generally decrease southwards as evapotranspiration increases. The coefficient drops from 0.6 in the tundra through 0.4 in the forest-tundra and 0.1–0.2 in the forest steppe to less than 0.1 in the dry steppe and semi-desert. The equivalent depth of runoff is also a useful indicator of the spatial variation in magnitude of surface water availability. In the tundra and taiga, the annual equivalent depth of river runoff ranges between 250 mm and 300 mm; in the steppe it varies between 20 mm and 100 mm and it approaches zero in the semi-desert. The equivalent depth also declines towards the centre of the continent. In mountainous regions, orographic increases in precipitation produce higher equivalent depths of runoff. This effect is demonstrated in the Alaska Range and St Elias and Coast mountains (Mackenzie and Yukon basins), Urals (Ob basin), the Altai and Sayan mountains (Ob basin), the Yenisei-Sayan-Baikal mountains (Yenisei and Lena basins), and the Verkhoyansk massif (Lena basin).

11.6.4 Nival River Regime

Streamflow regime is the average seasonal rhythm of river discharge. The most ubiquitous natural process is snowmelt, which controls the maximum flows of boreal rivers. Because of the strong climatic seasonality, most rivers within these basins are characterised by a highly uneven annual distribution of runoff. All five rivers have a nival regime with powerful spring floods and low summer and winter water levels. The spring flood regime dominates most of the region. Spring water levels are especially high on the rivers draining mixed forest, forest steppe and northern steppe zones. Further north, snowmelt is a crucial control (although in the tundra and the taiga, snow melts at a slower rate), and other factors such as rainfall and groundwater are also important. Therefore, the difference between spring and summer water levels is less pronounced. Runoff in smaller basins ceases completely in winter. The most striking feature of the Lena's discharge is the huge range from a minimum of 366 to a maximum of 241 000 $m^3\ s^{-1}$, the mean discharge being only 16 530 $m^3\ s^{-1}$(Figure 11.12).

Mackenzie River discharge has a pronounced seasonal rhythm, with low flow in the winter and spring high flow followed by a gradual flow decline in the summer that is often interrupted by rain-induced events (Stewart et al. 1998). Throughout the cold winter months (November–April), the river acquires an ice cover. Rivers in Mackenzie valley generally have a spring snowmelt dominated mean monthly flow pattern classified as a subarctic, nival flow regime. Ice jamming during the spring break up local obstructs to flow, and may produce flooding. Rivers draining the mountains rise and fall rapidly in response to rain storms and diurnal snow melt. Monthly mean runoff suggests a nival regime for the Mackenzie but it is actually the aggregate effect of a diversity of regimes. From south to north, the Mackenzie and its tributaries traverse cold temperate, subarctic and arctic environments with varying amounts of permafrost. The bulk of Mackenzie River runoff comes from the western mountains, with diminishing runoff contributions towards the northeast and its discharge pattern can be taken as an example of a large arctic river (Figure 11.13).

Figure 11.13 presents the runoff regimes of the major sub-basins and of Mackenzie River at Arctic Red River before it enters the delta (Table 11.5). In the south the Athabasca River flows mostly across cold temperate areas of low gradient but its headwaters lie in high mountains with glaciers. Thus, its seasonal runoff follows a proglacial regime. Due to the small size of the glacier area relative to the whole Athabasca basin, the late summer high flow becomes much less evident when, for example, Fort McMurray is reached. Despite a drainage area of more than 300 000 km^2 Athabasca River contributes only 17% of the Mackenzie runoff, whereas Peace River, which drains a similar area, provides 23%. This disparity arises because of the higher proportion of the Peace River basin being located in the Cordillera. Again, as Peace River is regulated for hydropower generation, extremes of discharge are modulated by water release in the winter and by water storage during the summer. The Liard River basin has an even

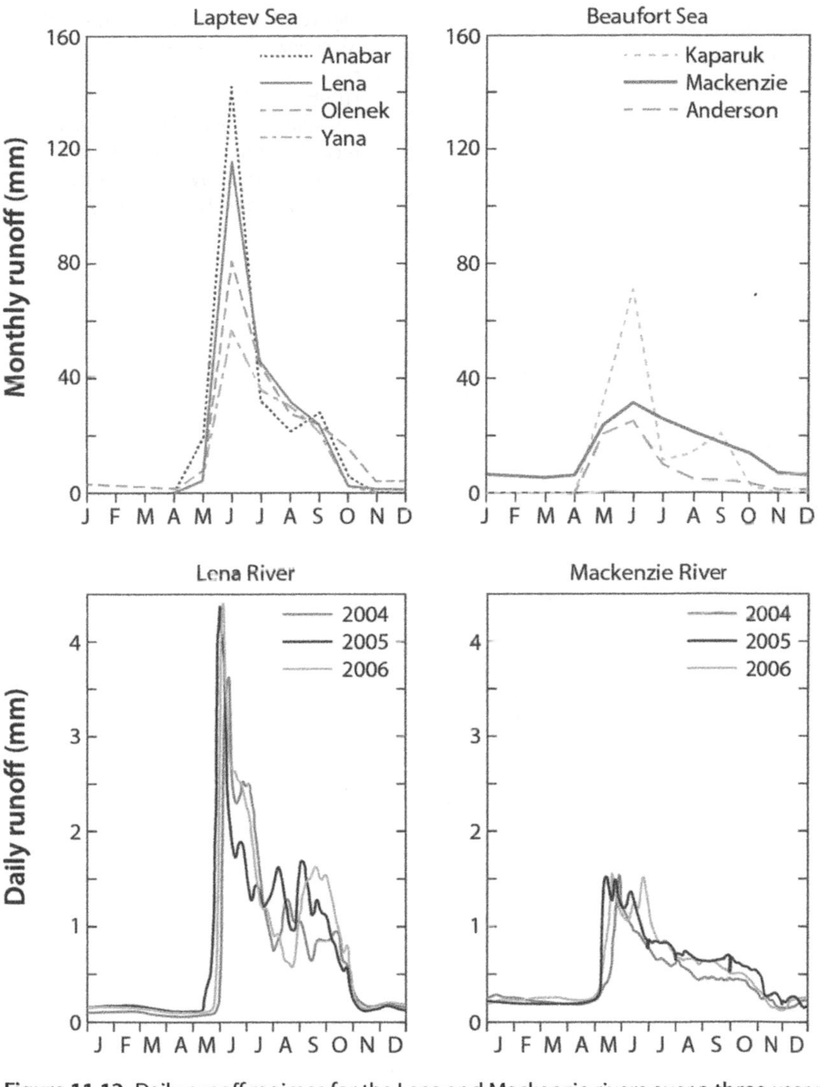

Figure 11.12 Daily runoff regimes for the Lena and Mackenzie rivers over a three-year period (2004–2006) (after McClelland et al. 2012).

higher proportion of its area in the Cordillera and provides 27% of the Mackenzie runoff. Discontinuous permafrost occurs in the northern and higher parts of the Liard basin, which also displays the most pronounced nival regime in the entire Mackenzie basin.

High specific discharge is also contributed by the smaller mountain basins. By contrast, tributaries that flow from the Canadian Shield and the Interior Plains produce lower specific discharges. The presence of numerous lakes and wetlands in discontinuous permafrost is critical for the Interior Plains rivers (Zoltai 1988). The Canadian Shield covers a third of the basin but produces only one fifth of the runoff, primarily due to low precipitation as well as a large number of lakes. The Great Bear River has a characteristic pro-lacustrine regime. The basin of the Mackenzie integrates the runoff of its sub-basins

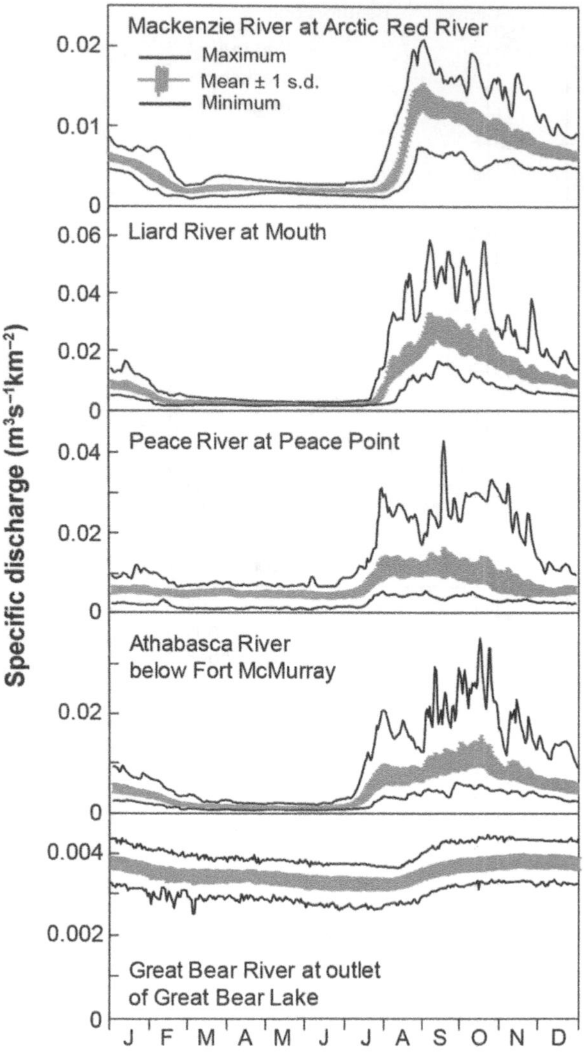

Figure 11.13 Runoff regimes of the Mackenzie River at the Arctic Red River station and its major sub-basins. Means and standard deviations of daily flow as well as recorded maxima and minima for each calendar day are presented (after Woo and Thorne 2003).

and therefore has a smoother hydrograph than most of its tributaries with the notable exception of those dominated by lake and reservoir storage.

In comparison, data from the Siberian rivers show a dominant nival regime but substantial modifications from widespread damming.

11.6.5 Lakes and Glaciers

The presence of lakes and glaciers causes variations in the standard subarctic nival regime (Woo 2012). The Tanana River, a tributary of the Yukon River in Alaska, is fed by the glaciers of the Alaska Range. The spring freshet is caused by snowmelt in May and June but discharge continues to increase during the summer from glacier melt. This resembles a proglacial regime. The presence of lakes can buffer streamflow from

Table 11.5 Sub-basins of Mackenzie River above Arctic Red River station: areas and mean annual runoff and percentage contribution to total flow (in square brackets) (after Woo and Thorne 2003).

Sub-basins	Area (km^2)	Annual runoff (mm) and percentage contribution
Northern Mountains	112 037	307 [10]
Liard	275 000	279 [27]
Peace	293 000	223 [23]
Athabasca	307 000	159 [17]
Great Bear	145 000	114 [6]
Low Plains	138 452	104 [6]
Great Slave	404 470	103 [14]
Mackenzie at Arctic Red River	1 680 000	169 [103]*

* Total contribution exceeds 100% because runoff is evaluated for individual sub-basins and their weighted sum does not necessarily match the measured flow at Arctic Red River station.

extremes. Lockhart River, a tributary of Mackenzie River in the North West Territory (NWT), flows through a chain of lakes in the Canadian Shield. By comparison with the Upper Liard, the Lockhart has a more subdued regime and late high flows. An extreme case of streamflow with a lacustrine regime is the Great Bear River, a tributary of Mackenzie River in NWT, which drains Great Bear Lake. This lake, being one of the 10 largest lakes in the world, completely dominates the drainage, and annual high flows are delayed until August and September, at least three months after the regional snowmelt.

The presence of myriad lakes in the Canadian Shield and in the Interior Plains, tends to dampen streamflow response to rainfall through their storage capabilities. Large lakes are notably influential in modifying discharge downstream. The reservoir on the Peace regulates its discharge according to the needs for hydropower production. The other three major lakes (Athabasca, Great Slave, and Great Bear), through their enormous storage capacities, usually retain the inflows for some time and then gradually release the water through their outlets. Great Slave Lake is notably effective in modifying the flow of the Mackenzie system (Figure 11.14).

Gibson et al. (2006) noted that three-quarters of the inflow to Great Slave Lake is derived from the Slave River which captures about 40% of the flow from the upper Mackenzie basin. The lake alters the rhythm of discharge from the Athabasca, the Peace and other minor tributaries that drain into the lake from the Shield and from the plains. Outflow from this lake is responsive to lake level variations and does not reflect the pattern of discharge from the upper basin. Baseflow is maintained by groundwater discharge and by releases from the large lakes, one of which is the reservoir on Peace River that usually provides over half of the total winter discharge of the Mackenzie system. Peters and Prowse (2001) noted that after the reservoir was built, winter flow increased by 250% when compared with its previous natural state.

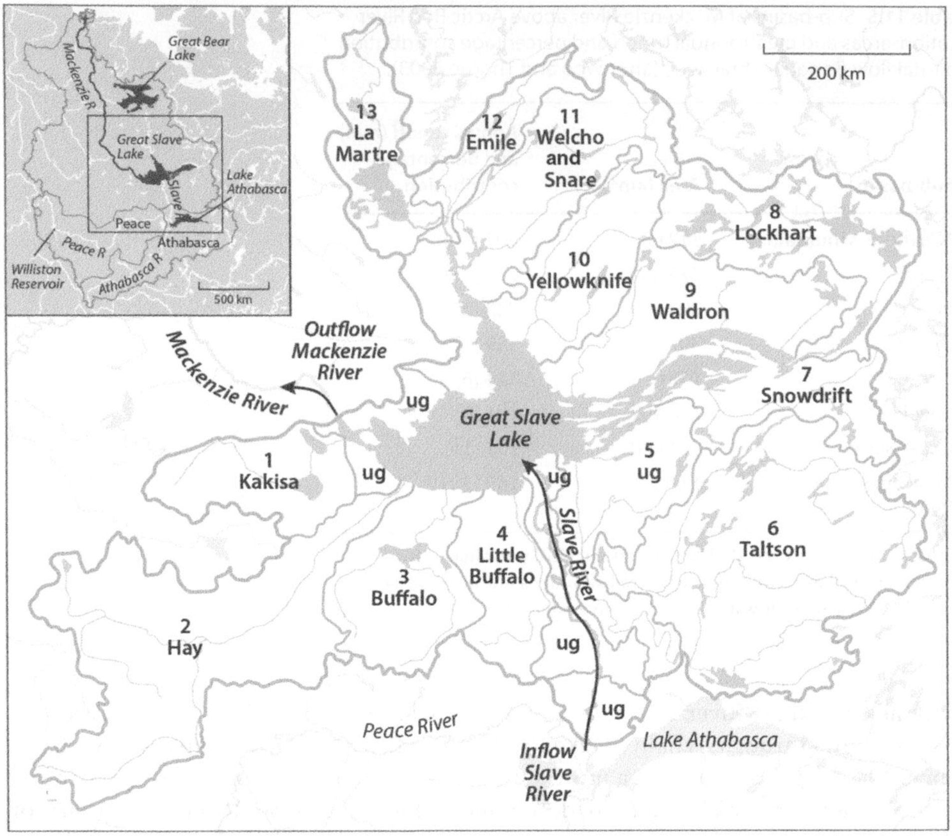

Figure 11.14 Great Slave Lake transforms the highly variable discharge of the Slave River to a more constant Mackenzie discharge (after Gibson et al. 2006).

Effects of glaciers on Yukon River flows are demonstrated by the Tanana and White River hydrographs with a second and extended high flow period corresponding to the later glacier melt maximum.

11.6.6 River Ice: Freeze and Break Up

Snowmelt arrives in May at most parts of the Mackenzie basin. Large amounts of snow, accumulated over the winter, are often released within weeks, sending floods down the Mackenzie River and its tributaries. The presence of a river ice cover hinders the downstream propagation of snowmelt runoff, creating ice jams that exacerbate flooding (Hicks and Beltaos 2008). Breakup begins in the south but is postponed for weeks at the lower Mackenzie. Summer rain may be of sufficiently high magnitude to produce rises in hydrograph in the sub-basins. An exceptional rainfall event can generate high flows and carry large sediment load, as exemplified by the mid-June 1974 high flow produced by 10–20 mm rain over the Liard basin that gave rise to a peak flow of 12 900 $m^3 s^{-1}$.

Autumnal rainfall, in a period of reduced evaporation, may cause hydrograph rises but such secondary peaks, if present, are generally of low magnitude. Freeze-up sets in as the flow continues to recede. The formation of river ice and the constriction of flow in the channels often produce an abrupt reduction in discharge due to hydraulic storage (Prowse and Carter 2002) but local flooding can result from ice blockage that raises the water level. The arrival of freeze-up in the north is usually over a month earlier than in the southern basin.

11.6.7 Scale Effects

The Liard basin is 275 000 km^2 in area. It is primarily underlain by discontinuous permafrost. Much of the basin lies in the mountainous Western Cordillera with high relief. The Liard River then flows through the Liard Plateau and the Liard Plain and debouches into the Interior Plains at Fort Liard before joining the Mackenzie at Fort Simpson. The southwestern part of the Liard basin lies in the Cassiar Mountains and has the highest precipitation. Woo (2012) noted that the key characteristics of spring freshet for a large basin depend on: (i) the amount of winter snow accumulation; (ii) time of arrival of warm spring; and (iii) gradual or rapid melt. However, high rainfall in July is able to build on the high recession flow of the spring to generate a second peak; in the case of July 1974 even a primary peak (Figure 11.15). The Upper Liard River has high flows in June. Its mountainous setting gives rise to altitudinal effects and snowmelt runoff is extended through July. Superimposed on the snow melt runoff is the streamflow response to rainfall events: (i) moderate rainfall in mid-June which did not affect the Liard at Upper Crossing; (ii) an exceptionally high rainfall event in mid-July which also bypassed the northwestern corner of the basin; (iii) moderate rain in early August the effect of which was most marked at the mouth due to localised high rainfall; and (iv) an isolated rain event in October.

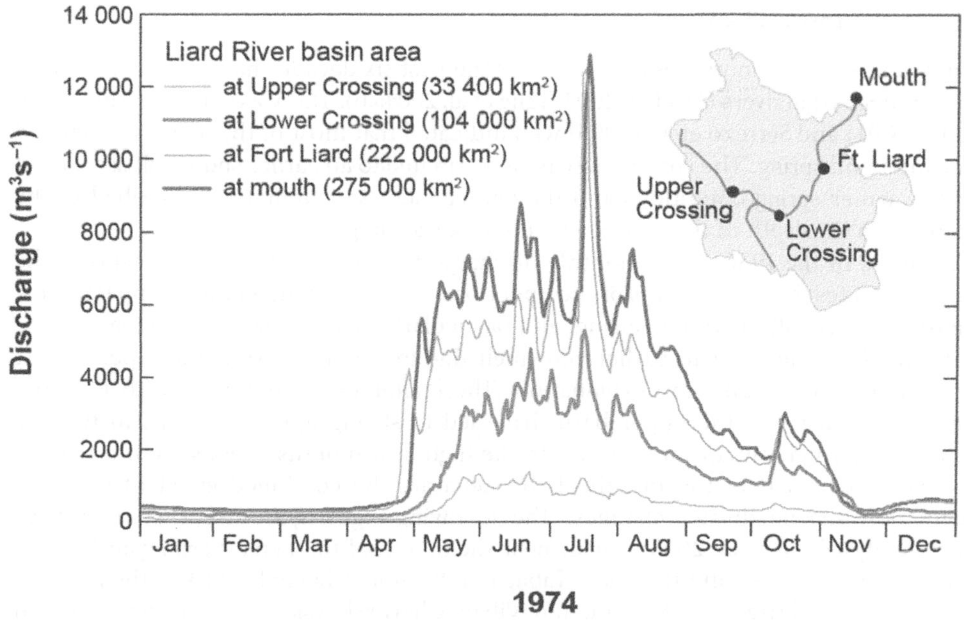

Figure 11.15 Discharge hydrographs of four stations along the Liard River in 1974, showing spring and summer high flow events (after Woo 2012).

Evidently, the spatial distribution, intensity, and timing of rainfall has a strong influence on streamflow response in this large subarctic basin.

11.6.8 Effects of River Regulation

The effects of flow regulation are most dramatically exemplified by the Peace River which is a tributary of the Mackenzie and the Vilyui River in the Lena basin. Before regulation by the Bennett Dam, Peace River had a typical nival flow regime. After 1968, the winter became the high flow period whereas the spring became the low flow period as flow releases were regulated to meet the demands for hydroelectric power. Artificial water releases can also produce extreme events. Woo et al. (2008) compared the flow regimes of Peace River before and after a dam was built in 1967 to regulate flow for hydroelectric production. A substantial reduction in spring high flow, accompanied by a large increase in winter low flow made the flow much more uniform through the year.

Vilyui River in the Lena basin also demonstrates the effect of damming on river regime. Originally a subarctic nival regime, this regime has been modified since damming at Chernyshevsky where the upstream basin area is 136 000 km^2. After the dam was built in 1967, winter flow increased by 400–600 $m^3 s^{-1}$ during November–April but decreased by 1400 $m^3 s^{-1}$ in May and by 2300 $m^3 s^{-1}$ in June. The July discharge increased by 300 $m^3 s^{-1}$ and little change occurred during August–October. About 350 km downstream (Suntan) similar changes in mean monthly flow occurred but 900 km downstream of the dam (Haryrik-Korno) the impact was much reduced because of unregulated tributary streams. In summary, summer high flows have been reduced by 55% at the mouth of the Vilyui and winter flows increased 30 times.

11.6.9 Historical Changes

Annual streamflow increases of 3–7% over the past six decades have been reported in the large Siberian rivers (Ye et al. 2003; Yang et al. 2004a,b). This was confirmed by Zhang et al. (2001) and Serreze et al. (2006) who indicated that most of the increase occurred in winter and spring. The spring increase is attributed to an earlier snowmelt associated with warmer spring conditions and the winter increase is tentatively ascribed to the increase in the depth of the active layer under a warming climate.

Analysis of the 1936–1999 monthly discharge data for the major sub-basins of the Lena indicates that the streams of the upper basin, without much human impact, show increases in runoff in winter, spring, and particularly in summer but a decrease in the autumn. A regime shift to earlier snowmelt due to climate warming and permafrost degradation is probably the explanation. These alterations combined with the documented changes in the upper basin have led to strong upwards trends in the low flow months and low upwards trends in the high flow months. Cold season discharge increases identified at the mouth of the Lena are the combined effect of reservoir regulation and headwater changes. The following eight hydrometric stations have been analysed (Ye et al. 2003): Kusur near the mouth of the Lena; Verkhoyansk at the junction of the Aldan and the Lena; Tabaga on the upper Lena; Hatyrik at the junction of the Vilyui and the Lena; Suntan, mid-Vilyui; Chernyshevsky, upper Vilyui; Malyakai tributary to Vilyui; and Suhana, on Olenek River. A reservoir was completed near Chernyshevsky in 1967. The long-term monthly discharge of major sub-basins of the

Table 11.6 Recorded hydrologic changes (1966–1996).

River	Temperature (°C)	Precipitation (mm)	Runoff (mm)	Permafrost % of 1966
Yenisei	+2.5	0	+27	71
Lena	+2.1	−24	+10	94
Ob	+2.2	−4	+1	19
Mackenzie	+1.4	−6	−5	55
Yukon	+2.2	+43	+13	90

Source: Data from ACIA 2005.

Ob basin (1936–1990) were analysed by Yang et al. (2004b). Differences between upper and lower parts of the basin were primarily a function of river regulation in mid-basin. Woo et al. (2007) showed that the Tanana and the Upper Liard had an increased winter low flow but a slight reduction in snowmelt runoff. The Great Bear River showed a general reduction in flow, especially for June–September but the Lockhart River had an increase in mean monthly flows, especially during the high flow months of July–October.

Dams appear to be the only unambiguous cause of general regime change in Siberia's large arctic rivers. Warming temperatures have had substantial effects on permafrost thaw in the discontinuous and sporadic permafrost belts but this effect has failed to influence large river regimes to the present date. All large arctic rivers, except the Mackenzie, have experienced increased runoff over the past 3–6 decades. Significant increases in discharge (10–20%) are anticipated by 2050 according to Intergovernmental Panel on Climate Change (IPCC) models after a doubling of CO_2 (CGC model) (Table 11.6).

11.7 Sediment Fluxes

11.7.1 Complications in Determining Sediment Fluxes Both Within Arctic Basins and to the Arctic Ocean

Sediment flux to the Arctic Ocean is influenced by a large number of factors: (i) magnitude, timing and spatial pattern of inputs of suspended sediment within arctic river basins; (ii) magnitude of inputs of suspended sediment to the Yukon, Mackenzie and Lena deltas and Yenisei and Ob estuaries; (iii) transport of bed material and shoaling of bed material and ice jams as obstacles to navigation; (iv) channel scour under ice and break-up conditions; (v) riparian and river bank permafrost; and (vi) extraction of bed material for use. These factors are difficult to evaluate as several problems are involved when attempting to establish the accuracy of sediment load estimates in northern and mountain rivers. In terms of discharge there are uncertainties during periods of extreme flows, and especially, immediately after spring break-up. Complicated backwater conditions are created by downstream ice jams in river channels and there are serious staff safety concerns. In terms of sediment concentration measurement, error is controlled largely by three factors: (i) too few measurements are made in

Table 11.7 Sediment in large arctic rivers.

River	Annual discharge (km^3)	Annual susp. load (10^6 t)	Annual dissol. load (10^6 t)	Specific total sediment yield ($t\ km^{-2}\ yr^{-1}$)	Specific susp. sed. yield ($t\ km^{-2}\ yr^{-1}$)	Specific dissol. load ($t\ km^{-2}\ yr^{-1}$)
Yukon	210	54	26	96	65	31
Mackenzie	310	100	64	90	55	35
Ob	390	16	34	18	6	12
Lena	520	21	60	33	9	24
Yenisei	620	5	43	19	2	17

Source: Data from Gordeev (2006) for Russian rivers and Milliman and Farnsworth (2011) for North America.

the month of May; (ii) errors are inherent in the use of sediment rating curves; and (iii) cross-sectional representativeness of the sampled concentrations is often unknown.

11.7.2 Flux of Suspended Sediment and Dissolved Solids

Ludwig and Probst (1998) and Milliman and Farnsworth (2011) provide generalised data for annual suspended sediment and dissolved sediment fluxes in the five largest arctic rivers (Table 11.7). The relatively high values for the Yukon and Mackenzie Rivers in specific suspended sediment yield are caused by the higher stream channel gradients, high relief tributaries in the lower parts of the basins, and a relative absence of engineered impoundments. The specific dissolved solids yield for each basin show a smaller range and a similar ranking. These data show that the dissolved solids are less controlled by dams than the suspended sediments. Perhaps the most interesting statistic is the ratio of suspended sediments to dissolved solids which shows Yukon (2.1), Mackenzie (1.5). Ob (0.5), and Yenisei and Lena (0.3). This clearly separates the mountain rivers (Yukon and Mackenzie) from the lowland rivers (Ob, Yenisei, and Lena).

Amounts of suspended sediment in the major arctic rivers show a distinct contrast between higher values in the North American rivers (Mackenzie and Yukon) and lower values in the Eurasian rivers (Lena, Yenisei, and Ob) (Figure 11.16).

This is primarily a consequence of damming of the Russian rivers and greater proportional runoff from mountainous terrain in the North American rivers (Holmes et al. 2002; Syvitski 2002). In comparison with other large world rivers, sediment concentrations in the Russian rivers are more similar to the Congo whereas the North American large rivers are intermediate between the Amazon and Changjiang.

11.7.3 Historical Changes in Water and Sediment Discharge in the Siberian Rivers

The discharges at the mouths of the Ob, Yenisei and Lena rivers tend to increase consistently in spite of dam and reservoir construction in their middle reaches. Temperature trends from 1883 to 1995 are unambiguous (Bobrovitskaya et al. 2003) for 10 out of the 13 long-term Siberian meteorological stations, and the continuing temperature increase is confirmed by ACIA (2005). There is, however, higher variability in the

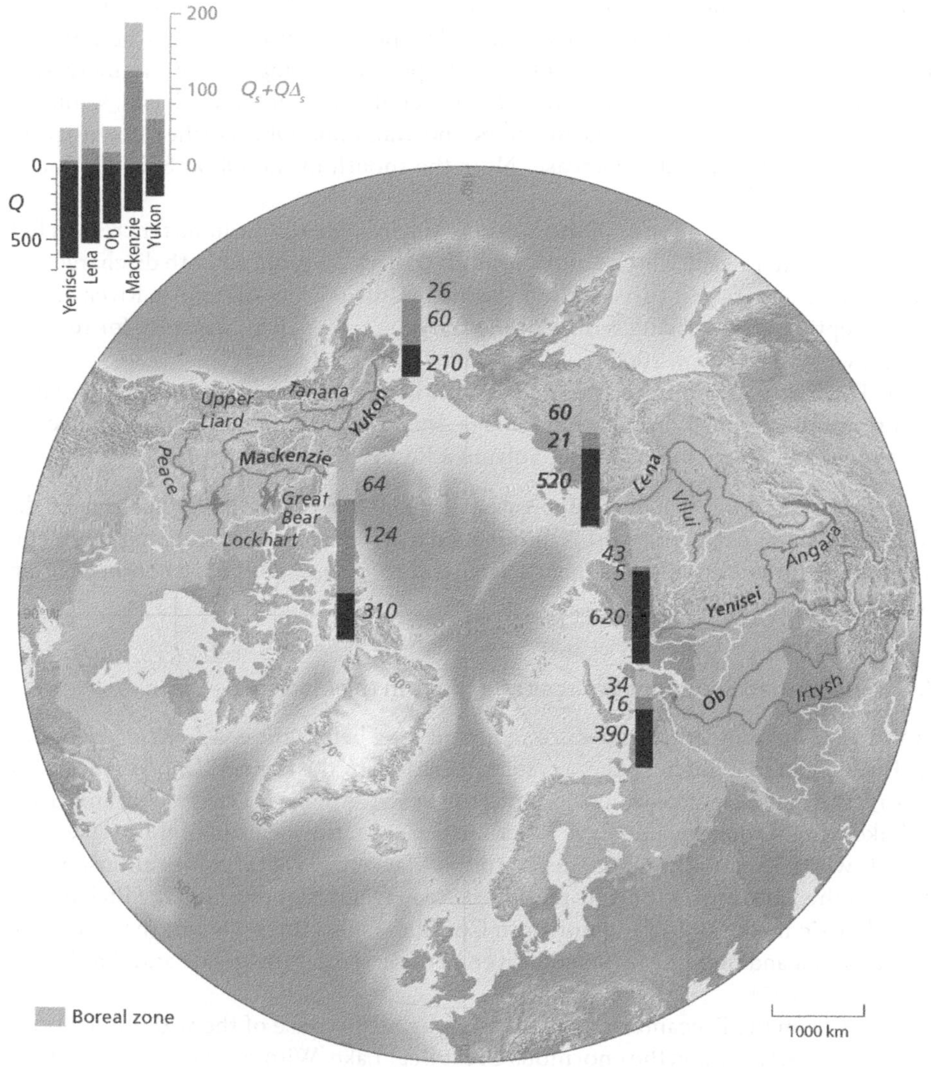

Figure 11.16 Water and sediment discharges in large arctic rivers. Black histograms are in km^3 yr^{-1}; grey histograms are in millions of tonnes of suspended sediments per year; white histograms are in millions of tonnes of dissolved sediments per year (after Milliman and Farnsworth 2011).

sediment flux record. Incidentally, the three long-term meteorological stations which show cooling as a departure from the common trend are all located in the coastal area of the Arctic Ocean.

Impoundments, however, greatly affect sediment yield, and a basin-wise summary (1960–1988) demonstrates the following:

(a) The Ob-Irtysh has increasing discharge at its mouth but not a corresponding change in sediment yield at Salekhard. Most of this increased discharge comes from the Irtysh where sediment yield is decreasing. At three stations on Ob River the discharge and sediment yield have dropped because of dams at Barnaul, Kolpashevo, and Tobolsk. One notable exception, at Belogorie, shows increasing sediment yield for reasons unknown (Meade et al. 2000) in the Ob basin, sediment yields increase between the headwaters and Kamenna Obi. Further downstream sediment yields gradually decrease. Near the mouth of the Ob at Salekhard, it is about 5.3 t km^{-2} yr^{-1}.

(b) The discharge at the mouth of the Yenisei-Angara-Selenga basin has increased but not the sediment yield which has decreased. Dams have reduced both discharge and sediment discharges at three stations in the middle reaches of Yenisei River. A single exception of increasing sediment discharge is recorded at Yeniseisk for reasons unknown (Meade et al. 2000).

(c) Lena River and its tributary, the Aldan, show increasing temperature and discharge but reduced sediment discharge at Tabaga. Increased sediment discharge, however, was recorded on the main stem at Solyanka Reservoirs.

Sediment in the lower reaches is deposited permanently on the millennial time scale, and the natural low sediment yield is diminished by large reservoirs trapping sediment. Changes in sediment discharge appear to depend more on dam construction than on climate change (McClelland et al. 2004).

11.7.4 Suspended Sediment Sources and Sinks in the Mackenzie Basin

Several major sinks intercept and store sediment long before the Mackenzie reaches its delta. These include Williston Lake, a reservoir created by the Bennett Dam on the upper Peace River; Athabasca Lake which contains the Peace–Athabasca river delta; and Great Slave Lake, which contains the Slave River Delta. Given the location of these lakes, the sediment, which eventually debouches onto the Mackenzie Delta, comes almost exclusively from the Liard and Peel rivers and short steep tributaries draining the Richardson and Mackenzie mountains of the northern Cordillera, plus the sediment that is eroded from the banks and bed of Mackenzie River between Fort Providence and Arctic Red River (Figure 11.17).

Sediment from the Precambrian Shield is negligible because of the widespread presence of wetlands, lakes, and the enormous Great Bear Lake. With respect to the sediment that is eroded from the banks and bed of the Mackenzie River, the reach between Fort Providence (outlet of Great Slave Lake) and Arctic Red River is the only one that is relevant for sediment supply to the Mackenzie Delta. The Mackenzie itself has a suspended load of only 2.5 million tonnes where it emerges from Great Slave Lake. The Great Bear, Great Slave, and Athabasca lake systems supply only 20% of the mean discharge of the Mackenzie River and an even smaller proportion of its annual sediment load. Delta growth in the Great Slave Lake and sediment starvation of the Peace–Athabasca Delta following the damming of Peace River are the main evidence of the dominant influence of lakes.

11.7.4.1 Sediment Yield in the Mackenzie Basin

When specific sediment yield is plotted against drainage basin area the usual Cordilleran relations are confirmed (Church et al. 1989; Figure 11.18)

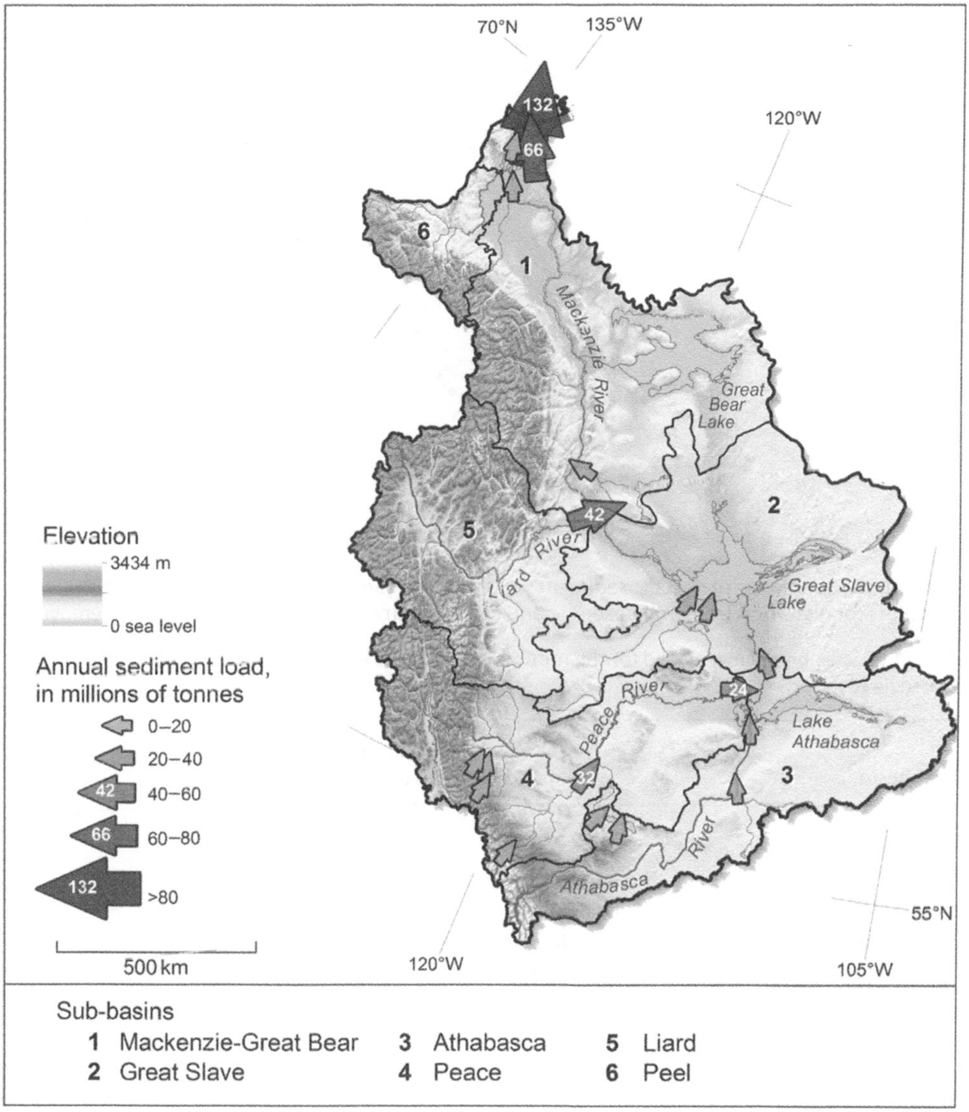

Figure 11.17 Cordilleran sediment sources (1974–1994) and lake sinks in the Mackenzie basin. Arrow stem width represents mean annual suspended sediment load in millions of tonnes (after Carson et al. 1998).

When considering sediment flux in the Mackenzie basin, it is important to consider (i) tributaries which have their origin within the Cordillera, (ii) the river banks, (iii) and the bed of the mainstem channel of Mackenzie River (Table 11.8).

11.7.4.2 West Bank Tributary Sources

Most of the west bank tributaries of the Mackenzie River do not have monitoring stations with long enough records to provide reliable estimates of annual sediment

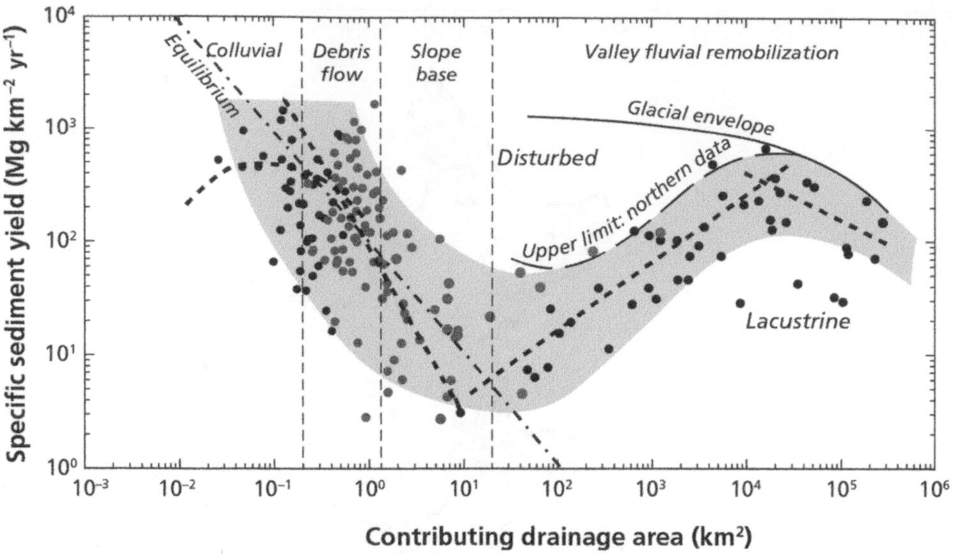

Figure 11.18 Correlation of specific sediment yield with discharge area (after Church et al. 1989). Datapoints for disturbed, lacustrine reservoir and glacial sediment yield regimes have been removed.

Table 11.8 Specific sediment yield for the mountain tributaries of Mackenzie River.

River basin	Area (km^2)	Specific sediment yield (t km^{-2} yr^{-1})
Peace River		
Peace at Peace River	186 000	179
Peace below Alces river	122 500	135
Peace at BCR Bridge	84 000	33
Beatton near Fort St. John	16 100	690
Pine near mouth	13 500	237
Halfway near mouth	9400	219
Murray near mouth	5620	262
Kiskatinaw near mouth	4370	476
Sukunka near mouth	2510	76
Murray above Wolverine River	2410	47
Moberly near mouth	1840	105
Sukunka above Chamberlain Creek	927	40
Dickebusch Creek near mouth	86	288
Quality Creek near mouth	30	30
Liard basin		
Fort Nelson below Muskwa River	43 500	347
Liard at lower Crossing	104 000	30
Liard above Beaver River	119 000	80

Source: Data from Church et al. 1989.

Table 11.9 Estimates of annual suspended load for main tributaries based on morphologic data (after Carson 1988).

River	Annual suspended load (10^6 t)
Peel	11.1
Keele	9.2
Redstone	8.9
Arctic Red	7.1
Root	6.8
Mountain	6.6
North Nahanni	5.2
Johnson	1.6
Dahadinni	1.0
Wrigley	0.6
Carcajou	0.5
Willowlake	0.3

Source: Data from Carson 1988.

transport. An estimate can be determined for a hydrologically homogeneous area by calculating the stream power index and a product of maximum basin elevation, basin area, and average stream slope for each tributary on the assumption that there is at least one monitoring station with reliable quantitative records. For the northern west bank tributaries, the Arctic Red River can be used as the most reliable and its power index calculated (Table 11.9).

The suspended sediment load for each tributary is determined by scaling the actual load of the Arctic Red River by an amount equal to the ratio of the power index of the tributary in question to the power index of the Arctic Red River. On this basis, it can be estimated that close to 60 million tonnes of sediment are contributed annually to the Mackenzie by the eleven largest west bank tributaries north of the Liard (Carson 1988; Figure 11.19a).

11.7.4.3 Bed and Bank Sources

By examining six sediment monitoring stations (Mackenzie River above Liard River, Liard River near its mouth; Martin River near its mouth; Redstone River near its mouth; Mackenzie River at Norman Wells; and Mackenzie River above Arctic Red River), Church et al. (1987) were able to isolate the inputs of sediment from the upper part of the Mackenzie basin from the sediment input from the Liard basin and from the river bank sediment sources of the lower Mackenzie main channel and left bank tributaries. They concluded that the Liard drainage is the major source of sediment input to the main stem as it is not controlled by large lakes and damming. Within the Liard basin, the Fort Nelson-Muskwa-Prophet river system and the upper Liard proper would seem to be the chief sediment sources. Severe summer storms produce high sediment yield from the west bank mountain tributaries (e.g. Sherstone 1983) so that the pattern of regional sediment yield may vary from year to year. Liard and the northern mountain

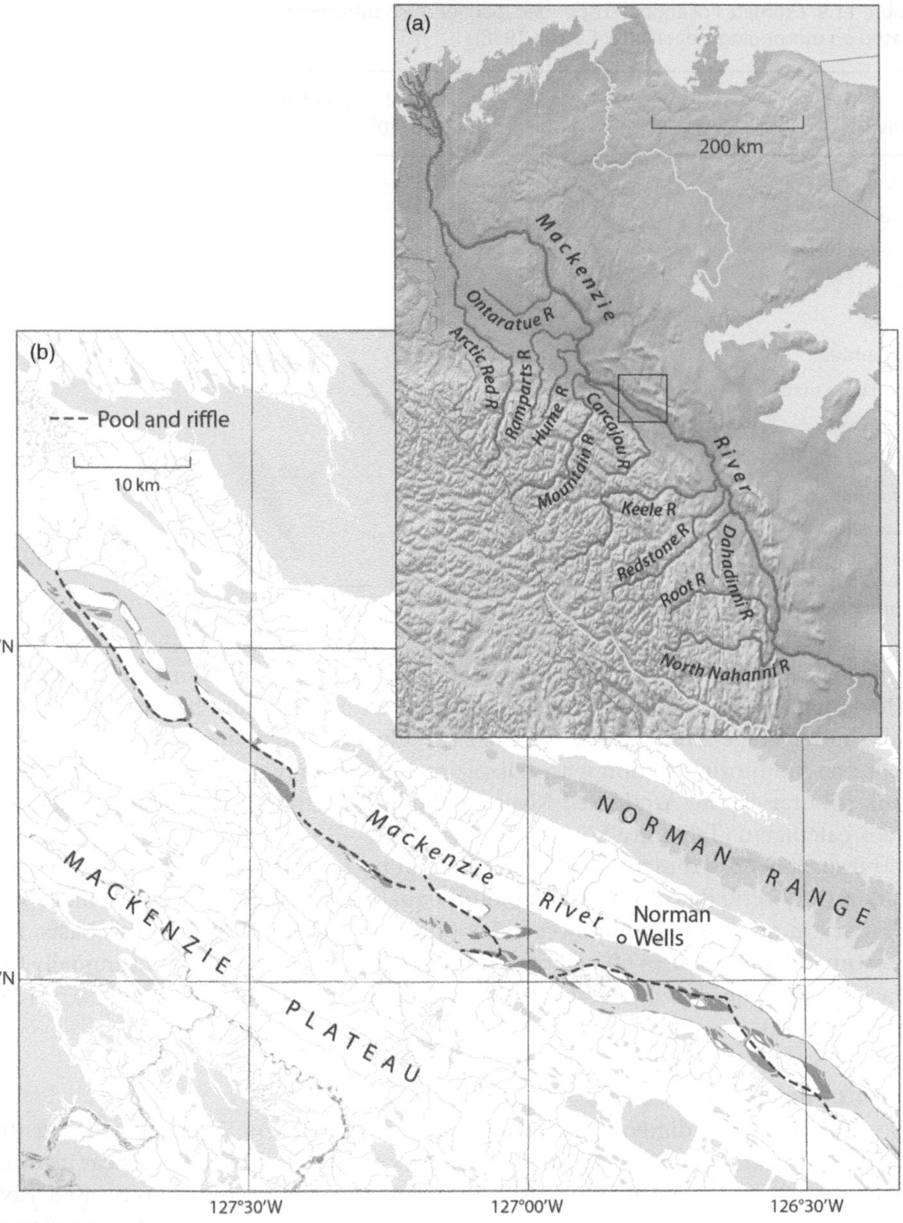

Figure 11.19 (a) The locations of eleven major west bank tributaries of the Mackenzie River. (b) The pattern of major pools and riffles in Mackenzie River near Norman Wells, NWT, Canada (after Church et al. 1987).

river systems which supply 77% of the mean discharge of the Mackenzie, can be expected to deliver an even higher proportion of the annual suspended sediment load to the Mackenzie River (Brooks 1996). The novelty of the Church et al. (1987) study was acquiring information on sediment transport from non-traditional sources, specifically historical planimetric and topographic surveys, aerial photographs, and hydrographic

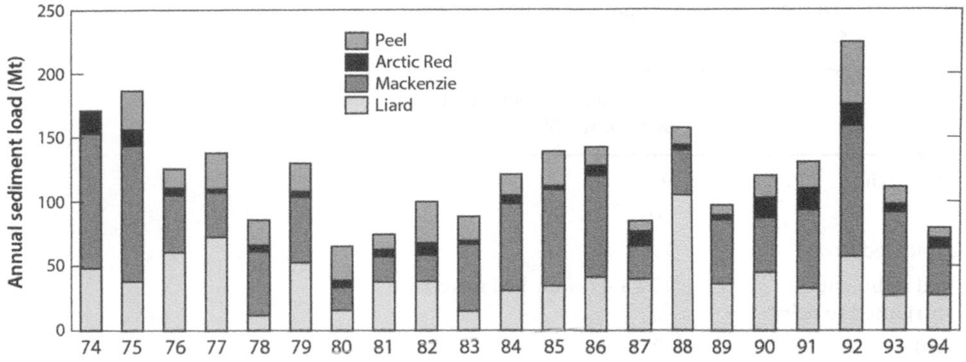

Figure 11.20 Annual fluctuations in fine sediment load inputs to the Mackenzie delta, 1974–1994. The 'Mackenzie' component refers to the Mackenzie delta input minus the Liard load (after Carson et al. 1998).

sounding, supplemented with limited observations of sediment transport. Erosion and sedimentation need to be understood on a reach-by-reach basis in order to assess river stability (Figure 11.19b). Rivers in the Mackenzie valley generally exhibit spring snowmelt dominated mean monthly flow pattern classified as a subarctic nival flow regime. During spring, ice jamming along the Mackenzie obstructs flow locally and elevates river stage to flood levels. Flow along the large tributary rivers draining the Mackenzie Mountains to the west can rise and fall rapidly in response to rain storms and diurnal snowmelt. The mean annual sediment supply to the delta is 128 million tonnes, of which about four is sandy bed material delivered by the Mackenzie River. More than 99% of this sediment is supplied to the delta during May–October. About 17% of the fine sediment load is supplied by the Peel River. Carson et al. (1998) determined the mean monthly fine sediment load, based on 1974–1994 data. The effect of incomplete monitoring in the month of May is unfortunate as both Arctic Red and Peel rivers deliver their largest monthly sediment load in that month and, in the case of the Mackenzie, May is the second most productive month (Figure 11.20; Tables 11.8 and 11.10).

Table 11.10 Mean monthly fine sediment loads (in 10^6 t) for delta-head rivers, North America.

Month	Mackenzie	Arctic Red	Peel	Total
May	21	3.1	10.1	34
June	34	2.4	8.2	45
July	21	0.6	1.5	23
August	14	1.0	0.7	16
September	4	0.1	0.3	4
October	2	0.1	0.0	2
Total	96	7.3	20.8	124

Source: Data from Carson et al. 1998.

Table 11.11 Estimates of mean annual suspended sediment load at four benchmark stations.

River	Mean annual suspended sediment (10^6 t)
Mackenzie	96
Peel	20.8
Arctic Red River	7.3
Total sediment contributed to delta	124 + 4 (sandy bed material)
Liard	41

Source: Data from Carson et al. 1998.

Carson et al. (1998) have discussed these sources of uncertainty in relation to the four benchmark stations on the Liard, Mackenzie, Arctic Red and Peel rivers. Three of these stations are referred to as the delta-head rivers (Mackenzie, Peel, and Arctic Red), as their aggregate sediment load is contributed to the Mackenzie delta (Table 11.8). The Liard River near its mouth is about 1100 km from the Mackenzie Delta (Tables 11.5 and 11.11; Figure 11.20).

In a break-up, river levels can rise to tens of metres above their normal banks, extending the floods above the river terraces and sending water and ice floes at high velocities (up to 5 m s^{-1}) down river when the ice jam breaks. Ice floes along rivers destroy riparian vegetation, ice shove armours the banks with boulder pavements, elevated river levels permit erosion of riparian zones high above the channel, and strong currents armed with ice floes can instantly reconfigure river channels and alter the morphology of flood plains. Riverside communities are also affected by these recurrent floods. The importance of this event is increased by the sinking of the delta simultaneously with a rising sea level due to climate change. Furthermore, the climate change driven thaw of permafrost may erode the delta itself. There is a need to know the rate of input of sediment over the delta surface in relation to rates of submergence.

Data from Siberian hydrometric stations (Figure 11.12) indicate a general tendency for discharge to increase and sediment concentrations to decrease over the past decades. The increasing discharge is consistent with increasing precipitation associated with warming trends. The decrease in sediment concentrations has more to do with the damming of many of the major tributaries of the Lena, Yenisei and Ob rivers (Table 11.12).

Specific sediment yield (or sediment mass per unit area per year) is derived from monitoring stations in the Mackenzie basin (Figure 11.18). Such data are important in investigating the effects of scale on sediment yield and Figure 11.18 demonstrates the presence of four different relations between area and specific sediment yield on colluvial slopes (<0.2 km^2), debris flow dominated areas (0.2–1 km^2), mountain slope bases (1–20 km^2) and the adjacent valleys (>20 km^2) (Table 11.8).

Most of the west bank tributaries of the Mackenzie River do not have monitoring stations with long enough records to provide reliable estimates of annual sediment transport. A first cut estimate can be determined for a hydrologically homogeneous area by calculating a stream power index, a product of maximum basin elevation, basin

Table 11.12 Changing water and sediment pattern in Siberian hydrometric stations.

Gauging station	Discharge over time	Sediment over time
Ob at Bernaul and Kolpashevo	Decreasing	Decreasing
Ob at Belogorie	Decreasing	Increasing
Irtysh at Khanty	Unknown	Decreasing
Irtysh at Tobolsk	Increasing	Decreasing
Ob at Salekhard (mouth)	Increasing	Constant
Yenisei at Kyzyl and Divnogorsk	Unknown	Decreasing
Yenisei at Yeniseisk	Decreasing	Increasing
Kan at Kansk	Unknown	Increasing
Angara at Tatarka	Increasing	Decreasing
Yenisei at Igarka (mouth)	Increasing	Decreasing
Lena at Solyanka	Increasing	Increasing
Lena at Tabaga	Increasing	Decreasing
Aldan at Okhotsky	Unknown	Increasing
Vilui at Ust	Unknown	Decreasing
Lena at Kuzur (mouth)	Unknown	Unknown

Source: Data from Bobrovitskaya et al. 2003.

area and average stream slope, for each tributary on the assumption that there is at least one monitoring station with reliable quantitative records. For the northern west bank tributaries, the Arctic Red River can be used, and its power index calculated. The suspended sediment load for each tributary is then calculated by scaling the actual load of Arctic Red River by an amount equal to the ratio of the power index of the tributary in question to the power index of Arctic Red River (Table 11.9).

11.8 Nutrients and Contaminants

Although the spring season lasts only two months, dissolved organic nitrogen and dissolved organic carbon (DOC) flux in that period exceed the flux during the entire six months of winter by more than 400% (Holmes et al. 2012) in the six rivers, including the Kolyma River with the large five. It is difficult to extrapolate such results to the whole arctic basin because almost all of the unmonitored rivers are small and totally northern.

11.8.1 Supply of Nutrients

The seasonal variations in continental runoff in the Arctic are accompanied by large seasonal variations in nutrient and organic matter inputs to the coastal ocean. Seasonal changes in water discharge alone account for much of the variation in constituent fluxes. In addition, seasonally varying concentrations of dissolved and particulate material contribute to variations in related fluxes. For example, organic matter concentrations increase dramatically during the spring freshet, while nitrate and silica concentrations show the opposite pattern. These concentration changes are tightly coupled to seasonal changes in water flow paths through the landscape (Gao et al. 2007). Frozen ground constrains water flow to organic-rich surface layers during the spring snowmelt period. As the ground thaws, an increasing proportion of the water moves through mineral soils along deeper flow paths. Where there is permafrost, flow paths are constrained to the seasonally thawed portion of the soil profile (active layer) and deep groundwater contributions are relatively small, e.g. via springs (Woo 2012).

Seasonal vegetation growth and microbial activity also contribute to variations in nutrient and organic matter concentration in arctic rivers. For example, growth of tundra vegetation on the North Slope of Alaska is nitrogen-limited during summer months and loss of inorganic nitrogen from soils to tundra streams is relatively low during the growing season (Williams et al. 2000). Increased microbial activity during summer months provides a source of inorganic nitrogen through mineralisation of organic matter but much of this nitrogen is taken up by the vegetation. Higher temperatures and longer residence times (in both soils and river networks) during summer base flow conditions combine to facilitate microbial decomposition of dissolved organic matter. In general, river waters entering the Arctic Ocean are rich in organic matter and depleted in inorganic nitrogen (Table 11.13).

A comparison of DOC and nitrate concentrations in major world rivers shows that nitrate concentrations are consistently low for the arctic rivers (Figure 11.21). Nitrate yields also are much lower, partly a function of the low discharge intensity. DOC concentrations by contrast are higher in the arctic rivers than elsewhere and DOC yields are about average because of the low runoff. The pattern for dissolved inorganic phosphorus is similar to that of nitrate and the pattern for dissolved organic nitrogen is similar to that of DOC. Dissolved silica concentrations in arctic rivers are not much different from other regions.

11.8.2 Transport of Contaminants

During the past 50 years, persistent organic pollutants (POPs), metals and radionuclides have been widely distributed into northern freshwater ecosystems by long range atmospheric transport (Macdonald et al. 2002).

Deposition from the atmosphere is augmented by industry, agriculture, and biotransport. There are two components of long-range transport pathways: transport to arctic freshwater basins; and processes within those basins. Both components tend to enhance bioaccumulation of contaminants.

POPs in arctic basins are transported through the atmosphere either as gases or adsorbed onto particulates. Global temperature increase will normally accelerate this cycling and increased wetness is likely to increase the capture of particulates and contaminants. Because much of the contaminant delivery to the Arctic occurs during late winter as 'arctic haze' or 'brown snow' events, sequestering by snow is an important process. Terrestrial organic carbon in soils and vegetation has a large capacity to store POPs. Many of the same concentrating processes for POPs are equally valid for mercury. However, mercury has one distinctive process called mercury depletion events. The process requires snow surfaces, solar radiation and the presence of sea salts (bromides and chlorides). Once mercury has been deposited, its entry into the hydrological and

Table 11.13 Baseline infomation for two of the four ACIA regions (Central Siberia and Western Canada/Chukchi/Bering).

Projected environmental changes	Central Siberia	Canada/Alaska
Mean annual temperature (°C)		
Baseline 1981–2000	−8 to +4	−8 to +12
2050 change from baseline	+0.5 to +4	0 to +4
Precipitation (mm month^{-1})		
Baseline 1981–2000	+10 to +70	+10 to +150
2050 change from baseline	−2 to +5	−4 to +1
Change in albedo (due to vegetation change)		
2050	−0.05 to +0.025	−0.10 to +0.025
*Ecosystem processes projected by the LPJ model**		
NPP (PgC yr^{-1})		
1960s–1980s (%)	62.3	63.8
NEP change in C storage		
1960–2080 (PgC)		
Vegetation	1.7	2.5
Soil	0.5	1.9
Litter	0.5	1.8
Total	2.8	6.2
*Landscape processes projected by the LPJ model**		
Increase in taiga area (%)		
1960–2020	6.1	4.2
1960–2050	9.4	8.2
Decrease in polar desert area (%)		
1960–2020	6.9	5.3
1960–2050	9.9	11.0

*Responses of a simulation model driven with outputs from four different climate models and forced with the B2 emissions scenario.

organic carbon cycles can be facilitated. Flooding of terrestrial landscapes releases mercury from submerged soils. Therefore, alteration of wetland distribution resulting from thawing permafrost is likely to increase mercury accumulation. Terrestrial organic carbon in soils and vegetation has a large capacity to store POPs, such as PCBs, DDT, HCH, and chlorobenzenes. Mercury deposition from the atmosphere is another major contaminant as there is now 2–3 times as much mercury cycling through the atmosphere and surface waters as compared with pre-industrial revolution times.

Biogeochemical export from rivers and linkages between river-supplied nutrients and productivity across the pan-arctic domain as a whole have been studied by Holmes et al. 2012 and Tank et al. 2011.

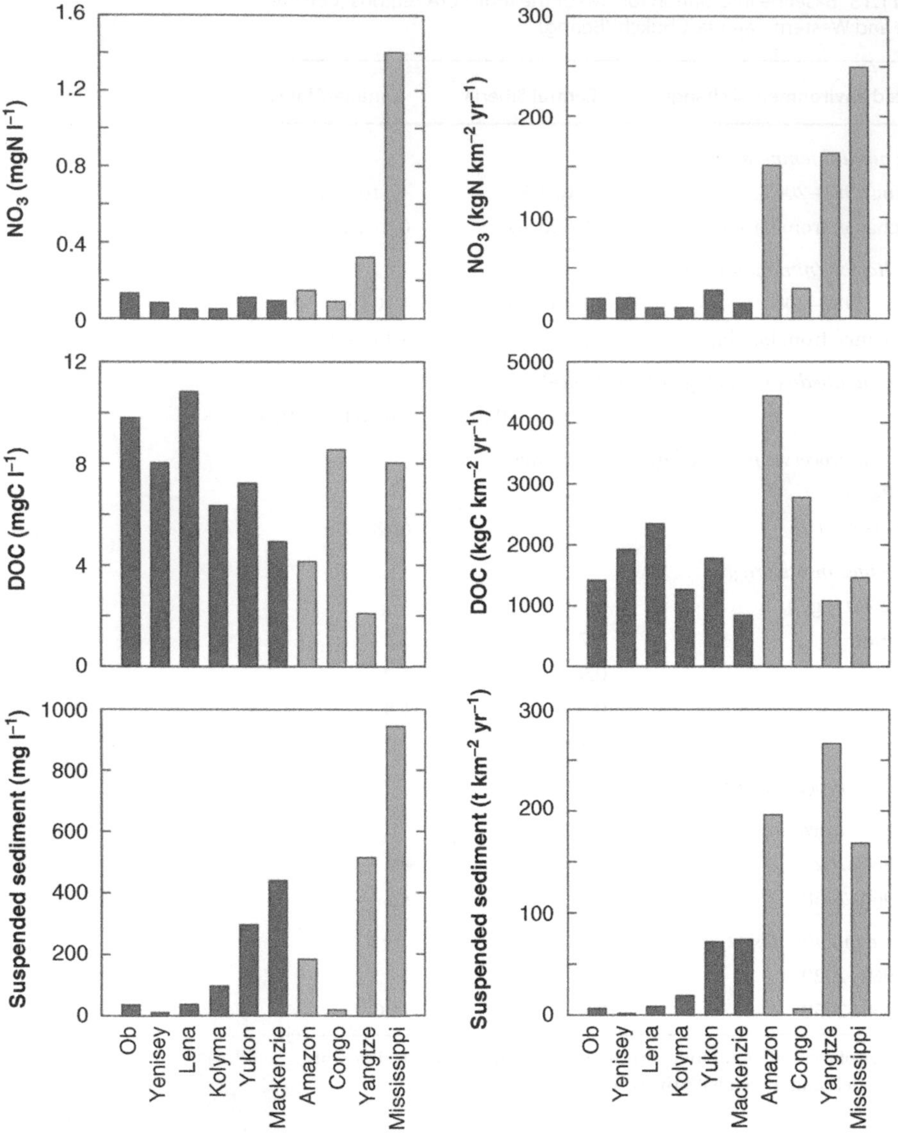

Figure 11.21 Concentrations (left columns) and yields (right columns) of nitrates, dissolved organic carbon (DOC) and total suspended sediments in the six largest rivers draining the pan-arctic watershed (Ob, Yenisey, Lena, Kolyma, Yukon, and Mackenzie) as well as several other major rivers around the world (Amazon, Congo, Yangtze, and Mississippi). Average annual discharge in km^3 yr^{-1}: Amazon (6590), Congo (1200), Yangtze (928), Yenisei (620), Mississippi (530), Lena (520), Ob (390), Mackenzie (310), Yukon (210), and Kolyma (130). Data from Holmes et al. (2002, 2012).

11.9 Mackenzie, Yukon and Lena Deltas

11.9.1 Mackenzie Delta

The classic study of the Mackenzie Delta region is provided by Mackay (1963). He notes that the maze of lakes and channels 'defies easy description'. The north–south length of the delta is 210 km; the width is about 62 km, and the total area covers 12 000 km^2. The delta is composite in origin, mostly built up by sedimentation from the Mackenzie River but the southwestern part receives sediment also from the Peel and Rat rivers. The dominant physical features are the lakes and river channels, many of which are interconnected. The river channels present an anastomosing network; they rarely meander and are best described as sinuous. Many lakes open on to the channels and many channels flow through lakes. The lake channels experience flow reversal with water flowing into the lakes at high water and out again at low water. The reversals occur most commonly after break-up of ice, but also follow pronounced rises in river level either from heavy precipitation or from surges accompanying storms. The channels in the northern part of the delta are rather stable by contrast with those affected by the Peel and Rat rivers in the south. Mackay recognised five lake types: (i) abandoned channel lakes; (ii) arcuate point bar lakes; (iii) floodplain lakes; (iv) thermokarst lakes; and (v) dammed lakes. These lakes can be described as the result of a contest between levee building, infilling, thermokarst melting, and lateral cutting. Mackay proposed that of these, levee building has a greater impact.

The Mackenzie River branches into many distributaries as it enters its delta (Burn 2010; Figure 11.22). The main channels follow levees that separate the river from the numerous delta lakes that, depending on the elevation of the sills that fringe the lakes, have continuous, intermittent (during floods) or no flow linkage with the river. In addition to flooding by the spring freshet, storm surges also produce a backwater effect that causes the river to overflow the sills and flood the delta wetlands and lakes. One such episode occurred on 15–18 September 1985, with backwater effect that could be traced up to Point Separation, over 200 km inland. The delta continuously progrades in the Beaufort Sea as the river brings in 100 million tonnes of sediment per year to extend its delta. According to Vermaire et al. (2013), arctic warming has led to an increase in coastal storm surges.

11.9.2 Lena Delta

The Lena River distributes water and sediment through four main channels and discharges about 515 km^3 of water annually. Over 1500 islands compose the 30 000 km^2 delta, and continuous permafrost underlies the area to 400–600 m depth (Boike et al. 2016; Figure 11.23). The majority of the channels freeze over in winter but the broad Bykov Channel to the east is seasonally navigable. This channel is the Lena's connecting link to the harbour city of Tiksi, the most important economic centre on the Northeast Passage. The islands which grow to form the delta rise 6–40 m above the sea level. In the delta, dry tundra occurs on polygon ridges, well-drained plateaux, and elevated polygon centres while wet tundra is found in depressed polygon centres, in water channels, and on collapsed ridges. The patterned terrain dominantly consists of ice wedge polygonal networks with depressed centres and thermokarst lakes. Water surfaces are classified as either overgrown water or open water with no vegetation.

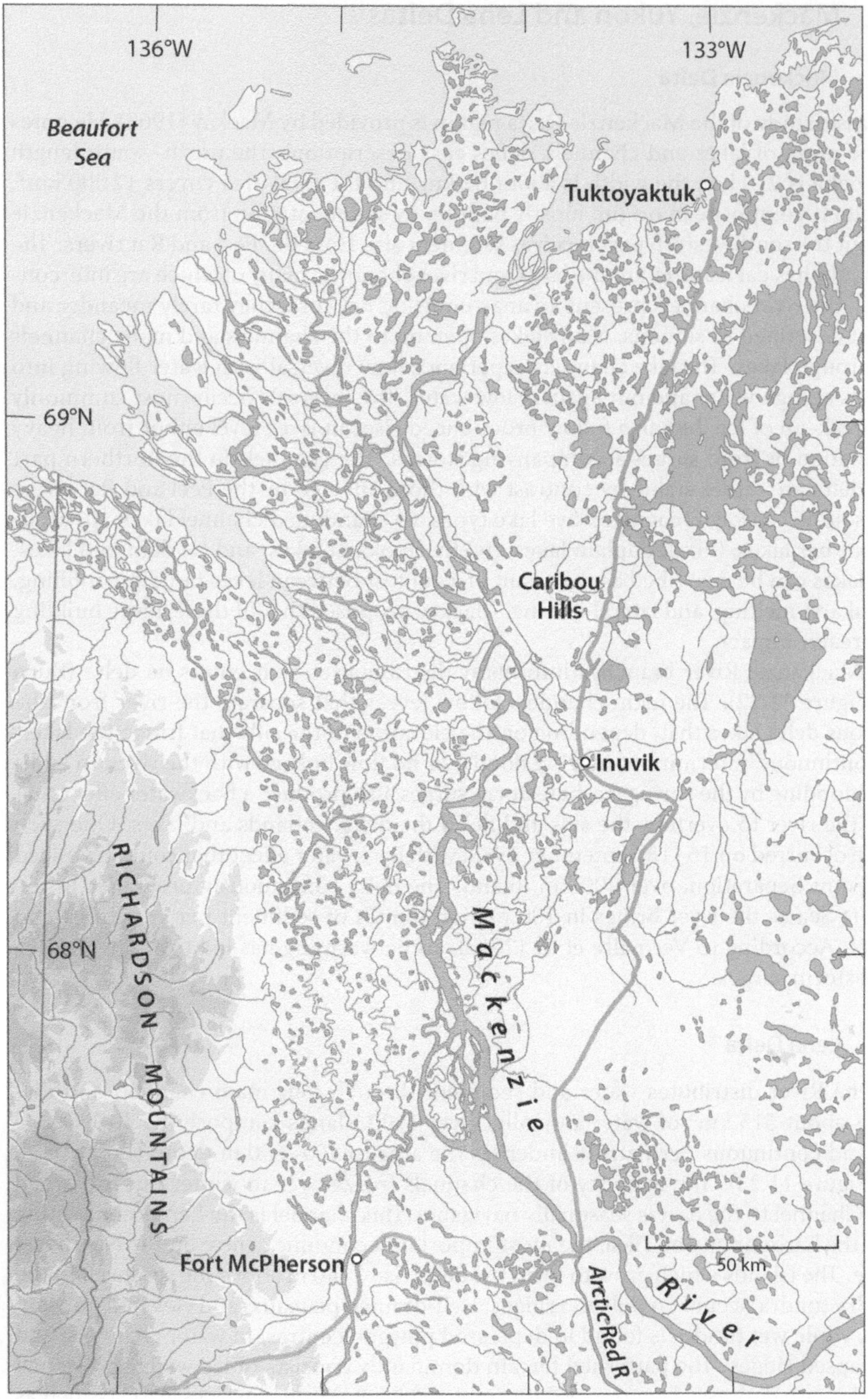

Figure 11.22 Map of the Mackenzie delta anabranching system (after Burn 2010).

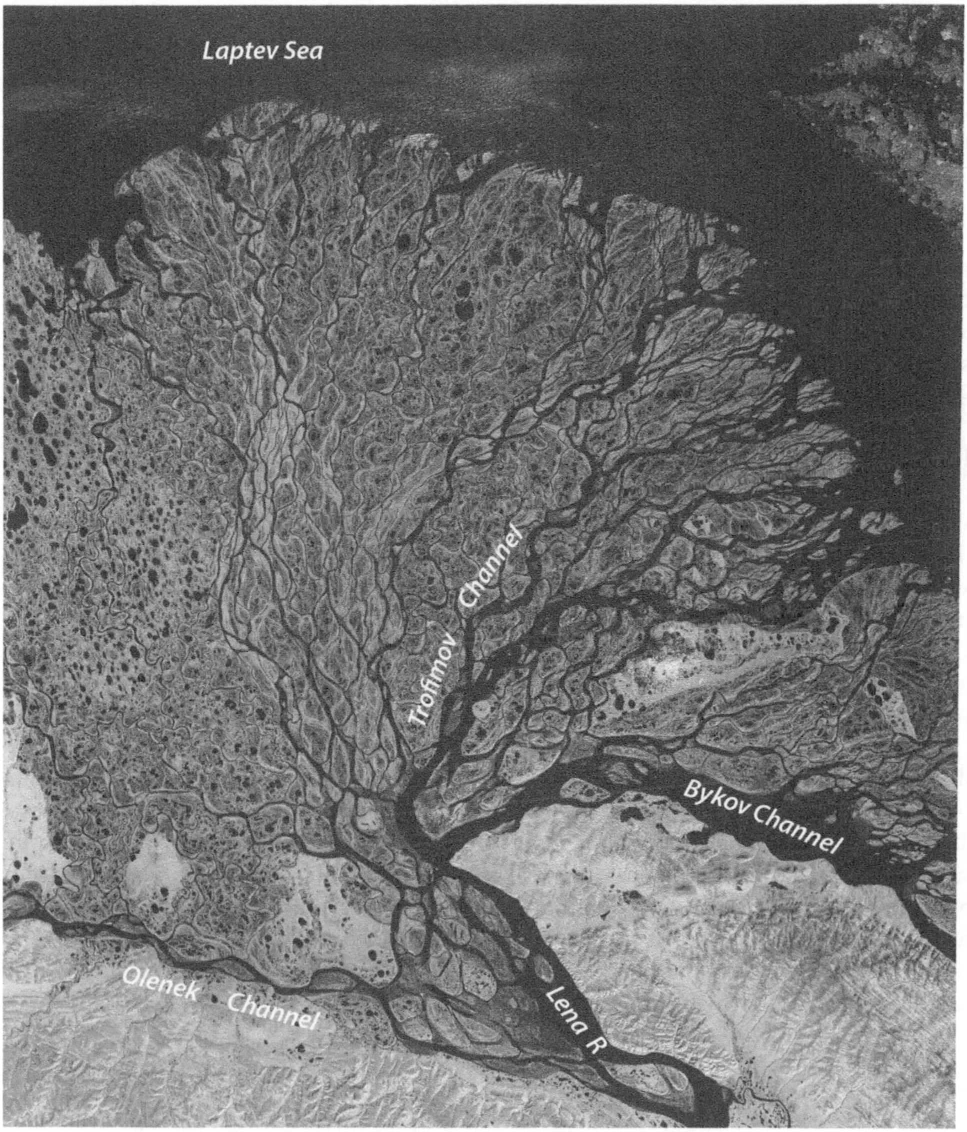

Figure 11.23 Lena–Olenek delta: to the west, the Olenek Channel that freezes over in winter and to the east, the Bykov Channel that is seasonally navigable. The latter channel connects the Lena River to the harbour city of Tiksi, the most important economic centre on the Northeast Passage through the Laptev Sea. The Trofimov Channel carries 70% of the water discharge. Source: Thematic Mapper, Landsat 7. US Geological Survey.

Overgrown water is found in troughs above ice wedges, in polygon centres with fluctuating water tables and in the shallow parts of ponds and lakes. Water bodies are dominated numerically by the polygonal ponds but in area by a relatively few thermokarst lakes. Polygonal ponds contribute 35%, thermokarst lakes 40%, and polygonal lakes about 15% of water surfaces. Polygonal lakes are transitional between polygonal ponds and thermokarst lakes. The shorelines of polygonal ponds are defined by ice wedge structures with steep flanks and flat bottoms. All soils are gelisols, greatly influenced by permafrost. The shorelines of the shallow ends of thermokarst lakes are highly irregular. Where deeper sections occur closer to the shore the shorelines are smooth, and the lakes have an oval shape.

11.9.3 Yukon–Kuskokwim Delta

In western Alaska, a single massive delta, over 23 000 km^2 in area, has been formed at the mouths of the Yukon and Kuskokwim rivers. This is almost as big as the Lena Delta and almost twice the size of the Mackenzie Delta (Thorsteinson et al. 1989). The delta is a depositional plain built since the sea reached its present level about 5000–6000 years ago (Figure 11.24).

The delta consists of several geomorphological units (Figure 11.25): (i) delta uplands, the highest delta environments above sea level with peatlands characterised by low-growing sedge and dwarf shrub hummocks, large thaw lakes, channels, and old streams; (ii) fluvial plain, gently sloping and displaying actively meandering and abandoned distributary channels and channel bars, natural levees, interdistributary marshes with peats and mud; a narrow band of forest (*P. mariana, P. glauca, Betula,* and *Populus*) along the Yukon River; (iii) a chenier plain, a narrow, low coastal beach or marsh that forms in response to variable sediment supply located south of the southernmost distributary where beach ridges alternate with thaw lakes; (iv) tidal flats; and (v) a sub-ice delta platform.

The modern Yukon Delta has formed within the past 2500 years after the river course shifted to its present location. Both the emergent and submerged portions of the delta contain three major distributaries which bifurcate as they approach the coast. These distributaries continue as far as 30 km beyond the current shoreline. There is rapid lateral migration of these distributaries. Although the delta is in the discontinuous permafrost zone, closed system pingos are actively forming on the delta, possibly further south than any others recorded. Permafrost is present, but discontinuous and relatively thin (2–3 m). Permafrost is not present along the larger channels and in the rapidly prograding shoreline.

11.10 Significance of Large Arctic Rivers

Large rivers as conveyors of sediments, dissolved solids, nutrients, and pollutants have long fascinated scientists. Four broad sets of questions are commonly raised regarding the major world rivers: (i) the quest to determine contemporary denudation rates; (ii) concerns about land degradation, the global extent of soil erosion and the significance of human impacts; (iii) the role of sediment in biogeochemical cycles; and (iv) inspiration for understanding fluvial stratigraphy. Large arctic rivers not only raise these questions but also queries regarding the sources and sinks of the rivers that drain half of the continental landmass of Russia, Canada, and Alaska. The city of Tomsk lies near the junction of the Ob and the tributary Tom River. When Charles Wenyon visited Tomsk in 1893 he noted 'Almost all the various industries of Siberia are represented in --- Tomsk; its rivers abound with fish, and its forests with game; in the hilly regions of

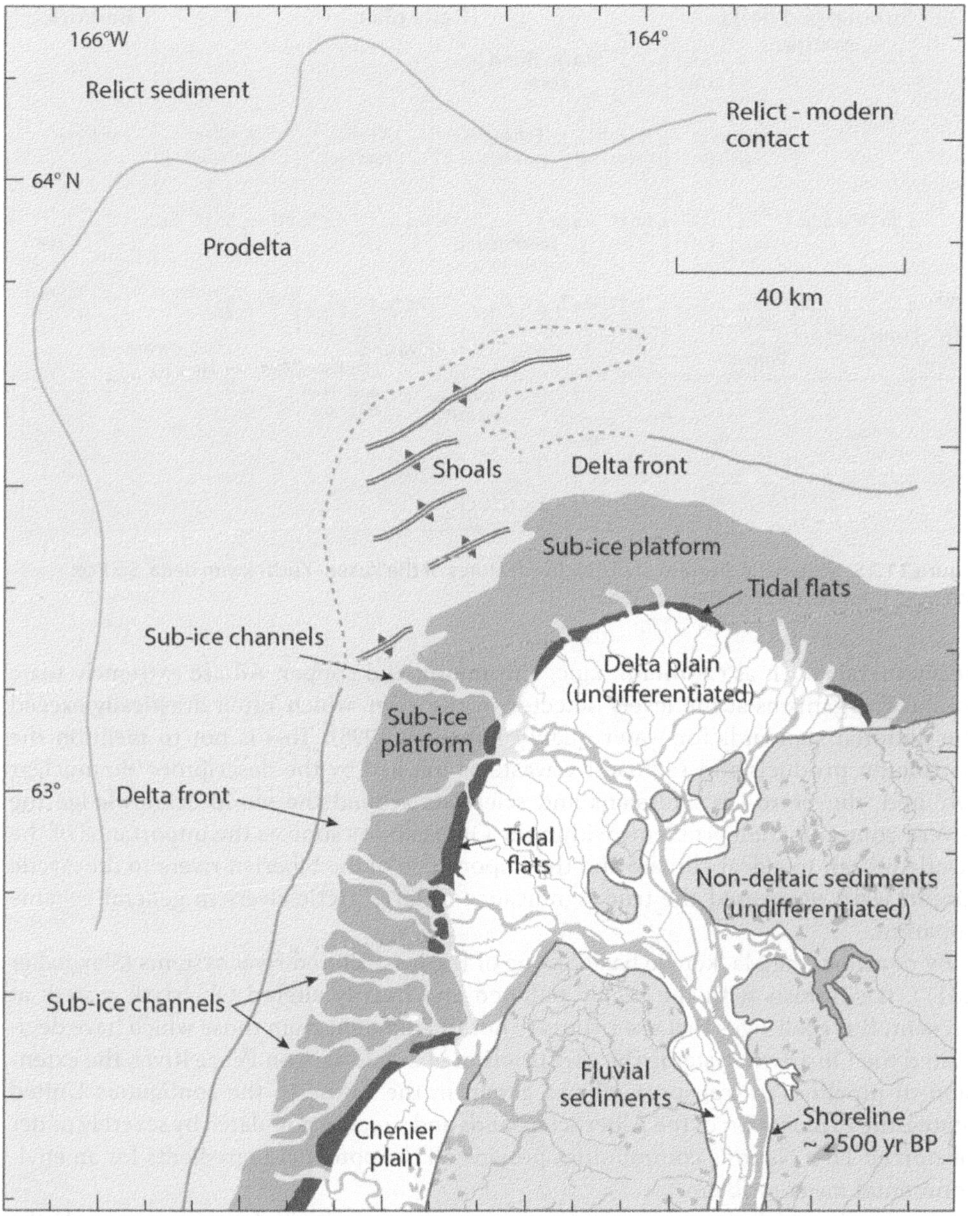

Figure 11.24 Depositional environments of the Yukon–Kushkokwim delta. Source: Adapted from Dupre 1980.

the south there is mining for gold and silver and precious stones; in the wide valleys farmers have tens of thousands of acres of fertile lands under cultivation; and other settlers rear great herds of horses and cattle on the plains' (quoted in Wohl 2011). By 1998, 'the river waters contain practically all types of pollutants (a variety of organic compounds, heavy metals and others)' (Vasiliev 1998). Major contaminants include petroleum compounds, phenols, formaldehyde, naphthalene, and its derivatives and

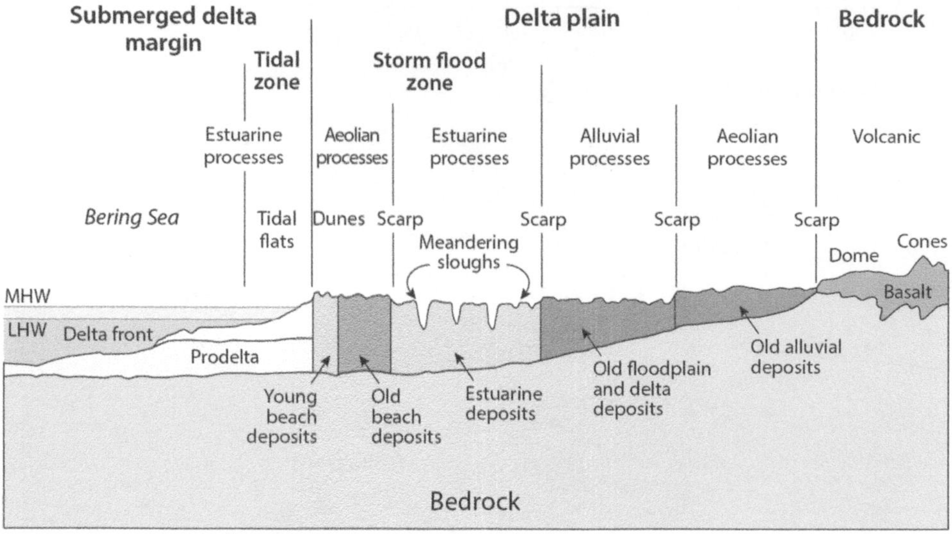

Figure 11.25 Schematic diagram of distinctive features of the Yukon–Kushokwim delta. Source: Adapted from Tande and Jennings 1986.

heavy metals such as cadmium, zinc, chromium, and copper. All are extremely toxic to living organisms at the levels detected in the Tom, which 'often drastically exceed the national standards for water quality' (Vasiliev 1998). This is not to mention the petroleum products and radioactive waste as implied by the descriptors 'the nuclear Ob' and 'the petro-Ob'. Russians and scientists around the world acknowledge the severe contamination problems within the Ob basin. As soon as the importance of the Arctic Ocean to global climate and the importance of the Siberian rivers to the Arctic Ocean were recognised the true significance of large arctic rivers in general became apparent.

By contrast, the Mackenzie basin is one of the least altered river systems (Slaymaker 2017). It contrasts with the highly polluted and heavily altered Ob-Irtysh system at present. But challenges that are rising are not too different from those which have decimated the Ob-Irtysh system. The construction of Site C dam on Peace River, the extension of pipelines to transport oil and gas from the Arctic to the contiguous United States, the exploitation of the Alberta tar sands and a region populated by severely under resourced First Nations communities provide all the potential ingredients for an environmental mega-problem.

The way in which these large arctic rivers are managed will determine not only the fate of the ecosystems within their basins but the fate of the oceans and the future of our planet.

Acknowledgment

All cartography in this chapter is original, was designed by Eric Leinberger, and is hereby acknowledged.

Questions

1 Consider the benefits of continuing warming trends in arctic environments. Do the benefits outweigh the disadvantages?

2 What are the limitations of applying a model of river behaviour developed for small basins to a large river basin?

3 What are the limitations of applying a model of river behaviour developed for temperate climate environments to arctic river basins?

4 Comment on the usefulness of applying the Schumm (1977) model of river behaviour to physiographic and geomorphological landscape regions.

5 Why do Siberian rivers show differing trends in their discharge and sediment concentration data over time?

6 How would you explain the scale effects demonstrated in the specific sediment yield data from the Mackenzie basin?

7 What are the main reasons for the low suspended sediment flux in large arctic rivers?

8 Why do the Mackenzie and Yukon rivers have higher suspended sediment flux?

9 Why do the Siberian rivers have higher dissolved solids than suspended sediment loads? This is a feature which is the reverse of the Mackenzie and Yukon rivers.

10 Why do the two priority issues recognised in 1988 remain unresolved today?

11 What are the main reasons why the Arctic Ocean is often described as an estuary?

12 Why are persistent organic pollutants a more serious problem in arctic than in temperate basins?

13 What are the main effects of climate change on arctic fish, fisheries, and aquatic wildlife as indicated by changing solute, nutrient and contaminant yield of the large arctic rivers?

References

ACIA (2005). *Arctic Climate Impact Assessment.* Cambridge, UK: Cambridge University Press.

Alisov, B.P. (1956). *Climate of the USSR.* Moscow: Moscow State University Press (in Russian).

Anisimov, O. and Fitzharris, B. (2001). Polar regions (Arctic and Antarctic). In: *Climate Change 2001: Impacts, Adaptation and Vulnerability* (eds. J. McCarthy, O.F. Canziani, N.A. Leary, et al.), 801–841. Cambridge, UK: Cambridge University Press.

Ashmore, P. and Church, M. (2001). *The Impact of Climate Change on Rivers and River Processes in Canada*, Geological Survey of Canada Bulletin 555. Ottawa, Canada.

Aylsworth, J.M. and Kettles, I.M. (2000). Distribution of peatlands. In: *The Physical Environment of the Mackenzie Valley, Northwest Territories: A Baseline for the Assessment of Environmental Change*, Geological Survey of Canada Bulletin 547 (eds. L.D. Dyke and G.R. Brooks), 49–55. Ottawa, Canada.

Aylsworth, J.M., Duk-Rodkin, A., Robertson, T., and Traynor, J.A. (2000). Landslides of the Mackenzie valley and adjacent mountainous and coastal regions. In: *The Physical Environment of the Mackenzie Valley, Northwest Territories: a Baseline for the Assessment of Environmental Change*, Geological Survey of Canada Bulletin 547 (eds. L.D. Dyke and G.R. Brooks), 167–176. Ottawa, Canada.

Baker, V.R. (2008). Greatest floods and largest rivers. In: *Large Rivers: Geomorphology and Management* (ed. A. Gupta), 65–74. Chichester, UK: Wiley.

Bobrovitskaya, N.N., Kokorev, A.V., and Lemeshko, N.A. (2003). Regional patterns in recent trends in sediment yield of Eurasian and Siberian rivers. *Global and Planetary Change* 39: 127–146.

Boike, J., Grau, T., Heim, B. et al. (2016). Satellite-derived changes in the permafrost landscape of central Yakutia, 2000-2011: wetting, drying and fires. *Global and Planetary Change* 139: 116–127.

Bostock HS (2014) Physiographic regions of Canada. Geological Survey of Canada, Ottawa, Canada. 'A' Series Map 1254A.

Brooks, G.R. (1996). *The Fluvial Geomorphic Character of the Lower Reaches of Major Mackenzie River Tributaries, Fort Simpson to Norman Wells*, Geological Survey of Canada Bulletin 493. Ottawa, Canada.

Brown, J., Ferrians, O.J., Heginbottom, J.A., and Melnikov, E.S. (eds) (1997) Circum-Arctic map of permafrost and ground-ice conditions. US Geological Survey in cooperation with the Circum-Pacific Council for Energy and Mineral Resources. Circum-Pacific Map Series CP-45, Washington, DC. Scale: 1:10 000 000.

Burn, C.R. (2010). The Mackenzie Delta: an archetypical permafrost landscape. In: *Geomorphological Landscapes of the World* (ed. P. Migon), 1–12. Dordrecht, Netherlands: Springer Verlag.

Carson, M.A. (1988) An assessment of problems relating to the source, transfer and fate of sediment along Mackenzie River, NWT. Inland Waters Directorate, Environment Canada, Ottawa and Yellowknife, Canada. Report KE144-7-4155/01-SS.

Carson, M.A., Jasper, J.N., and Conly, F.M. (1998). Magnitude and sources of sediment input to the Mackenzie Delta, 1974–1994. *Arctic* 51: 116–124.

Church, M.A. (1977). River studies in northern Canada: reading the record from river morphology. *Geoscience Canada* 4: 4–12.

Church, M. and Ryder, J.M. (1972). Paraglacial sedimentation: a consideration of fluvial processes conditioned by glaciation. *Geological Society of America Bulletin* 83: 3059–3072.

Church MA, Miles M, Rood K (1987) Sediment transfer along Mackenzie River: a feasibility study. Environment Canada, Ottawa, Canada. Report IWD-WNR(NWT), WRB-SS-87-1.

Church, M., Kellerhals, R., and Day, T.J. (1989). Regional clastic sediment yield in British Columbia. *Canadian Journal of Earth Sciences* 26: 31–45.

Cubasch, U., Hasselman, K., Hock, H. et al. (1992). Time dependent greenhouse warming computations with a coupled ocean-atmosphere model. *Climate Dynamics* 8: 55–69.

Derksen, C. and Brown, R. (2012). Spring snow cover extent reduction in the 2008-2012 period exceeding climate model projections. *Geophysical Research Letters* 39: L19504. https://doi.org/10.1029/2012GL053387.

Duk-Rodkin, A. and Lemmen, D.S. (2000). Glacial history of the Mackenzie region. In: *The Physical Environment of the Mackenzie Valley, Northwest Territories: A Baseline for the Assessment of Environmental Change*, Geological Survey of Canada Bulletin 547 (eds. L.D. Dyke and G.R. Brooks), 11–20. Ottawa, Canada.

Dupre, W.R. (1980) Yukon Delta processes study. Final Report of the Environmental Assessment of the Alaska Continental Shelf, Environmental Research Laboratory, Boulder, Colorado: 1–78.

Dyke, L.D. (2000). Climate of the Mackenzie Valley. In: *The Physical Environment of the Mackenzie Valley, Northwest Territories: A Baseline for the Assessment of Environmental Change*, Geological Survey of Canada Bulletin 547 (eds. L.D. Dyke and G.R. Brooks), 21–30. Ottawa, Canada.

Dyke, A.S. (2004). An outline of North American deglaciation with emphasis on central and northern Canada. In: *Developments in Quaternary Science* (eds. J. Ehlers and P.L. Gibbard), 373–424. Netherlands: Elsevier.

Gao, L., Peng, C.-L., and Macdonald, R.W. (2007). Mobilization pathways of organic carbon from permafrost in arctic rivers in a changing climate. *Geophysical Research Letters* 34: L13603.

Gibson, J.J., Prowse, T.D., and Peters, D.L. (2006). Hydroclimatic controls on water balance and water level variability in Great Slave Lake. *Hydrological Processes* 20: 4155–4172.

Gordeev, V.V. (2006). Fluvial sediment flux to the Arctic Ocean. *Geomorphology* 80: 94–104.

Grosswald, M.G. and Hughes, T.J. (1995). Paleoglaciology's grand unsolved problem. *Journal of Glaciology* 41: 313–332.

Hicks, F.E. and Beltaos, S. (2008). River ice. In: *Cold Region Atmospheric and Hydrologic Studies, the Mackenzie GEWEX Experience, vol. 2: Hydrologic Processes* (ed. M.K. Woo), 281–305. Berlin: Springer.

Holmes, R.M., McClelland, J.W., Peterson, B.J. et al. (2002). A circumpolar perspective on fluvial sediment flux in the Arctic Ocean. *Global Biogeochemical Cycles* 16: 1–45.

Holmes, R.M., McClelland, J.W., Peterson, B.J. et al. (2012). Seasonal and annual fluxes of nutrients and organic matter from large rivers to the Arctic Ocean. *Estuaries and Coasts* https://doi.org/10.1007/s12237-011-9386-6.

Koronovsky, N. (2002). Tectonics and geology. In: *The Physical Geography of Northern Eurasia* (ed. M. Shahgedanova). Oxford, UK: Oxford University Press.

Lammers, R., Shiklomanov, A., Vorosmarty, C.J. et al. (2001). Assessment of contemporary Arctic river runoff based on observational discharge records. *Journal of Geophysical Research* 106: 3321–3334.

Lemmen, D.S., Duk-Rodkin, A., and Bednarski, J.M. (1994). Late glacial drainage systems along the northwest margin of the Laurentide Ice Sheet. *Quaternary Science Reviews* 13: 805–828.

Ludwig, W. and Probst, J.L. (1998). River sediment discharge to the oceans: present-day controls and global budgets. *American Journal of Science* 298: 265–295.

Macdonald, R.W., Mackay, D., and Hickie, B. (2002). Contaminant amplification in the environment: revealing the fundamental mechanisms. *Environmental Science and Technology* 36: 457A–463A.

Mackay, J.R. (1963). *The Mackenzie Delta Area, N.W.T.*, Geographical Branch Memoir 8. Ottawa, Canada: Canada Department of Mines and Technical Surveys.

McCarthy, J.J. and Martello, M.L. (2005). Climate change in the context of multiple stressors and resilience. In: *Arctic Climate Impact Assessment*, 946–988. Cambridge, UK: Cambridge University Press.

McClelland, J.W., Holmes, R.M., Peterson, B.J., and Stieglitz, M. (2004). Increasing river discharge in the Eurasian Arctic: consideration of dams, permafrost thaw and fire as potential agents of change. *Journal of Geophysical Research* 109: D18102. https://doi.org/10.1029/2004JD004583.

McClelland, J.W., Holmes, R.M., Dunton, K.H., and Macdonald, R.W. (2012). The Arctic Ocean estuary. *Estuaries and Coasts* 35: 353–368.

Meade, R.H., Bobrovitskaya, N.N., and Babkin, V.I. (2000). Suspended sediment and freshwater discharges in the Ob and Yenisey rivers, 1960–1988. *International Journal of Earth Science* 89: 578–591.

Milliman, J.D. and Farnsworth, K.L. (2011). *River Discharge to the Global Oceans.* Cambridge, UK: Cambridge University Press.

Peters, D.L. and Prowse, T.D. (2001). Regulation effects on the lower Peace River, Canada. *Hydrological Processes* 15: 3181–3194.

Prowse, T.D. and Carter, T. (2002). Significance of ice-induced storage to spring runoff: a case study of the Mackenzie River. *Hydrological Processes* 16: 779–788.

Schumm, S.A. (1977). *The Fluvial System*. New York: Wiley.

Serreze, M.C., Barrett, A.P., Slater, A.G. et al. (2006). The large scale freshwater cycle of the Arctic. *Journal of Geophysical Research* 111 https://doi.org/10.1029/2005JC003424.

Shahgedanova, M. (2002). Climate at present and in the historical past. In: *The Physical Geography of Northern Eurasia* (ed. M. Shahgedanova), 70–102. Oxford, UK: Oxford University Press.

Sherstone, D.A. (1983). Sediment removal during an extreme summer storm: Muskwa River, northeastern BC. *Canadian Geotechnical Journal* 20: 329–335.

Shiklomanov, I.A., Shiklomanov, A.I., Lammers, R.B. et al. (2000). The dynamics of river water inflow to the Arctic Ocean. In: *The Freshwater Budget of the Arctic Ocean* (ed. E.L. Lewis), 281–296. Dordrecht, Netherlands: Kluwer.

Slaymaker, O. (ed.) (2017). *Landscapes and Landforms of Western Canada*. Dordrecht, Netherlands: Springer Verlag.

Smith, D.G. (1994). Glacial Lake McConnell: paleogeography, age, duration and associated river deltas, Mackenzie River basin, western Canada. *Quaternary Science Reviews* 13: 829–843.

Stewart, R.E., Leighton, H.G., Marsh, P., and Kochtubajda, B. (1998). The Mackenzie GEWEX study. The water and energy cycles of a major North American river basin. *Bulletin of the American Meteorological Society* 79: 2665–2683.

Strahler, A.H. and Strahler, A.N. (1994). *Modern Physical Geography*, 4e. Chichester, UK: Wiley.

Syvitski, J.P.M. (2002). Sediment discharge variability in arctic rivers: implications for a warmer future. *Polar Research* 21: 323–330.

Tande, G.F. and Jennings, T.W. (1986). *Classification and Mapping of Tundra near Hazen Bay, Yukon Delta National Wildlife Refuge, Alaska*. Anchorage, Alaska, USA: US Fish and Wildlife Service.

Tank, S.E., Manizza, M., Holmes, R.M. et al. (2011). The processing and impact of riverine nutrients and contaminants in the near and off-shore Arctic Ocean. *Estuaries and Coasts* https://doi.org/10.1007/s12237-011-9417-3.

Teller, J.T., Boyd, M., Yang, Z. et al. (2005). Alternative routing of Lake Agassiz overflow during the Younger Dryas: new dates, paleotopography and a reevaluation. *Quaternary Science Reviews* 24: 1890–1905.

Thorsteinson, L.K., Becker, P.R., and Hale, D.A. (1989). *The Yukon Delta: a Synthesis of Information. NOAA/National Ocean Service*. Anchorage, Alaska, USA: US Department of Commerce.

Tumel, N. (2002). Permafrost. In: *The Physical Geography of Northern Eurasia* (ed. M. Shahgedanova), 149–168. Oxford, UK: Oxford University Press.

Vasiliev, O.F. (1998). Water quality and environmental degradation in the Tom River basin (western Siberia): the need for an integrated management approach. In: *Restoration of Degraded Rivers: Challenges, Issues and Experiences* (ed. D.P. Loucks). Dordrecht, Netherlands: Kluwer.

Velichko, A.A., Kononov, Y.M., and Faustova, M.A. (1997). The last glaciation of Earth: size and volume of ice sheets. *Quaternary International* 41/42: 43–51.

Vermaire, J.C., Pisaric, M.F.J., Thienpont, J.R. et al. (2013). Arctic climate warming and sea ice declines lead to increased storm surge activity. *Geophysical Research Letters* https://doi.org/10.1002/grl.50191.2013.

Walsh, J.E. (2005). Cryosphere and hydrology. In: *Arctic Climate Impact Assessment*, 184–242. Cambridge, UK: Cambridge University Press.

Water Resources Institute (2003). *Water Resources eAtlas*. Washington, DC, USA: Water Resources Institute.

Williams, M., Eugster, W., Rastetter, E.B. et al. (2000). The controls on net ecosystem productivity along an arctic transect. *Global Change Biology* 6: 116–126.

Wohl, E. (2011). *A World of Rivers: Environmental Change on Ten of the World's Great Rivers*, :42–69 (Ob River) and :297–327 (Mackenzie River). Chicago, USA: University of Chicago Press.

Woo, M.-K. (2012). *Permafrost Hydrology*. Dordrecht, Netherlands: Springer Verlag.

Woo, M.-K. and Thorne, R. (2003). Streamflow in the Mackenzie basin, Canada. *Arctic* 56: 326–340.

Woo, M.-K., Mollinga, M., and Smith, S.L. (2007). Climate warming and active layer thaw in the boreal and tundra environments of the Mackenzie Valley. *Canadian Journal of Earth Sciences* 44: 733–743.

Woo, M.-K., Thorne, R., Szeto, K., and Yang, D. (2008). Streamflow hydrology in the boreal region under the influences of climate and human interference. *Philosophical Transactions of the Royal Society B* 363: 2251–2260.

Wrona, F.J., Prowse, T.D., and Reist, J.D. (2005). Freshwater ecosystems and fisheries. In: *Arctic Climate Impact Assessment*, 354–452. Cambridge, UK: Cambridge University Press.

Yang, D.Q., Ye, B.S., and Shiklomanov, A. (2004a). Stream flow changes over Siberian Yenisei River. *Journal of Hydrology* 296: 59–80.

Yang, D.Q., Ye, B.S., and Shiklomanov, A. (2004b). Discharge characteristics and changes over the Ob River watershed. *Journal of Hydrometeorology* 5: 595–610.

Ye, B.S., Yang, D., and Kane, D.L. (2003). Changes in Lena River streamflow hydrology: human impacts versus natural variation. *Water Resources Research* 39: 1200–1214.

Zhang, X., Harvey, K.D., Hogg, W.D., and Yuzyk, T.R. (2001). Trends in Canadian streamflow. *Water Resources Research* 37: 987–998.

Zlotin, R. (2002). Biodiversity and productivity of ecosystems. In: *The Physical Geography of Northern Eurasia* (ed. M. Shahgedanova), 169–190. Oxford, UK: Oxford University Press.

Zoltai, S.C. (1988). *Wetlands of Canada*, National Wetland Working Group: Ecological Land Classification Series No. 24. Montreal, Canada: Sustainable Development Branch, Environment Canada, Ottawa and Polyscience Publications Inc.

12

Climate Change and Large Rivers

12.1 Introduction

The form and function of a river depend, as we have seen, primarily on the physiography of its drainage basin and the ambient climate. For example, a series of changes in climate and sea level happened during the Quaternary, forcing the rivers to alter their form and function. The current climate change is also likely to impact present-day rivers. Rivers may change their forms, alter their courses, and lose stationarity (Blum 2007; Milly et al. 2008). Stationarity is the concept that measures of natural systems, within a time period, tend to fall within an expected range of variability. It is a statistical concept much used in water engineering. As settlements and economic enterprises tend to be located on river plains and deltas, loss of stationarity impacts both the local environment and the lifestyle of residents. We may have to re-learn how to live near rivers.

The impact of climatic change along a large river will be complex (Milly et al. 2008), and may vary amongst its different parts: headwaters, megafans, sections on rock, sections on alluvium, and delta. The impact of climate change may also vary from river to river depending on the river's geographical location. There are many different types of impact. Many studies have reviewed climate change and its impact on various aspects of the environment and people's lifestyles but only a limited number have projected its effects on large rivers and their deltas. The impact is difficult to project, as the effects of El Niño Southern Oscillation (ENSO) and similar events, and various anthropogenic modifications are often superimposed on climate change, the final result reflecting all factors. It has been said that the only certainty in predicting the effect of climate change is uncertainty.

Impacts can be determined by (i) observations, (ii) modelling, and (iii) studying past analogues, usually from the Late Quaternary. Observations are the best, but so far we have had very few opportunities for measuring time-based changes on rivers. Models provide a generalised forecast, often at a large-scale, and not always at the localised level we want. Construction of scenarios that may arise from climate change, using cases in the Late Quaternary, is helpful but perhaps too generalised. We may be able to identify changes in precipitation at a subcontinental scale but possibly not foretell places of future levee disruption. We can, however, assume that due to global warming leading to climate change, a river may behave differently in the future than it does at present. This may modify the regional physiography and utilisation of river water.

Introducing Large Rivers, First Edition. Avijit Gupta.

12.2 Global Warming: Basic Concept

The shortwave high temperature radiation energy from the Sun heats the surface of the Earth, which then radiates low-temperature thermal energy of long wavelengths back through the gaseous atmosphere to space. The escape through the atmosphere into space occurs through gaps in the atmosphere used by certain wavelengths of radiation not covered by constituent atmospheric gases. Such gaps are called windows. The two types of radiation should be in balance. When the temperature of the surface of the Earth increases, and so does the atmospheric temperature, an imbalance is created. This is global warming (Figure 12.1).

Nitrogen and oxygen make up about 99% of the atmosphere but take no part in global warming. Instead, several minor constitutional gases – water vapour, carbon dioxide, and others – heat the atmosphere in spite of their low volume. They do it by partially blocking the windows of wavelength at which Earth's thermal radiation escapes through the envelope of atmosphere to space. This happens naturally, raising the average temperature of the atmosphere from -6 to 15 °C. Such gases are called greenhouse gases because the process partly resembles the heating inside a greenhouse. The phenomenon is also seen in our neighbouring planets, Mars and Venus, where the process is entirely natural. The Martian atmosphere consists almost entirely of CO_2, which produces a greenhouse effect. The atmosphere of Venus, thick with CO_2, is a very effective barrier

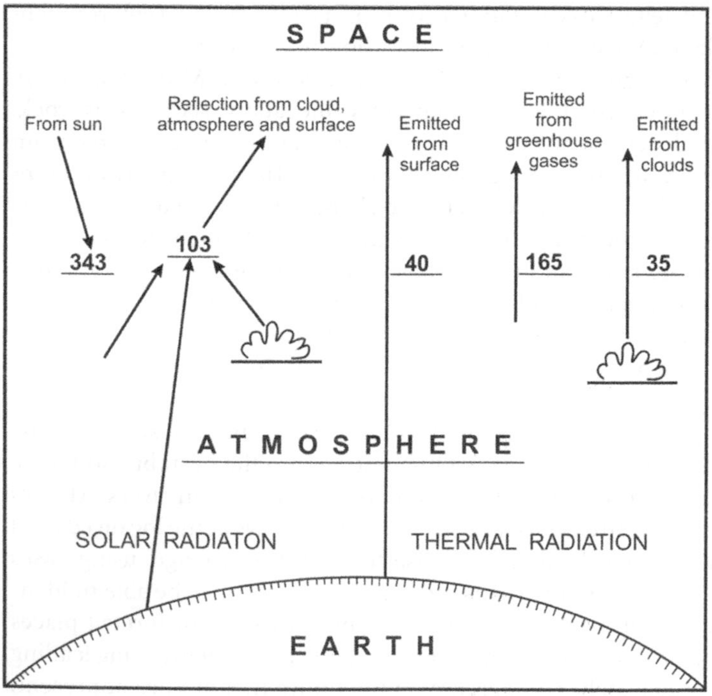

Figure 12.1 Solar and thermal (terrestrial) radiation entering and leaving the atmosphere and balancing its radiation budget. Components of the radiation given in watts per square metre. Data taken from Houghton (2004).

Table 12.1 Common gases that produce enhanced greenhouse effect.

Gas	Major sources
Carbon dioxide	Burning of fossil fuels; deforestation; the upper part of the sea water which contains dissolved CO_2; biological activities
Methane	Decomposition of organic material; leakage from gas pipelines and oil wells; generation from rice fields; fermentation from cattle
Nitrous oxide	Chemical industry; deforestation; agriculture
Chlorofluorocarbons	Used in industrial applications; recently they have been controlled as they destroy stratospheric ozone
Ozone	From other gases by reactions

Water vapour is a very effective greenhouse gas but its presence in the atmosphere is naturally determined, not anthropogenically enhanced. There are also several short-lived gases such as sulfur dioxide and carbon monoxide.

to thermal radiation from the planet, and the greenhouse effect raises its temperature close to 500 °C.

As the greenhouse gases increase in the atmosphere, they progressively block thermal radiation from escaping to space and thus heat the atmosphere. This happens, for example, by burning of fossil fuels which release additional CO_2 into the air. This is an anthropogenic phenomenon and known as the *enhanced greenhouse effect*. Table 12.1 lists the gases that give rise to the enhanced greenhouse effect.

In brief, raising the temperature of the lower atmosphere due to an enhanced greenhouse effect is generally referred to as global warming. The effect of global warming is difficult to determine, as it is also affected by feedback from other factors. Such feedback can be both positive and negative. Feedback comes mainly from two sources: atmospheric particles; and episodic climatic patterns, such as ENSO.

Fine particles present in the atmosphere tend to reduce the greenhouse effect directly, by absorbing incoming solar radiation or scattering it to space, and indirectly, by enhancing clouds. Fine particles are also referred to as aerosols. Aerosols come from a variety of sources: windblown particles from agricultural fields; deserts; sea sprays; volcanic eruptions, etc. Aerosols are mainly sulfates, organic carbon, black carbon, nitrates, and dust. Complications also arise from the episodic occurrence of climatic drivers such as ENSO which modifies the effect of climate change across the face of the Earth. Other such patterns are the North Atlantic Oscillation (NAO), the Northern Annular Mode (NAM), the Southern Annular Mode (SAM), the Pacific–North American (PNA) pattern, and the Pacific Decadal Oscillation (PDO). These climatic patterns modify temperature and rainfall, both positively and negatively, over and beyond the alterations caused by climate change.

The most important greenhouse gas is CO_2, which continues to increase in the atmosphere because of human activity. At present, excellent records of atmospheric CO_2 as a time-series are available from the Mauna Loa Climate Observatory in Hawaii. The record is due to the foresight of Roger Revelle, who was instrumental in starting the project during the International Geophysical Year, 1957–1958. The volume of earlier CO_2 content in the atmosphere is determined from air bubbles in ice cores from Antarctica and Greenland. Before the industrial revolution started, around 1750, the content of CO_2 was stationary at about 280 ppm by volume. It reached 379 ppm in 2005, easily exceeding any other concentration of CO_2 over the last 650 000 years (IPCC 2007), and even higher after that (IPCC 2014). The increased amount was released in the atmosphere mainly by the burning of fossil fuels and deforestation. Exchange of CO_2 between the atmosphere and the dissolved CO_2 in the top 100 m of the ocean adds to global warming. Biological effects are also significant in determining the CO_2 flux.

The estimated model emissions of CO_2 in different conditions, known as scenarios, have been determined by the Intergovernmental Panel on Climate Change (IPCC) and the World Energy Council (WEC). One of the scenarios of the IPCC is *IS 92*, known as the business-as-usual scenario because it assumes a doubling of the global population, moderate economic growth, and no active reduction of the emission of the gas – a reasonable assumption of the future. This scenario shows a nearly three-time rise in total emission towards the end of the current century. The increase in emission is expected mostly from energy use. Other scenarios similarly demonstrate future projections of CO_2 under different conditions. In general, the scenarios reflect what we expect to happen: an increase in the use of fossil fuels, release of CO_2 emission to the atmosphere, and enhanced global warming.

Methane (CH_4) is another greenhouse gas, which is known as marsh gas, as it is naturally generated by decomposition of organic material in wetlands. About 2000 years ago, its atmospheric concentration was about 0.7 ppm by volume. Its concentration increased to 1803 ppb by 2011 (IPCC 2014). It is in a much lower concentration than CO_2 but has a strong greenhouse effect. Its amount in the atmosphere is increased by leakage from gas pipelines and oil wells, generation from rice fields, melting of permafrost, and fermentation from cattle. Nitrous oxide (N_2O) also is a greenhouse gas; its natural volume is enhanced by chemical industry, deforestation, and agriculture. Its presence in the preindustrial atmosphere was 270 ppb but by 2011 it was 324 ppb (IPCC 2014). The combined radiative forcing of CO_2, CH_4, and N_2O is 2.30 W m^{-2} and this rate probably has previously never happened during the entire Holocene (IPCC 2007).

Chlorofluorocarbons (CFCs) are anthropogenically created chemicals which enhanced the greenhouse effect in the second half of the twentieth century, when they were much in use. They were originally designed as safe gases to be used in industry, such as in refrigeration, but CFCs have been banned under international agreements, mainly due to their harmful destruction of ozone in the atmosphere. Ozone is another constituent of the atmosphere that has a greenhouse effect.

We may expect a world of enhanced global warming, unless the release of the volume of such gases to the atmosphere is restricted by international agreements as discussed recently in Paris and Katowice (Poland).

At present, records of land temperature are collected and maintained at numerous weather stations. The Sea Surface Temperature (SST) is collected by ships sailing across oceans. Global temperature measurements are compiled through the World Weather Watch (WWW) system. The surface temperature of the Earth is also determined and recorded by satellites with designated sensors. Geographically compiled temperature records have been available from the middle of the nineteenth century. The earlier land data were mostly from the northern temperate areas and urban settlements. The earlier ocean temperatures were approximate, as on travelling vessels the water sample used to be collected in a bucket first and then measured, unlike the present practice of measuring temperature near the entrance to the water intake pipe of ships. Past temperature is also determined from air bubbles trapped in the ice of ice cores. Examination of this air reveals three pieces of information: (i) the composition of the atmosphere at the time when the bubble was trapped; (ii) the date of the ice layer which contained the bubble; and (iii) the temperature of that time determined by calculating the ratio between isotopes of oxygen in the air bubble.

All investigations tend to show a time series with a steep increase in global temperature from the later decades of the twentieth century, a curve very similar to the one showing the presence of CO_2 in the atmosphere. Climatic information indicates that the last 50 years have been unusually warm. The previous time when the polar regions were even warmer than the present for an extended period happened about 125 000 years ago, in the last interglacial (MIS 5e). It is likely that at that time the global sea level was 4–6 m higher than present (IPCC 2007). It is possible that the enhanced warming effect of the greenhouse gases would have been even more effective but for the feedback from aerosols and various atmospheric effects. Uncertainties exist in understanding the level of these feedbacks.

The Paris Climate Conference of December 2013 unanimously agreed to hold the increase in average global temperature within 2 °C of the pre-industrial level with an aspirational target of 1.5 °C. However, Wang et al. (2017) have shown that extreme El Niño frequency increases linearly with rising temperature and nearly doubles even at 1.5 °C warming. This increasing frequency is expected to continue for a hundred years even after the mean global temperature has stabilised. This would severely influence patterns of weather and associated environmental problems for a considerable time. On the other hand, their models indicate that there will be no change for La Nina with 1.5 or 2 °C increase in temperature.

Warming has not been globally uniform but the projected scenarios present geographical climate patterns for the last several decades. IPCC (2007) expects warming to be greater on land, significantly higher over the high northern latitudes, and the lowest over the Southern Oceans and part of the North Atlantic. Average temperature from the Arctic increased at nearly twice the global mean rate for the past 100 years and with high decadal variations (IPCC 2007). According to the IPCC scenarios, snow cover should contract. The current satellite images show conspicuous reduction in both the Arctic (Figure 12.2) and Antarctic sea ice. On land, temperatures at the top of the permafrost layer have generally increased since the 1980s. The seasonally frozen ground also has decreased in extent.

There have been other changes in global climate. Precipitation has increased in different regions at the same time. Rising temperature increases the water-holding capacity of the atmosphere, and it is not surprising that the frequency of extreme precipitation has increased over most land. Mid-latitude westerly winds have strengthened in both hemispheres. Intense and long droughts have been observed, mainly in parts of the tropics and subtropics. An increase in the strong tropical cyclone activities has been observed in the North Atlantic. Although no clear trend has been discerned regarding tropical cyclone activities, in general, an increase in the frequency of intense tropical cyclones has been noticed (IPCC 2007). According to various models, such tropical cyclones would have greater wind speed at their peaks and heavier precipitation.

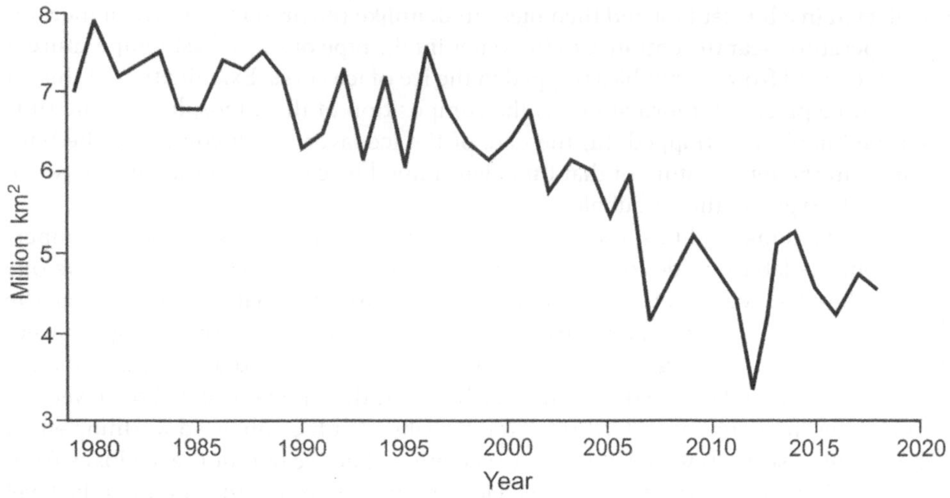

Figure 12.2 Changing extent of Arctic sea ice in recent years (from satellite imagery). Source: NSIDC/NASA.

12.3 A Summary of Future Changes in Climate

We can list the major results of enhanced global warming as follows:

1. Each of the last decades since 1950 has been successively warmer than the preceding one. IPCC (2014) considers that it is virtually certain, with more than 99% probability, that global warming has occurred since the middle of the twentieth century. All enhanced scenarios show increasing future temperatures, although at different rates.
2. Higher temperature, more hot days, and an increase in heatwaves will occur in the twenty-first century.
3. IPCC considers it to be virtually certain that the upper ocean (0–700 m) warmed up between 1971 and 2010.
4. The ice sheets of Greenland and the Antarctica had lost mass by the beginning of the twenty-first century; other glaciers have shrunk worldwide; and since the 1950s a very substantial part of the waters of the Arctic has opened up.
5. In general, the mean sea level rose by about 0.19 m (range 0.17–0.21) between 1901 and 2010. It is expected to rise continuously through the twenty-first century. However, rates may vary across the world. The rising is due to (i) thermal expansion of seawater with rising temperature and (ii) melting of glacial ice on land. Chen et al. (2017) examined the increasing rate of global mean sea level (GMSL), mainly from satellite altimetric data, and calculated an increased rate of rise from 2.2 ± 0.2 (1993) to 3.3 ± 0.3 mm (2014) per year.
6. A continuous emission of greenhouse gases will continue to cause more atmospheric warming and associated changes in the components of the climate system.
7. Changes will affect the global water cycle but not uniformly. The range between the wet and dry seasons, and between wet and dry areas, will increase. However, regional exceptions are likely, which suggests large regional uncertainties which in turn would hamper planning of efficient strategies for adaption to climate change.

Extreme precipitation is anticipated to increase by about 7% for each degree Celsius of warming (Lenderink and Fowler 2017).
8. Larger and more intense precipitation events are expected.
9. Intense summer droughts will strike certain localities.
10. Increased rainfall and wind velocity will mark tropical storms. Tropical cyclones of categories 4 and 5 will be more frequent.
11. Larger droughts and floods will be associated with ENSOs.
12. It is likely that the variability of the Asian monsoon will increase (Houghton 2004; IPCC 2014).

This is a summary of expected major changes in climate due to enhanced greenhouse effects that are happening now, and which will continue to happen in the future. There will be other changes but lesser in scale. The enhanced greenhouse warming is likely to affect many aspects of human practice such as urban development or intensive agriculture. Geomorphological processes, such as mass movement on slopes or flooding as a channel-maintaining process, also will be modified. The expected impact of climate change on large rivers will be reviewed in the following section.

A recent report on the state of the ocean and cryosphere in a changing climate has been produced by IPCC (2019) on occasion of the 51th session of the IPCC in Monaco. Such changes directly affect communities living in coastal environments, small islands, polar areas and high mountains. Ice sheets and glaciers have lost mass. For example, the Greenland ice sheet has lost ice mass at an annual average rate of 278 +/− 11 GT between 2006 and 2015. This is equivalent to about 0.77 mm global sea level rise annually. The total rise from all sources are necessarily higher. The total sea level rise between 1902 and 2015 is likely to be 0.16 m (range 0.12–0.21 m). Precipitation, winds, and events associated with large tropical cyclones have increased. A large river therefor would be impacted at its mountainous source, along its course, and also at its base level. The impact would not only affect the morphology and behaviour of a big river, physical structures on the river may also be impacted (IPCC 2019).

12.4 Impact of Climate Change on Large Rivers

Changes involving water cycle and sea level will affect hydrological and geomorphological processes and subsequently modify river morphology (Best et al. 2007; Gupta 2010; Pickering et al. 2014). It should be emphasised that changes would not be the same along the course of a single long river, and would vary amongst multiple rivers because of their different geographical location. The channel, floodplain, and the rest of the drainage basin will all be affected, but differently. The following is a list of major changes that are likely to be seen in large rivers:

1. Retreat of mountain glaciers and early melting of ice and snow in a calendar year. This will alter annual river hydrographs, with peaks arriving earlier in spring.
2. Changes in amount of annual rainfall with regional variation.
3. Enhanced seasonality of precipitation, increased peakedness in the hydrograph separated by longer periods of low base flow. Increased rainfall from extreme events, leading to bigger floods, for example, the flow in the Nile River is expected to increase

by about 13% with a 50% greater standard deviation indicating a strong variability (Conway 2017).
4. Increase in the strength of very large storms (categories 4 and 5 of tropical cyclones); increased summer cyclonic activities in the North Atlantic.
5. Increased temperature and possible movement of surface and subsurface water may enhance frequency and scale of mass movements on slopes, leading to greater and possibly coarser amounts of sediment reaching the river.
6. Alluvial floodplains may increase in size and also different types of bars.
7. A rise in sea level, varying from place to place.
8. Major changes in the Arctic. Average annual temperature is expected to increase by 1–3 °C; most of the enhanced warming will occur during winter (ACIA 2005). Significant changes, for example the melting of river ice, will influence the hydrology of the Arctic rivers. For details of warming of the Arctic and changing major rivers, see Chapter 11.
9. Melting of permafrost. This will result in enhanced movement of groundwater and sediment, leading to bank collapse and alteration of deltas at a large scale. The entire Lena Basin is under permafrost, and so is 80% of the Yenisei Basin. Permafrost soils contain a huge pool of organic carbon and future thawing of permafrost may be of considerable concern because the released CH_4 will have a much higher potential for global warming, compared with CO_2 (Bridgham 2017).
10. Rising frequency of landslides in the Arctic and high mountains because of melting and debuttressing of glacial ice.
11. Parts of middle and lower latitudes becoming drier, other areas wetter.
12. Rain falling on fewer days but very intensely.

The dimensions, appearance, and behaviour of rivers are controlled by a number of variables, of which four are of prime importance: water, volume of sediment load, texture of load, and channel slope (Lane 1955). The impact of climate change may alter one or more of these robust variables, causing rivers to adjust to new environmental conditions. Generalisation is difficult as climate change is expected to modify multiple characteristics of the drainage basin of a large river. A list of these characteristics would include factors such as land use, mass movements in the headwater mountains, variation in discharge, variation in amount and nature of sediment, varying adjustments amongst the tributary basins, etc. These would impact floodplains and the main river channel. Blum (2007) has demonstrated from past climatic changes that the nature of change is difficult to predict as the relationship between the forcing mechanisms of climate change and the response of the river is probabilistic and non-linear. Furthermore, the adjustment may take a long period of time. The upper and lower parts of the large river may not be in phase, and climate-driven changes in discharge and sediment may operate on the channel at different scales. Local impacts may vary following perturbations in the physiography. Even the largest river, the Amazon, changes form and adjusts to regional geological variations (Mertes and Dunne 2007).

Regional perturbations are expected along large rivers which cover long distances. The Irrawaddy traverses three rock-cut gorges, separated by flood basins. The Mekong has a long rocky course, followed downstream by a lower shorter section in alluvium. Large rivers would show different adjustments on rock, on alluvium, and at the contact between different sections. The anthropogenic impoundments and reservoirs will complicate the pattern of adjustment. The rates of erosion and sedimentation will change, possibly giving rise to a new channel pattern and bar formation in a large river. The types and intensity of climate change will impact differently on the upper and lower courses of the river but as a river is a connected source-to-sink system, the results of the impact should reflect this integration. Given time the river would adjust to physiography, climate change, and a rising sea level.

12.5 Climate Change and a Typical Large River of the Future

A large river of the future, from mountainous headwaters to the sea, may need to cope with the retreat of glaciers, a changed annual hydrograph, increased seasonality in discharge, and enhanced sediment supply in its upper course. It is reasonable to expect additional sediment derived from its headwater mountains under increased frequency of slope failures, possibly giving rise to a bar-filled (braided?) channel with coarse sediment and building of large fans at the highland–lowland contact (Goodbred Jr 2003). Further downstream, the upper and middle parts of the river may experience seasonality in water and sediment fluxes. If this occurs, aggradation in the channel and episodic flooding of the channel and floodplain could be common. The river would need to adjust to changes in water and sediment discharges simultaneously, on the given channel base. Similarly, the lower course of the river may experience accelerated delta-building, deposition of fine grains over channel sand, bursting of levees, and flooding of low areas. It may appear to be a different river!

The possible detailed outcomes of climate change along the *upper and middle* courses of a large river could be anticipated as:

- Greater dependence on rainfall and less on snowmelt. The rain could be from short intense falls.
- Lengthened duration of low flow. It is not yet clear how the baseflow would change.
- Mean discharge value may increase. Small changes in mean rainfall may increase the size of large floods (Knox 1993).
- A changed pattern of transfer and storage of sediment in coupling (Harvey 2002) between slopes and streams and at river confluences. Building of fans of different sizes.
- Deviation from the general pattern during ENSO and other atmospheric phenomena (Aalto et al. 2003).
- A wider, shallower, and more flood prone river with a pattern of episodic sediment transfer and long-term storage of sediment, provided accommodation space is available in the valleys.

The *lower courses and deltas* of large rivers may adjust differently to the changed pattern of water and sediment arriving from upstream and from polyzonal sub-basins. They also need to adjust to the regional effects of climate change. A rising sea level will provide more accommodation space for transported sediment, increase coastal erosion, and cause saline intrusion. Depending on the geographical location, the lower plains will experience enlarged tropical cyclones and more efficient basin penetration by storms of all sizes. Storms, moving inland, will cause floods. Levees will break, crevasse splays will form, sediments will fill flood basins and wetlands, and avulsions and abandonment of distributary channels will mark low flat deltas.

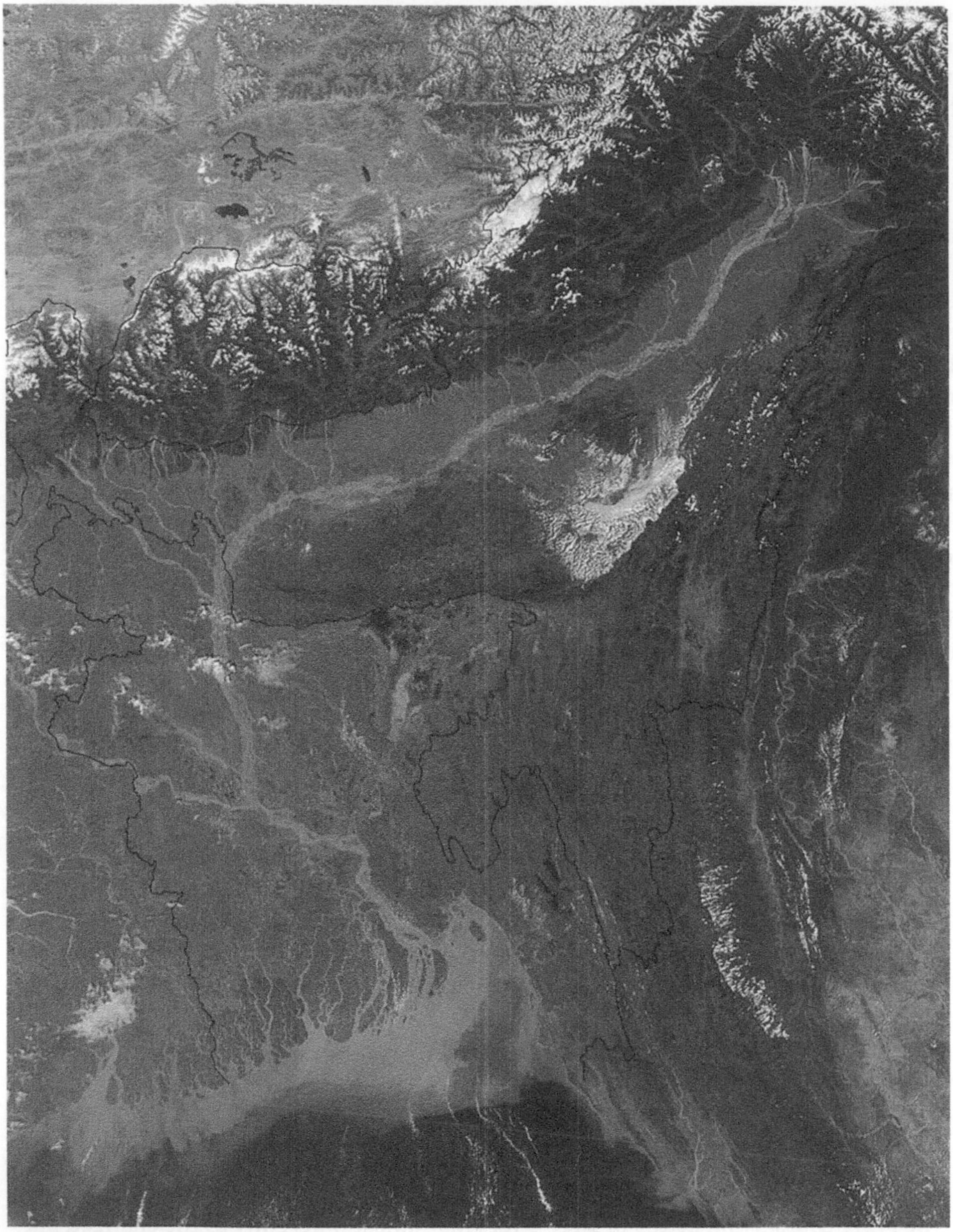

Figure 12.3 MODIS image of lower Ganga and Brahmaputra valleys and the delta. Source: NASA.

Figure 12.3 is a MODIS image of the Brahmaputra from the Himalaya to the Bay of Bengal, displaying the large valley of Assam and the delta area with the Ganga joining the Brahmaputra from the west. If we hypothetically raise the sea level by 1 m and superimpose the type of changes in a flooded delta as opined by Blum and Aslan (2006), we may expect the landform as suggested in Figure 12.4. The map of the lowland and the

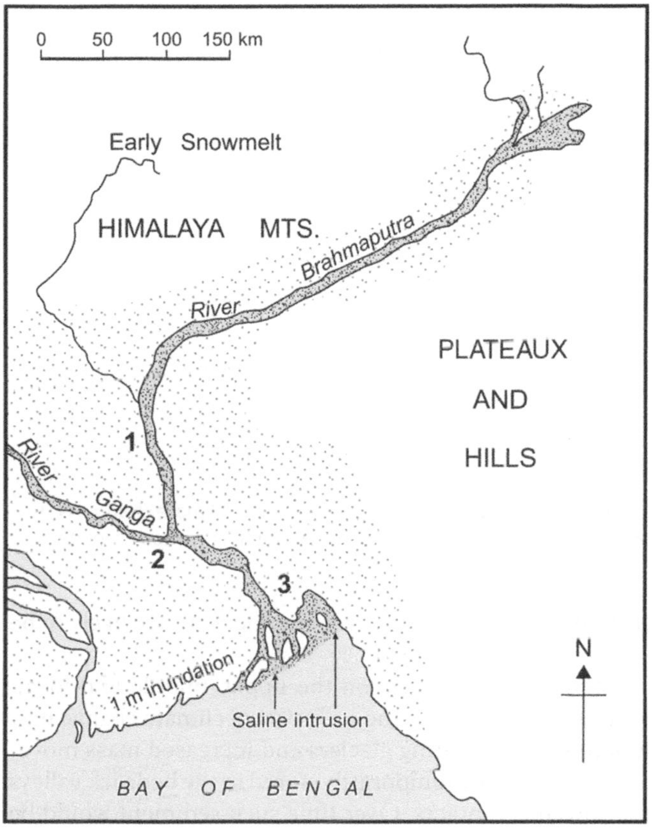

Figure 12.4 Possible effects of climate change in the Ganga-Brahmaputra Delta following approximate sea-level rise of 1 m. 1, Upper Delta changes, mostly crevasses splays, avulsions, flood basin filling; 2, Lower Delta changes, mostly fines deposited over channel sand; and 3, accelerated saline intrusion, increased damage and changes from tropical cyclones, coastal erosion. Source: Adapted from Gupta 2010.

delta would show a different physiography, ecology, and population concentration than those of the present time.

Based on past sedimentary records of the Texas Gulf Coast, Blum and Aslan (2006) modelled the changing style of avulsion in deltas over time. In the early stage, a rising sea level will result in valley-filling, channel occupation, and low aggradation. This is followed by high aggradation in channels, frequent crevassing, and filling of flood basins by avulsion. The old channels are reoccupied in the final stage and the aggradation will decrease in nearly filled valleys. Presumably, the speed and efficiency of such developments would depend on the rate of supply of water and sediment from the upper basin.

Such changes, apart from creating a modified lower course and a new delta, will have significant repercussions on the life of the people of the delta where often (i) villages are situated on levees, (ii) transport follows a network of rivers, and (iii) flood basins are farmlands. Abandonment of channels, avulsion of rivers, passage of most of the upstream discharge into a small distributary thereby enlarging it, and siltation of the

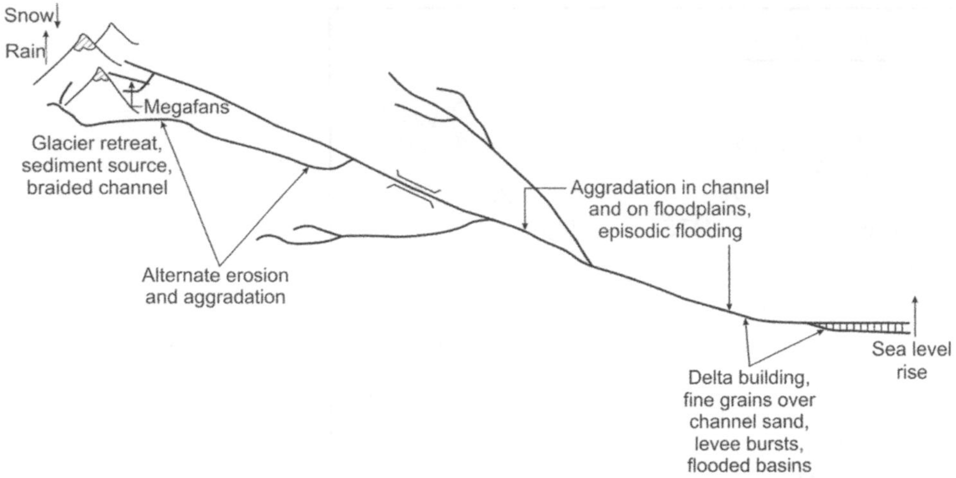

Figure 12.5 Probable geomorphic changes in a hypothetical large river due to climate change. Note the difference between this figure and Figure 4.2 which represents a natural river without climate change.

former major channel are likely to happen. The lower river and the delta are likely to be a dynamic landscape.

A river is a source-to-sink system and the effects on the upper, middle and lower courses should be seen in integration (Figure 12.5). The early part of climate change may produce huge volumes of sediment from retreating glaciers and increased mass movements in the headwaters. The sediment may be temporarily stored in the bedrock valleys and as alluvial fans in highland–lowland contacts. Over time such sediment would be gradually transferred and stored downstream along mixed bedrock – alluvium and alluvial valleys and on floodplains. The effect of climate change on sediment distribution and storage will influence the channel pattern and bar forms along the river. As time goes by, the river may look somewhat metamorphosed and behave differently.

A reference to the description by Goodbred (2003) of changing adjustments for the Ganga over approximately the last 50 000 years is relevant. During MIS 3 (48–24 ka ago), the regional climate was cool and moderately wet. The Himalayan glaciers advanced, and aggradation occurred in the valleys. The discharge of the river was moderate, and the sediment load was moderate to high. The advancing sediment led to the formation of mountain-front megafans and sedimentation in the lower valleys beyond. Incision and aggradation happened in the delta, shelf, and deep sea Bengal Fan. When the climate turned cold and dry (MIS 2, 24–18 ka ago), the river was low in both discharge and sediment load. Glaciers advanced a limited distance because of the dryness, and very little activity took place in the river basin. The climate warmed up during Late MIS 2 and the Early Holocene (18–7 ka ago) when it was also very wet. Both water and sediment discharges were high. Widespread erosion happened in the megafans and alluvial valleys. Rapid deposition occurred further down the basin, in the delta, shelf, and the Bengal Fan. Since mid-Holocene, the climate has been warm and wet. Discharges of water and sediment have been high, the Himalayan glaciers have retreated, and both valley aggradation and incision have occurred in the upper valleys. A pattern of slow aggradation was common in the lower alluvial plain, and on the delta and the fan. Current climate change, with melting glaciers, altered annual rainfall, and enhanced large storms, appears to be another step in this sequence of events for the Ganga and the Brahmaputra, a step we need to understand as it will involve the quality of life of the high population in the river valleys and deltas. Very large changes could be involved.

12.6 Conclusion

The general tenets of climate change are known and accepted but we have limited knowledge regarding its specific impact on large rivers. It is likely that the impact of climate change on large rivers will metamorphose their channels – making them wider, more seasonal, and more flood prone, except where influenced by regional geological structures. The change may be different in different parts of the river. Channel avulsion and flood basin sedimentation would become common in megafans and deltas. Large rivers, and probably most of the drainage network of the Earth, would change because of the climate, and the rivers of the future may differ from their present appearance, regarding both form and behaviour. As the form and function of rivers change, their statistical parameters and recurrence intervals of discharges need to be revised. Parameters for large river management have to be redesigned. We need to study the major rivers to estimate a set of rules for understanding and managing large rivers as such changes would impact the lifestyles of a very large number of people. Rivers of the future are likely to differ from the present ones, a possibility which we need to be explore. The analogue of a future large river could very well be another big river of the present time at a different location or the Quaternary version of a present river as discussed by Goodbred (2003). Discovering such changes will be fascinating.

Questions

1 Climate change is likely to modify present rivers. Why?

2 Would the changes be significant?

3 Will all the rivers of the future be flood prone? What effect would such a change have on the form and function of rivers?

4 Describe sequentially the changes that a large river may be expected to undergo from the source to sea. How integrated would be the effect of changes?

5 Would the effect of climate change be comparable irrespective of the geographical location of rivers?

6 What is the implication of the stationarity of rivers being changed? How could it be coped with?

7 Why are the deltas of large rivers expected to function as dynamic landscapes following climate change?

8 If a scenario for a changing large river is constructed, how reliable would the expectations be?

9 On what information such scenarios should be based?

10 Select a large river with a delta. Comment on the extent and intensity of change due to enhanced global warming.

References

Aalto, R.E., Maurice-Bourgoin, L., Dunne, T. et al. (2003). Episodic sediment accumulation on Amazonian floodplain influenced by El Niño/Southern Oscillation. *Nature* 425: 493–497.

ACIA (2005). *Arctic Climatic Impact Assessment*. Cambridge: Cambridge University Press.

Best, J.L., Ashworth, P.J., Sarker, M.H., and Roden, J.E. (2007). The Brahmaputra-Jamuna River, Bangladesh. In: *Large Rivers: Geomorphology and Management* (ed. A. Gupta), 395–433. Chichester: Wiley.

Blum, M.D. (2007). Large river systems and climate change. In: *Large Rivers: Geomorphology and Management* (ed. A. Gupta), 627–659. Chichester: Wiley.

Blum, M.D. and Aslan, A. (2006). Signatures of climate vs. sea-level change within incised valley fill successions: Quaternary examples from the Texas Gulf Coast. *Sedimentary Geology* 190: 177–211.

Bridgham, S.D. (2017). Methane origins. *Nature Climate Change* 7: 477–478.

Chen, X., Zhang, X., Church, J.A. et al. (2017). The increasing rate of global mean sea-level rise during 1993–2014. *Nature Climate Change* 7: 492–495.

Conway, D. (2017). Future Nile River flows. *Nature Climate Change* 7: 319–320.

Goodbred, S.L. Jr., (2003). Response of the Ganges dispersal system to climate change: a source to sink view since the last interstade. *Sedimentary Geology* 162: 83–104.

Gupta, A. (2010). The effect of global warming on large rivers and deltas. In: *Developing Countries Facing Global Warming: A Post-Kyoto Assessment* (eds. M. De Dapper, D. Swinne and P. Ozer), 125–138. Brussels: Royal Academy for Overseas Sciences.

Harvey, A. (2002). Effective timescales of coupling in fluvial systems. *Geomorphology* 44: 175–201.

Houghton, J. (2004). *Global Warming: The Complete Briefing*. Cambridge: Cambridge University Press.

IPCC (2007). *Climate Change: The Physical Science Basis*. Cambridge: Cambridge University Press.

IPCC (2014). *Climate Change 2013: The Physical Science Basis*. Cambridge: Cambridge University Press.

IPCC (2019). Summary for Policymakers In: IPCC Special Report on the Ocean and Cryosphere in a Changing Climate (eds. H.-O. Portner, D.C. Roberts, V. Masson-Delmotte, et al.) In press.

Knox, J.C. (1993). Large increases in flood magnitude in response to modest changes in climate. *Nature* 361: 430–432.

Lane, E.W. (1955). The importance of fluvial morphology in river hydraulic engineering. *Proceedings of the American Society of Civil Engineers* 81: 1–17.

Lenderink, G. and Fowler, H.J. (2017). Understanding rainfall extremes. *Nature Climate Change* 7: 391–393.

Mertes, L. and Dunne, T. (2007). Effects of tectonism, climate change, and sea-level change on the form and behaviour of the modern Amazon River and its floodplain. In: *Large Rivers: Geomorphology and Management* (ed. A. Gupta), 117–144. Chichester: Wiley.

Milly, P.C.D., Betancourt, J., Falkenmark, M. et al. (2008). Stationarity is dead: whither water management? *Science* 319: 573–574.

Pickering, J.L., Goodbred, S.L., Reitz, M.D. et al. (2014). Late Quaternary sedimentary record and Holocene channel avulsion of the Jamuna and Old Brahmaputra River valleys in the upper Bengal delta plain. *Geomorphology* 227: 123–136.

Wang, G., Cai, W., Gan, B. et al. (2017). Continued increase of extreme El Niño frequency long after 1.5°C warming stabilization. *Nature Climate Change* 7: 568–572.

Index

4000 Islands 201

a
Abies balsamiea 222
Above Ground Woody Biomass (AGWD) 75, 76
adaptive management 178
aerosols 267
Africa 156
African Plate 13, 138
Akosombo Dam 155
Alaknanda 45, 133
Alaska 216
Alaska Range 217, 229, 296
Alaskan Cordillera 216
Alcan Highway 229
Alder 238, 242
Allahabad 45
alluvial fan 37–38
Altai 126, 216, 226, 229
Amazon 1, 3, 8, 9, 12, 13, 17, 21–23, 26–29, 32, 35, 37–41, 48, 56, 57, 62, 64, 70, 125, 127, 128, 137–141, 167, 181, 184, 240, 272
Amazonia 74, 109
Atlantic Multidecadal Oscillation (AMO) 21
Amu Darya 167
Andes 1, 13, 14, 17, 18, 20, 26, 27, 31, 41, 42, 57, 137–139, 192
Angkor 193, 206
Annamite Mountains 195, 197, 198, 200–202
Antarctica 267
Appalachian Mts. 25
Aquatic Terrestrial Transition Zone (ATTZ) 60, 74, 85
Araguaia 60, 73, 74, 85, 88
Aral Sea 126, 218
Arctic Climatic Impact Assessment (ACIA) 213
Arctic Ocean 212, 214, 253
Arctic Red River 232, 242, 245, 247, 248
Arctic rivers 211–258
Aricha 133
Arkansas 25
Ashworth and Lewin 52
Asian Development Bank 193
Assam 133
Assosa 156
Aswan High Dam 141, 153–155, 167
Atbara 26
Athabasca basin 218, 249
Atlantic 13, 20, 27, 29, 39, 125, 131, 140
Australia 31
avulsion 51

b
Baikal 217
Bananal 56, 82
Bangladesh 36, 45
Barind 113
Barnaul 215, 228, 242
baroclinic 20–25
barotropic 16–20
baseflow 211
Bassac 186
Bay of Bengal 13, 109, 114, 133
beach ridges 109

Introducing Large Rivers, First Edition. Avijit Gupta.

Beaufort Sea 156, 212, 213, 254, 255
Belogorie 242
Bengal Basin 136
Bengal Deep Sea Fan 112
Bennet Dam 238, 242
Bering Sea 213, 214
Bering Strait 211, 212, 213, 218, 253
Betula spp. 222
Bhagirathi 44, 133
Big Bend 152
biodiversity 68
Biological Oxygen Demand (BOD) 177
blackwater 57, 60, 69, 82
Blue Nile 8, 26, 124, 141, 155, 157
Blum 129, 130
Blum and Törnquist 130
bog 224
Bonneville 151
Borneo 125
Brahmaputra 15, 22–24, 30, 32, 35, 36, 44, 46, 128, 133, 157, 225, 277
Brazil 27, 42
Brazil Shield 39, 41, 57, 137

c

Ca Mau 198, 202
Cabo Norte 43
Camaleao Lake 61, 71, 77, 78
Cambodia 193, 194, 198, 201, 205, 207
Campbell 174, 205–207
Can Tho 198
Canada 224
Canadian Cordillera 216
Canadian Shield 217
Canyon Ferry 152
Cardamom Hills 196
Caribbean Sea 57
Caspian 126, 218
Cassiar Mountains 237
Central Asian Orogenic Belt 216
Cerrado 62, 70, 72
Chan 82
Chandina 113
Chang Saen 197, 205
Changjiang 10, 13, 18, 22, 24, 25, 32, 110, 119, 141, 147–149, 155, 156, 158, 173, 178, 179, 240
Channel maintenance flow 177
Chao Phraya 157
Chengdu plains 149
chenier ridges *see beach ridges*
Chenla 193
Chernyshevsky 238
Chiang Khan 197
Chiang Saen 1
China 128, 148, 156, 193, 194, 195, 202, 204, 205
Chindwin 133
Chittagong 112
Chukchi Sea 213
Cicih 87
Clean Water Act 55
clearwater 57, 60
climate change 265–277
climate change and large rivers 271–273
Coast Mountains 229
Colorado River 1, 10, 25, 151, 156, 160, 163
Columbia River 126, 151, 152, 167
Comilla 113
Congo 22, 29, 35, 167, 240
Cuando 120

d

Danube 7, 9, 32, 59
Datong 18, 24, 27, 166, 169
decomposition 76–77
delta morphology 110–112
delta 103–115
Devprayag 44, 133
Dietrich 49
discharge 17–18, 21–26
Dneiper 126, 152, 179, 183
Doce River 70
Dolphin 207
Don 196
Dongting 24, 25
Dujiangyan Irrigation System 148
Dunne 40, 41

e

East China Sea 24
East Siberian Sea 218
East St. Louis 10
Eastern Syntaxis 133

Ebro 158
Egypt 26, 148
El Niño 21, 39
El Niño Southern Oscillation (ENSO) 21, 23, 39, 48, 140, 265, 267, 273
Elephant Hills 196
English Channel 125
enhanced greenhouse effect 267
environmental flow 177
environmental flow release (EFR) 177
Enzel 128
Equator 20, 21, 23
Erie canal 149
Ethiopia 26, 141, 153, 157
Euphrates 55, 147, 148
Eurasian Plate 13, 119, 194
Everglades 85, 87, 88

f

Fairbanks 229
Farakka Barrage 137
felling cycles 81
fen 224
fish 55, 59, 67, 160, 207–208
flood pulse 51, 193
Flood Pulse Concept (FPC) 55, 60, 64
floodplain 43, 55–89
floodplain, management 80–82
floodplain, biogeochemical cycle 73–80
floodplain, fishing 81
floodplain, migration 87
floods 29, 57, 62, 63, 67, 105, 106, 213, 232, 248
Fly 49, 111
foodweb 79–80
Forest Code 83
forest steppe 220, 231
Fort McMurray 220, 232
Fort Nelson-Muskwa-Prophet river system 245
Fort Peck 152
Fort Providence 242
Fort Rendall 152
Fort Simpson 237
Fort St. John 229
Front Ranges 150
future changes in channel 270–271

g

Ganga 9, 12, 13, 22, 24, 30, 32, 37, 44–48, 55, 119, 128, 178, 179, 181, 277
Ganga delta 109, 110, 112–115
Ganga Sagar 46
Ganga-Brahmaputra System 133–137
Gangasagar 30
Gangotri 44
Gangow Dam 49
Garrison 152
Gilbert 111
glacial stage 124–125
Glen Canyon Dam 163, 164
global warming 211, 238, 265–270
Godavari 109, 119
Goodbred 134, 276
Grand Canal 148
Grand Coulee 151
Grand Ethiopian Renaissance Dam (GERD) 141, 155–156
Grant et al. 163
gravity anomalies 49
Great Bear Lake 217, 218, 236
Great Bear River 233, 235, 239, 248
Great Plains 25
Great Slave Lake 213, 218, 242
greenhouse gases 243, 266
Growth Oriented Logging (GOL) 81
Gulf of Martaban 10
Gulf of Mexico 119, 129–131
Gurapá 138
Guri Dam 156
Guyana Coast 109
Guyana Current 109
Guyana Shield 18, 39, 41, 57, 137

h

Han 25
Hankow 24
Harappan Civilisation 148
Haridwar 45
Haryrik 238
Haryrik-Korno 238
Himalaya 11, 14, 20, 24, 30, 31, 44, 47, 110, 119, 134, 194
Hoa Bin Dam 160

Holocene 42, 109, 110, 114, 121, 127–128, 131, 132, 139, 140
Hoover Dam 10, 151
Hovius 1
Huanghe 1, 105, 147, 148, 157, 158
Hudson Bay 218
Hudson R. 149
Hugli 133, 137

i

Igapos 60, 69, 74, 76
impoundments 159–166
India 30, 128, 133
Indian Ocean 24
Indian Plate 13, 119
Indus 1, 26, 55, 147, 148, 160
Integrated Water Resources Management (IWRM) 182
Interglacial stage 127
Intergovernmental Panel on Climate Change (IPCC) 239, 268, 269
Interior Cordillera 217
International Basin Arrangements 180
International Commission on Large Dams (ICOLD) 155
Intertropical Convergence Zone (ITCZ) 20, 25, 39, 140
Inuvik 23
Iquitos 138
Irkutsk 228
Irrawaddy 7, 10, 23, 32, 49, 50, 105, 133, 272
Irtysh 126, 213, 241
Itaipu 56

j

Jamuna, *see Brahmaputra*
Japurá 56
Java 125
Jialing 165
Juruá 56
Jutai 27, 42, 138

k

Kailash 35
Kalahari 121
Kale 128
Kalimantan 111
Kamenna Obi 242
Kara Sea 213
Katowice 268
Katrina 132
Kazakhstan 212
Kemerovo 228
Kessel 132
Khartoum 26
Kinsahasa 23
Kirgizistan 126
Knox 131
Koblenz 185
Kolpashevo 242
Kolyma R. 249
Kong 196, 197, 201, 202, 205
Korat Plateau 195, 197, 202
Kosi 12, 37–38
Krasnoyarsk 228
Kuehl et al. 140
Kummu and Varis 205
Kuskokwim River 256
Kusur 238

l

La Niña 21, 23, 140, 269
Lake Agassiz 130, 218
Lake Bonneville 126
Lake Maracaibo 138
Lake McConnell 218
Lake Mead 151
Lake Powell 163
Lake Victoria 26
Lancang Cascade 168, 187, 205, 206
land cover changes 157–159
Lao PDR 193, 194, 196, 202, 205
Laptev Sea 213
large rivers, source to sink 35–37
large rivers, morphology 48–52
large rivers, sediment 26–32
Large-scale transfer of river water 166–167
Larix dahurica 220
Larix laricina 222
Last Glacial Maximum (LGM) 57, 125, 134
Latrubesse and Stevaux 157

Lena 1, 6, 8, 18, 25, 213, 215–218, 226, 228, 238, 239, 253, 272
Li 87
Li Bing 149
Liard basin 236, 237
Liard River 237, 244, 245, 246, 247
Limpopo 119–121
Liones dos M 85
Llanos Beixes 85
Llanos dos Moxos 85
Lockhart River 235, 239
Luang Prabang 198
Luanga 120

m

Mackenzie 6, 22, 25, 26, 126, 213–215, 217–220, 232, 235, 236, 239, 244, 245, 248, 253, 256, 258, 274
Madeira 27, 41, 42, 56, 139
Madhupur 113, 114
Magdalena 81
Mahakam 110, 111
Malacca Strait 125
Malay Peninsula 125
Manaus 12, 43, 57, 71
Mangroves 110
mantle plume 8
Marajó Rift 39
Marchantaria 71
Marine Isotope Stage (MIS) 121–122
Martin River 245
Mato Grosso 56, 80, 88
Mauna Loa Climate Observatory 267
MDBC 185
Meade 1, 26, 38
Mediterranean Sea 141, 218
megafan 37
Meghna 44, 46, 136
Mekong 1, 4, 6, 9–11, 15, 22, 23, 32, 49, 109, 119, 156, 158, 159, 168, 181, 183–188, 193–201, 207–208
Mekong Committee 194, 204
Mekong River Commission (MRC) 179, 196, 204
Mekong River–degradation of aquatic life 207–208
Mekong River Strategic Environmental Assessment (SEA) 207–208
Mertes and Dunne 35, 38, 42, 139, 140
Mesopotamia 148
Milankovich 134
Millenium Ecosystem Assessment (MEA) 88
Milliman 31
Milliman and Farnwell 132
Milliman and Syvitski 39, 41, 46
Min River 149, 165
Miranda River 76
Mishmi hills 24
Mississippi 1, 5, 8, 10, 14, 25, 31, 32, 35, 55, 69, 103, 107, 119, 125, 126, 128–133, 141, 147, 149–150, 178, 179, 181, 212
Missoula 126
Missourie 1, 25, 31, 131, 141, 150, 152, 167, 178, 212
Mohawak River 149
Momir 80
Mongolia 127, 212
Monsoon 48, 110, 112, 115
Monte Allegro Intrusion 138
Mun River 201, 207
Murray-Darling 26, 119, 127, 173, 179
muskegs 211
My Tho 198
Myanmar 193, 194
Myitkyina 49

n

Nam Ngum 197, 202, 205
Nam Ou 202
Nam Theun 197, 202
Namche Barwa 24, 133
Nargis 105
Narmada 7, 128
NAZCA Plate 39
Negro 9, 43, 56, 57, 64, 138
Net Primary Production (NPP) 73, 75, 76
New Orleans 132
Niger 8, 25, 63
Nile 1, 5, 13, 14, 22, 25, 26, 55, 103, 127, 141, 147, 148, 153, 160, 167, 179, 184
Nile Basin Initiative 184–185

Nitrogen cycle 77–78
Noakhali 112
Norman Wells 245
North Annular Mode (NAM) 267
North Atlantic 123
North Atlantic Oscillation (NAO) 21, 267
North Brazil Current 48
Northeast Passage 253
Novokuznetsk 228
Novosibirsk 228
nutrient 78–79
nutrient cycles 88
nutrients and contaminants 249–252

o
Oahe 152
Ob 6, 8, 18, 25, 126, 158, 213, 215, 217, 220, 223, 229, 240, 258
Óbidos 18, 27, 41, 42, 57, 138
Oc Eo 193
Ohio 8, 13, 25, 57, 129–131, 152
Okavango 85, 87, 120
Olenek River 238
Orange 8
Orinoco 5, 17, 19, 21, 26, 27, 29, 43, 57, 138
orographic uplift 20
Oubangui River 23
Oymyakon 220

p
Pacific Decadal Oscillation (PDO) 21, 267
Pacific Ocean 24, 123
Pacific-North American Pattern (PNA) 267
Padma *see Ganga*
Pakistan 26
Pakse 197, 198, 205
Paleofloods 128
Paleo-vérzea 52, 58
Panama 123
Pangasianodon gigas 207
Pangea 25, 119
Pantanal 56, 58, 61, 63–65, 70, 73, 74, 75, 76, 82, 85
Papua New Guinea 49
Paraguay River 56, 58, 82
Paraná 32, 56, 61–63, 70, 184
Paris Climatic Conference 269–270
Peace River 228, 232, 235, 238, 242, 258
Peat 224
Peel Plateau 218
Peel River 246, 247, 253
Peninsular India 45, 46
permafrost 25, 212, 214, 218, 224–227, 232, 238, 248, 250, 272
Phapheng Falls 201, 207
Phnom Penh 193, 196, 198, 202
Picea glauca 220
Picea mariana 220
Pinus banksiana 222
Pinus contorta 220
Pinus silvestris 220
Plantations 82
Platte 132
Pleistocene 9, 112, 114, 121, 124, 127, 134, 139, 218
Point Separation 253
polders 82
Populus tremuloides 220
Potter 1, 8, 12, 13
Poyang 24, 25
Probarbus julllieni 207
product respiration ratio (P/R) 59
Proto Primavera Dam 56, 88
Punjab rivers 134
Purus 56, 138

q
Qinghai 164
Quaternary 1, 48, 52, 58, 121, 123, 126, 130, 141

r
Rajmahal 45, 133
Ramsar Convention 62
Rat River 253
Red River *see Sông Hóng*
Redstone River 245
Relative Growth Rate (RGR) 75
Revelle 267
Rhine 69, 125, 183
Rhodinia 8
ria 57, 82

Rio Grande 8, 119
River Continuum Concept (RCC) 59
river ice 236–237
river management 176
river restoration 17
River-Floodplain Systems (RFS) 56, 57, 59, 69, 71
Rivers and Harbors Act 131, 149
Rocky Mountains 25, 31, 129, 130, 178, 217
Roraima 86
Rosieres Dam 157
Rupununi 59

s

Sacramento River 151
South Alantic Convergence Zone (SACZ) 21, 39
Sagaing Fault 50
Saigon River 202
Sakha 220
Salekhard 158, 215, 241
Salinas 61
Salween 10, 17, 32, 35, 49, 119
Sambor 206
San 196, 197, 201, 202, 205
San Francisco River 56
Sanmenxia 157
satellite imagery 159
savanna 64
Savannakhet 1, 198
Save 120
Sayan Mountains 226, 229
Scablands 125
scenarios 269
Schumm 10, 214
Se San 2 Dam 205
Sea Surface Temperature (SST) 268
Sediment 239–243
Serageldin 173
Shaba Highlands 23
Shasta 151
Shire 120
Siberia 16, 213, 214, 217, 218, 220, 226
Sichuan 25
Siem Reap 206
Sioli 60
Sirajganj 36
Snake River 126
Social and political management 178–179
Solimoés 138
Solyanka 242
Sông Hóng 13, 119
South America 26, 39
South Atlantic Convergence zone (SACZ) 140
South Atlantic Plate 39
South China Sea 125
Southern Annular Mode (SAM) 21, 267
Srepok 196–198, 201, 202, 205
St. Elias Mountains 232
St. Lawrence 32
St. Louis 150
Stationarity 265
Steppe 220, 224, 232
Stung Treng 197, 201
Sudd 26
Suhana 238
Sumeria 148
Sunda Shelf 125
Suntan 238
Surinam 59
sustainable development 173
Swatch of No Ground 112, 114, 134
Sylhet 112, 115
Syr Daria 167

t

Ta et al. 206
Tabaga 242
Taiga 220, 231, 232
Talik 226
Tanana 234
Tapejós 56, 57
Taq River 58
Teays-Mahomet-proglacial valley system 129
TECCONILE 184–185
Tefe 57, 80
Tennessee 131, 152
Tennessee Valley Authority (TVA) 153
Texas Gulf Coast 275

Thailand 193–195, 197, 205
Three Gorges Dam 11, 24, 27, 141, 155, 156, 164–166, 187, 205
Tibetan Plateau 24, 110, 119, 164, 193
Tigris 55, 147, 148
Tiksi 253
timber extraction 81
Tobolsk 242
Tocantins 56, 140
Tom River 256
Tomsk 229, 256
Tonle Sap Lake 168, 181, 196, 205, 207
transition stage 125–127
Triple junction 18
tropical cyclones 20, 29
tropical storms 20
Tsanpo 13, 119, 133
Tundra 220, 224, 231, 232

u
Uatuma River 76
United Nations 194
United States 31
United States Army Corps of Engineers (USACE) Hydrologic Engineering Center 178
Upper Crossing 237
Urals 31, 226, 229

v
Vannote 59
várzea (*also see floodplain*) 69, 70, 72, 74, 75, 81, 86, 87
Vaupes Arch 57
Verkhoyansk 232, 238
Vientiane 14, 194, 196, 198, 200
Vietnam 193, 194, 202, 205
Vilyui River 238
virtual tivers 176
vision 179
Volga 126, 152, 167

w
Wagga 46
Walling 203
Walling and Fang 152
Washington 126
Wasson 128
Water Framework Directive (WFD) 55
Western Cordillera 237
wetland 83–90
White Nile 26
White River 236
whitewater 69
Williams and Wolman 101
Williston Lake 242
Wilson Cycle 8
Wohl 176
wood decay 77
World Bank 147
World Energy Council (WEC) 268
World Weather Watch (WWW) 268
Wrangell-St. Elias Mountains 216
Wu River 165

x
Xiaowan Dam 156, 205

y
Yakutsk 229
Yamuna 45, 133
Yangtze *see Changjiang*
Yarlung 133
Yellowknife 229
Yenisei 6, 8, 18, 25, 125, 213, 215, 216, 217, 220, 232, 239
Yeniseisk 242
Yichang 24, 27, 158, 165, 168
Yukon 6, 8, 25, 211, 213–215, 217, 220, 229, 231, 234, 239, 256
Yunnan 195, 197

z
Zambezi 1, 8, 119–121
Zhujiang 32

Printed and bound by CPI Group (UK) Ltd, Croydon, CR0 4YY
27/10/2024
14580312-0003